PROBABILITY & STOCHASTIC PROCESSES

PROBABILITY & STOCHASTIC PROCESSES

Editor

Hermenegild Salzwedel

scitus
academics

Probability & Stochastic Processes

Edited by **Hermenegild Salzwedel**

This Book is Printed on Acid-Free Paper. ♾

ISBN: 9781681179650
Library of Congress Control Number: 2017949104

www.scitusacademics.com

Notice

Printed in United States of America.

Preface

In probability theory and associated fields, a stochastic or random process is a mathematical object usually defined as a collection of random variables. In the past, the random variables were allied with or indexed by a set of numbers, typically viewed as points in time, giving the explanation of a stochastic process representing numerical values of some system randomly changing ultimately, such as the growth of a bacterial population, an electrical current fluctuating due to thermal noise, or the movement of a gas molecule. Stochastic processes have played a significant role in various engineering disciplines like power systems, robotics, automotive technology, signal processing, manufacturing systems, semiconductor manufacturing, communication networks, wireless networks etc. Among the above engineering applications of stochastic processes are extensively used as mathematical models of systems and phenomena that appear to fluctuate in a random manner.

This volume 'Probability & Stochastic Processes' is concerned with stochastic processes and their applications in the modeling, analysis and optimization of stochastic systems, i.e. processes characterized both by temporal or spatial evolution and by the presence of random effects. It deals with all aspects of stochastic systems analysis, characterization problems, stochastic modeling and identification, optimization, filtering and control and with related questions in the theory of stochastic processes. With an emphasis on applications in engineering, applied sciences, business and finance, statistics, the book provides several practical examples that demonstrate how random phenomena take place in nature and how to employ probabilistic techniques to precisely model these phenomena. This book is oriented towards a broad spectrum of mathematical, scientific and engineering interests.

Table of Contents

CHAPTER 1

A Novel Approach to Probability

Oded Kafri

Kafri Nihul Ltd., Tel Aviv, Israel

ABSTRACT

When P indistinguishable balls are randomly distributed among L distinguishable boxes, and considering the dense system **P**□ **L**, our natural intuition tells us that the box with the average number of balls P/L has the highest probability and that none of boxes are empty; however in reality, the probability of the empty box is always the highest. This fact is with contradistinction to sparse system **P**□ **L** (i.e. energy distribution in gas) in which the average value has the highest probability. Here we show that when we postulate the requirement that all possible configurations of balls in the boxes have equal probabilities, a realistic "long tail" distribution is obtained. This formalism when applied for sparse systems converges to distributions in which the average is preferred. We calculate some of the distributions resulted from this postulate and obtain most of the known distributions in nature, namely: Zipf's law, Benford's law, particles energy distributions, and more. Further generalization of this novel approach yields not only much better predictions for elections, polls, market share distribution among competing companies and so forth, but also a compelling probabilistic explanation for Planck's famous empirical finding that the energy of a photon is $h\nu$.

Keywords: Probability, Statistics, Benford's Law, Zipf's Law, Planck's Law, Configurational Entropy

1. INTRODUCTION

Probability theory is an important branch of mathematics because of its huge economic significance. Since our knowledge about the future is limited, the most effective decisions rely on statistics.

Usually it is assumed that similar underlying events should have a similar resultant expectation value. From the law of large numbers [1] we expect that a large number of trials will yield the average value. From the central limit theorem [2] we expect that the scattered deviations from average will be distributed as in a Gaussian curve (the normal distribution).

Accordingly, when pollsters try to predict the results of a poll, i.e. elections, they treat each party individually. They find its expected number of seats from a small sample. Then they combine the results of all the parties together and obtain the expected distribution of the seats among the parties in the election. The a priori assumption of any pole is that the distribution of the seats among the parties, if the people vote randomly, is equal.

Based on the second law of thermodynamics, we suggest an alternative assumption that should be made, namely, that any configuration of seats among the parties should have an equal probability. When we calculate the distribution of seats under this assumption, we obtain an unequal distribution of seats among parties which is very similar to that obtained in real elections, see Figure 3. Nevertheless, each of the parties has an equal a priori probability to win each of the unequal number of seats of the distribution.

To generalize the analysis we call each party a box, and the available seats in parliament, balls. In statistical physics a box is called a state, and a possible configuration of the balls in the boxes is called a microstate. The logarithm of the number of microstates is called entropy [3] . The second law of thermodynamics states that a system in equilibrium has maximum entropy. Maximum entropy exists when all the microstates have an equal probability and all the states have an equal probability.

We distinguish between two kinds of statistical systems:

Dense systems, namely, systems in which the number of balls is bigger than the number of boxes.

Sparse systems, in which the number of balls is much smaller than the number of boxes and/or no more than one ball can exist in a box.

For example, elections are dense systems, since there are more seats than parties. Similarly, a best sellers' list is dense as the number of readers is greater than the number of the books. In wealth distribution, there are more dollar bills than people, there are more surfers per second than sites in the internet, and more listeners than songs, etc. It seems that many of the human's social activities can be described by dense systems statistics.

On the other hand lotteries are a sparse systems, since the number of winners is smaller than the number of tickets sold. The sparse systems statistics are very useful in describing distributions in space, because in many cases two objects cannot occupy the same space, i.e. two buildings cannot coexist in the same area. Similarly, two atoms or molecules cannot coexist in the same state.

In order to visualize the difference between the classic approach and the new approach to probability let us look at 12 balls distributed randomly in 3 distinct boxes. The average is 4, therefore most people will assume that the best guess is to find 4 balls in any randomly chosen box. The second best guesses are 3 and 5 balls in a box. Yet, in reality, if you have to gamble your best bet should be an empty box.

The guess of 4 balls is based on the law of large numbers, and the guess of 3 and 5 balls are based on the central limit theorem. However, the probability to find 4 balls in a box is only 10%, 3 balls is 11%, 1 ball is 13%, and the most probable is an empty box which has 14% probability. These probabilities were calculated according to the new approach and it reflects the known fact that there is a higher probability to be poor than to be rich.

The reason for that counterintuitive result is found in the microstates of this system. For example: there are only 3 different configurations in which boxes with 12 balls exist; namely, 12, 0, 0; 0, 12, 0; and 0, 0, 12. For these configurations, there are only 3 full boxes with 12 balls, while the number of empty boxes is double that, namely 6. Empty boxes are found, in addition, in many other configurations as well. This is the intuitive explanation why the expectation of the number of the boxes having n balls is monotonically increasing as n decreases.

It is not easy to count the number of appearances of the boxes with n balls in all the microstates, as it requires tedious combinatorial calculations because the number of microstates is exploding very fast. Moreover, the numbers are specific to any values of balls and boxes. However, the techniques of maximizing the entropy, which was first suggested by Planck yields a proven good approximations. The reason why entropy maximization yields the best solution to these problems is because the entropy is defined in equilibrium, in which the number of microstates is maximum because each microstate has an equal probability.

From the requirement of an equal probability to each microstate, we can derive the relative number of boxes $\varphi(n)$ that have n balls. We call $\varphi(n)$ frequency, and the balls distribution function is calculated by $\rho(n) = n\varphi(n)$

The relation between n and $\varphi(n)$ is derived by maximizing the entropy, using the Lagrange multipliers technique.

We can extend the theory more by giving the balls a "property". This is done by letting the balls have energy, assets, links etc.; In this case the property distribution calculations are similar to the calculations of the energy distribution among particles in statistical mechanics.

These kinds of probabilities are of great importance in statistics, econometrics, information theory and physics. Here we address the calculation of these probabilities in a unified way. In statistical mechanics this kind of analysis is called maximum entropy [4] , in information theory - Shannon limit [5] , and in classical physics - thermodynamic equilibrium [3] .

As is shown hereafter, this analysis yields for integer number of balls in a box: Zipf's law [6] [7] , Benford's law [8] - [10] , Elections votes distribution, Planck's law [11] , Bose-Einstein distribution [12] [13] , Fermi-Dirac distribution and Maxwell-Boltzmann Distributions [14] - [16] .

When the balls have a property, a chemical potential usually appears in the distribution functions.

It should be noted that maximum entropy techniques were used in probability theory and in econometrics [17] [18] . However, the entropy's expression that was used in these publications, namely, Boltzmann-Gibbs entropy [19] is approximated for infinitely small P/L , as opposed to the examples discussed in this paper in which $P \square L$. The Boltzmann-Gibbs entropy is applicable to sparse systems and yields the Maxwell-Boltzmann distribution in which the boxes with average values of balls have the highest probabilities. For example, in ideal gas the highest number of molecules have the average energy. Similarly the highest number of people have average height. This is with contradistinction to the example of 12 balls distributed in 3 boxes in which the highest probability is to find an empty box, similarly to the wealth distribution in which is it easier to find a poor man than to find a rich man.

2. ENTROPY, FREQUENCY AND DISTRIBUTION FOR INTEGER NUMBER OF BALLS

The model consists of identical indistinct balls and identical distinct boxes. The balls have no interaction between themselves and are distributed randomly among the boxes in a way that maximizes the entropy of the system under various constraints. We show how various statistical laws and distributions are derived using this model.

3. DENSE SYSTEMS

Here there are no inherent limitations to the number of the balls in a box. Namely, it can be any number up to the total number of balls or any limit that we artificially impose.

We consider two constraints: the first is the total number of balls P in the boxes, and the second is the total number of the boxes. When the balls have a property, the constraint on the number of balls is replaced by a constraint on the total amount of property.

When the balls are pure logical entities like numbers, we say that the balls have no chemical potential. A reader of a book or a seat in election has no chemical potential. When the balls consist of a measurable quantity, the balls have chemical potential. For example, balls of a given weight or a given amount of energy have chemical potential. Similarly, balls with assets, have chemical potential.

Suppose that we have P identical indistinct balls and L identical distinct boxes. The number of the distinguishable ways to arrange the balls in the boxes (microstates) is given by the number of the configurations as was first suggested by Planck,

$$W = \frac{(L+P-1)!}{(L-1)!P!} \tag{1}$$

We designate $n = P/L$ and apply the Stirling formula $\ln n! \cong n\ln n - n$, we obtain [11] an expression for the entropy S,

$$S = \ln W \cong L\left[(n+1)\ln(n+1) - n\ln n\right] \tag{2}$$

We designate the relative number of the boxes with n balls by $\varphi(n)$ and therefore the number the different groups of boxes having n balls is $N = L/\varphi(n)$, or,

$$L = N\varphi(n) \text{ and accordingly the number of balls is } P = Nn\varphi(n) \tag{3}$$

where L and P of Equation's (3) are the two constraints of the system. The left one is the constraint on the total number of boxes L, and the right one is the constraint on the total number of balls P.

Example 1 The Distribution of Balls in the Boxes: Zipf's Law and Planck-Benford Law

To write the Lagrange function we substitute the entropy from Equation (2) and add the constraints for P and L from Equation (3). The Lagrange equation $F(n)$ is,

$$F(n) \cong L\left[(n+1)\ln(n+1) - n\ln n\right] + \beta'\left[P - Nn\varphi(n)\right] + \alpha\left[L - N\varphi(n)\right] \quad (4)$$

where β' is a Lagrange multiplier of the constraint of the number of balls, and α is the Lagrange multiplier of the constraint of the number of boxes. The maximum entropy solution is obtained when

$\partial F(n)/\partial n = 0$.

Since $L\partial\left[(n+1)\ln(n+1) - n\ln n\right]/\partial n = L\ln\frac{n+1}{n}$

and $-\beta' N\partial\left[n\varphi(n)\right]\partial n = -\beta' N\varphi(n)$, and $\alpha N\partial\left[\varphi(n)\right]\partial n = 0$,

we obtain from Equation (4),

$$n = \frac{1}{\exp(\beta\varphi(n)) - 1} \quad (5)$$

where we designate $\beta' \equiv \beta L/N$.

Equation (5) is the probabilistic Planck equation. It should be noted here that L is canceled out in the derivation and we remain with knowledge only about the boxes with balls. Where $\beta\varphi(n) \square 1$, then $\exp(\beta\varphi(n)) \cong 1 + \beta\varphi(n)$ and we obtain, $n\varphi(n) \cong \beta^{-1}$,

which is Zipf's Law. From Equation (5) it is seen that Zipf's law is obtained when the number of balls in a box n is large. Zipf's law is known in its outcome that,

$$\frac{\varphi(1)}{\varphi(2)} = \frac{\varphi(2)}{\varphi(4)} = \frac{\varphi(4)}{\varphi(8)} = \cdots = 2$$
,

which was first found in texts in many languages [6] , namely, the most frequent word appears in long texts twice as many times as the second most frequent word, and the second most frequent word appears twice as many times as the fourth most frequent word and so forth.

Equation (5) can be written also in a different way, namely,

$$\varphi(n) = \beta^{-1} \ln\left(1 + \frac{1}{n}\right) \tag{6}$$

To obtain the normalized relative frequency of a box with n balls, we divide the frequency by the total number of boxes with balls,

$$\sum_{n=1}^{N} \varphi(n) = \beta^{-1} \sum_{n=1}^{N} \ln\left(1 + \frac{1}{n}\right) = \beta^{-1} \ln(N+1)$$

Which yields,

$$\phi(n, N) = \ln\left(1 + \frac{1}{n}\right) \Big/ \ln(N+1) \tag{7}$$

Choosing the value of N reflects the difficulties arises from the subjective nature of the entropy. If we go back to our example of 12 balls distributed in 3 boxes, N may be 3 or 12, depending on the question that we ask. If we want to know the distribution of particles in 3 boxes, than $N = 3$ and the equilibrium distribution of the balls between the boxes is 50%, 29%, 21%. However a box may have a probability to have any number of balls up to 12. In this case, when we substitute $N = 12$ in Equation (7), we calculate the probability of finding a given possible number of balls. For example, the probability of finding a box with 5 balls is $\ln(1.2)/\ln(13)$, which is about 7.1%. The confusion between the these two kind of probabilities is the reason why most people will conclude that if million balls distributed randomly in 3 boxes the best bet is to find 333,333 balls in a box, while the probability to find exactly 333,333 balls in a box is only about $2.17 \times 10^{-5}\%$.

We call Equation (7) Planck-Benford law [3] [20] . It is worth noting that in Equation (7) the Lagrange multipliers were cancelled out in the normalization. This makes this law universal, as it independent in the total number of balls and the total number of boxes. This is the reason why Benford's law and Zipf's law were discovered so many years ago. However, the Lagrange multiplier β is not cancelled out in the density distribution function - given by

$$\rho(n,\beta) = n\varphi(n) = \beta^{-1} n \ln\left(1+\frac{1}{n}\right) \tag{8}$$

In Figure 1, the density distribution function $\rho = n\varphi(n)$ of 20 balls and $\beta = 1$ that was calculated from Equation (8) is presented by the upper orange curve. As is seen, the plateau of Zipf's law in which $n\varphi(n) \approx 1$ appears pretty fast. The lower black curve is the normalized relative frequency of the boxes having n balls which were calculated from Equation (7). The boxes with a smaller number of balls have higher frequencies.

Zipf found his ratio of the frequencies of words in 1949 [6] [7] . Zipf attributed the phenomenon to psychological argumentations of efficiency. It is shown here that the reason is probability.

Example 2 Applications of Planck-Benford Law: Benford's Law

When we substitute $N = 9$ in Equation (7) we obtain Benford's law.

$$\phi(n) = \ln\left(1+\frac{1}{n}\right) \Big/ \ln 10 = \log\left(1+\frac{1}{n}\right) \tag{9}$$

Namely, Benford's law is the maximum entropy distribution when the boxes (digits) may have 1, 2, 3, 4, 5, 6, 7, 8 or 9 balls in a box, see Figure 2. The Planck-Benford distribution excludes the empty boxes (the digits zero) as was explained previously and this is the reason why Benford's law is attributed erroneously to the first digit which is never 0. It is worth noting that Zipf's law and Benford's law are about normalized frequencies and are not density distributions. Usually we are interested in the density distribution function $\rho(n,\beta)$ which is the distribution of the balls, namely the fraction of the boxes having n particles multiplied by n. The distribution function is plotted in the upper curve of Figure 1.

It should be noted that N is the number of the distinguishable boxes that have different number of balls. The number of the boxes in the ensemble which is the total number of digits in the file is not relevant. Benford's law

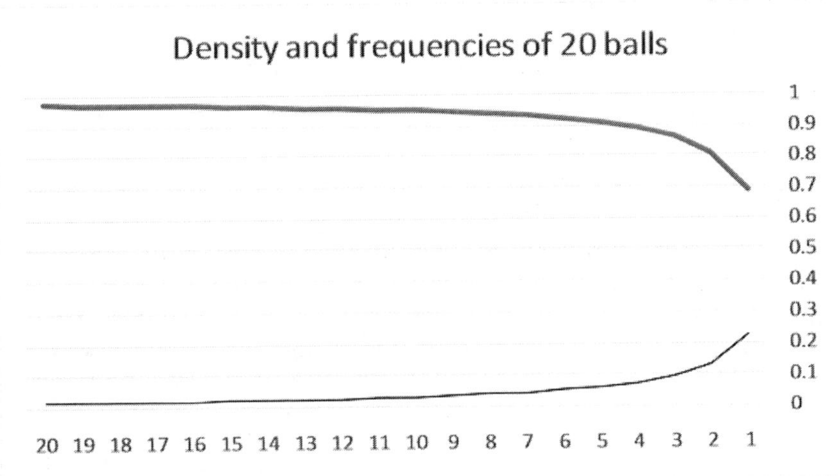

Figure 1. The distribution of 20 balls in 20 boxes. The X axis is the number of balls in a box and the Y axis is the probability/density. The lower black curve is the normalized relative probability of the boxes with n balls [Equation (7)]. It is seen that the probability to find a small number of balls in a box is higher for small numbers. The upper orange curve is the density distribution $\rho = n\varphi(n)$ and $\beta^{-1} = 1$, namely the probability to find a box with n balls. The plateau is Zipf's law.

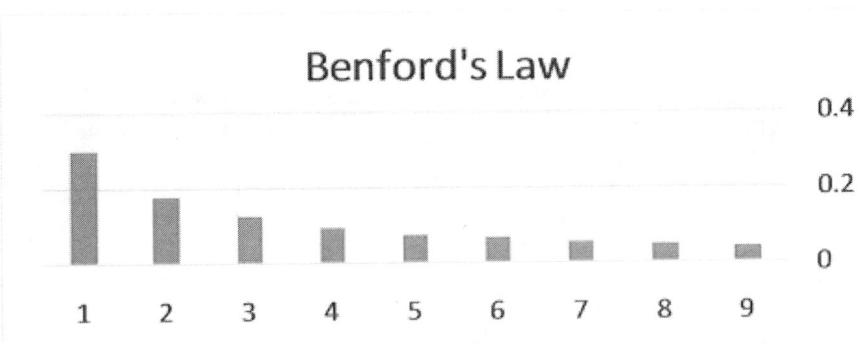

Figure 2. The normalized relative frequencies of boxes with n balls where $1 \le n \le 9$.

was found in the normalized relative frequencies of occurrences of the first digits of logarithmic tables by New-comb who also guessed empirically Equation (9) in 1881 [8] . Later it was discovered to be correct in many random decimal data by Benford in 1938 [9] .

Example 3 Applications of Planck-Benford Law: Prior Distributions of Polls

Equation (7) may be useful for predicting the prior distribution of polls. For example, in the Israeli parliament there were (eventually) 10 parties competing for 120 seats in the March 2015 election. We can use Equation (6) to find the equilibrium distribution of seats among the ten parties. This problem is different from Benford's law. Here, the number of the different boxes is equal to the number of boxes. In Benford's law, we ask what the normalized frequency of boxes having n balls. In polls we ask how the seats will be divided among the parties. Here we take $\phi(n)$ of Planck-Benford and divide the seats according to the relative normalized frequency of the parties. Therefore, the number of seats of the n's party is calculated by

$$M(n) = \text{Round}\left[120\frac{\ln\left(1+\frac{1}{n}\right)}{\ln(11)}\right].$$

Figure 3 shows that the distribution of the seats in the Israel parliament obeys Planck Benford's law, which shows that no major forgery was made. It is worth noting that no free parameter was used in this calculation.

Example 4 Planck Law and Planck-Einstein Assumption

We have derived the probabilistic Planck equation, Equation (5), for pure logical entities that we call balls. Now let us assume that each ball has an amount of energy h' , and that all the balls are moving at the same velocity c, where the velocity has units of boxes per second. In this case the spatial frequency of the appearance of the balls φ will be replaced by the temporal frequency $\nu = c\varphi$, and the energy of the system becomes $E = Lnh'$. We obtain from the number of balls $P = Nn\varphi(n)$ the energy $E = Nnh\nu$ where $h \equiv h'/c$. Now the number of balls constraint of Equation (4) becomes an energy constraint and the Lagrange equation is,

$$F(n) \cong L\left[(n+1)\ln(n+1) - n\ln n\right] + \beta'(E - Nnh\nu) + \alpha(L - N\nu/c) \quad (10)$$

which yields the famous Planck equation,

$$n = \frac{1}{\exp\left(\beta h v(n)\right) - 1} \tag{11}$$

The present probabilistic outcome of Planck-Einstein assumption that $E(n) = E/N = nh\nu$, which means that the higher the energy of a photon, the higher its frequency, is a direct outcome of the constraint on the number of balls of Equation (3), namely, $P = Nn\varphi(n)$. If we view a single oscillator as the group of all the boxes with the same number of balls n, then N is the number of the (longitudinal) modes of the radiation. In the Zipf's law regime all modes have the same energy, however, this is not true for $n < 1$, in the quantum limit. In the quantum limit, the occupation number is less than one and has an exponential probability decay like in ideal gas.

When the modes have the same energy, a smaller number of balls in a box means more energy per ball, and since boxes with a smaller number of balls appear more frequently, the frequency of the radiation is higher. It means that Planck's empirical assumption that the higher the frequency, the higher the photon energy, that was empirically invoked by him to obtain the Blackbody Radiation formula, namely $E(n=1) = h\nu$, is an outcome of Zipf's law. Planck's equation and his assumption were published by Planck in 1901 [11] .

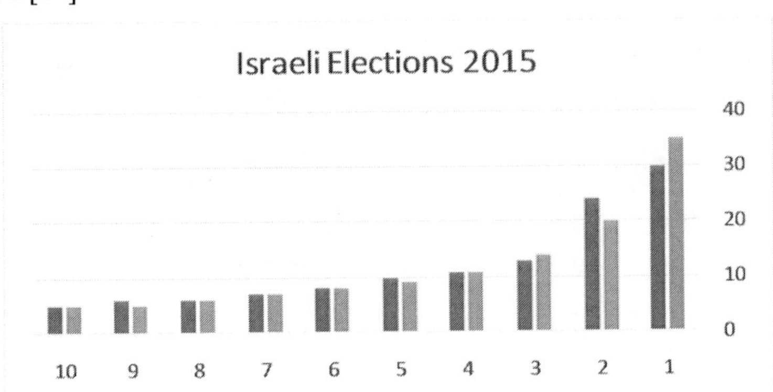

Figure 3. The distribution of 120 seats among 10 parties. The right blue bars are the Planck-Benford distribution and the left orange bars are the actual distribution of seats.

Example 5 Balls with Property: Bose-Einstein Distribution

Now let us assume that the balls are static and have a simple property ε. Now the constraint on the number of balls and the constraint on the energy are simply $P = Ln$ and $E = \varepsilon Ln$. The Lagrange equation becomes,

$$F(n) \cong L[(n+1)\ln(n+1) - n\ln n] + \beta(E - \varepsilon Ln) + \alpha(P - Ln) \tag{12}$$

We ask that $\partial F(n)/\partial n = 0$.

and obtain

$\ln(n+1) - \ln n - \beta\varepsilon - \alpha = 0$, which yields

$$n(\varepsilon,\alpha,\beta) = \frac{1}{\exp(\beta\varepsilon + \alpha) - 1}$$

and the property distribution function $\rho(\varepsilon) = \varepsilon n(\varepsilon)$ is,

$$\rho(\varepsilon,\alpha,\beta) = \frac{\varepsilon}{\exp(\beta\varepsilon + \alpha) - 1} \tag{13}$$

Equation (13) is the Bose-Einstein distribution. We saw that photons that are eventually pure energy behave like balls in boxes. However objects with a chemical potential in which the number of balls and their energies are independently constrained have chemical potential.

In Figure 4 we provide three plots of Equation (13), for $\rho(\varepsilon) = \varepsilon n(\beta = 1, \alpha = 0)$ which is a Planck function, and $\rho(\varepsilon) = \varepsilon n(\beta = 1, \alpha = 0.1)$ and $\rho(\varepsilon) = \varepsilon n(\beta = 1, \alpha = 1)$ which are Bose Einstein functions.

In the plots the functions are multiplied by an arbitrary number-10, namely the curves are $\rho(x) = \frac{10x}{\exp(x+\alpha) - 1}$ with $\alpha = 0$, which is the Planck equation, below we set $\alpha = 0.1$ (Bose Einstein), and the lowest is the blue with $\alpha = 1$.

The upper plot of Figure 1, the upper plot of Figure 4 and the plot of Blackbody radiation that appears in the textbooks look entirely different.

These differences are worth an explanation. In Figure 1 we plotted $\rho(n,\beta) = n\varphi(n)$. Namely the density of balls versus the number of balls in a box. We see that the density of the balls at high n's is independent of n. In Figure 4 the energy density $\rho(\varepsilon,\beta) = \varepsilon\varphi(\varepsilon)$ is plotted versus the energy for high energies $\varepsilon > 1$ which also means $n \gg 1$. In this region the density decays exponentially, but these energies are not shown in Figure 1 at all. At a very small photon energy where n is large, the density becomes, according to Equation (8), $1/\beta$ as $\lim \rho_{\varepsilon\to 0}(\varepsilon,\beta) = 1/\beta$. It should be noted that at high energies of the balls in which $n \leq 1$ all the distributions in nature decay exponentially and most of the balls have the average "property" value. However, this is not true for $n \gg 1$ in the Zipf's law regime.

In most textbook the energy flux of the radiation is plotted and therefore the density function is multiplied by the number of the radiation modes per a given volume. The number of modes N in a given volume is proportional to λ^{-3}, where λ is the wavelength of the radiation. Therefore N is proportional to ν^3 or ε^3. When we multiply the density $\rho(\varepsilon,\beta)$ by ε^3, namely $\varepsilon^4\varphi(\varepsilon)$, the bell-like blackbody radiation curve that in common in the literature is obtained.

In the lower graphs of Figure 4 we see that the chemical potential has a dramatic effect that causes the Planck monotonically decreasing function to be similar to Maxwell-Boltzmann function.

Bose-Einstein distribution was suggested by Bose in 1924 [12] and by Einstein in 1925 [13] .

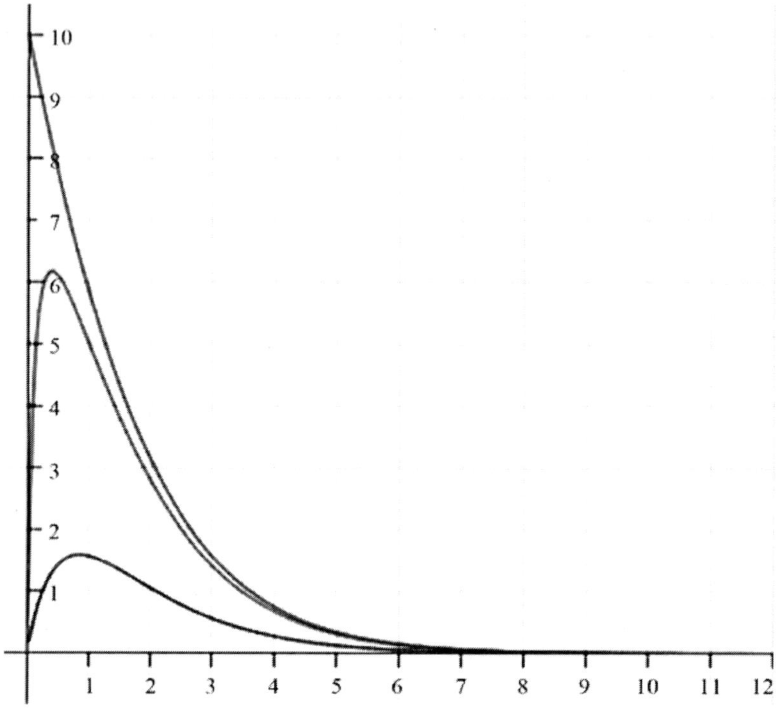

Figure 4. The upper red curve is the Planck distribution and the purple and blue below are the Bose-Einstein distributions with finite chemical potentials.

4. SPARSE SYSTEMS

Example 6 Fermi-Dirac and Maxwell-Boltzmann Distributions

Now we look at a system in which no more than one ball can exist in a box. The number of the configurations is given by

$W = L!/(L-P)!P!$ and $S = \ln W \cong -L\left[n\ln n + (1-n)\ln(1-n)\right]$

with $n < 1$. Equation (12) becomes,

$$F(n) \cong -L\left[n\ln n + (1-n)\ln(1-n)\right] + \beta(E - \varepsilon L n) + \alpha(P - Ln) \tag{14}$$

We ask that $\partial F(n)/\partial n = 0$.

We obtain $\ln\frac{1-n}{n}=\beta\varepsilon+\alpha$ or, $n=\frac{1}{\exp(\beta\varepsilon+\alpha)+1}$ at the limit where $\beta\varepsilon$ is large, one obtains that $n\propto \mathrm{e}^{-\beta\varepsilon}$ which is called the canonical distribution. The density distribution function is given by $\rho=\varepsilon n$ namely,

$$\rho(\varepsilon,\alpha,\beta)=\frac{\varepsilon}{\exp(\beta\varepsilon+\alpha)+1} \tag{15}$$

When $\alpha=0$ and $\beta\varepsilon$ is high, Equation (15) is called the Maxwell-Boltzmann distribution and when $\alpha\neq 0$ it is called Fermi-Dirac distribution. In the examples below we chose the values

$\rho(x)=10x\exp(-x)$, $\rho(x)=\frac{10x}{\exp(x)+1}$ and $\rho(x)=\frac{10x}{\exp(x+1)+1}$.

We see in Figure 5 that the three curves are very similar and can easily be confused.

Maxwell-Boltzmann Distribution was first suggested by Maxwell in 1860 [14] . Fermi-Dirac distribution was suggested by Fermi and Dirac in 1926 [15] [16] .

5. DISCUSSION

The formalism of maximizing the entropy using Lagrange multipliers technique is widely used in statistical mechanics to find thermal equilibrium energy distributions i.e. [4] . Ensemble of a fixed amount of particles, energy and volume is called a microcanonical ensemble. Here we strip the energy and volume off their physical identities and remain only with balls and boxes obeying the statistical second law, namely, equal probabilities to the boxes and equal probabilities to all the configurations of the balls in the boxes. An equal probability for any box yields an equal distribution of particles in the boxes. However, an equal distribution is not a maximum entropy solution. In nature, a statistical system tends to reach the equilibrium distribution in which entropy is maximum. To maximize the entropy one needs to add an equal probability requirement to the microstates.

Zipf's law was found by Zipf in the frequency of words in texts in many languages. Benford's law which is known also as the "first digit law" was found in the first digit of logarithmic tables (Benford's law is true for any digit in random decimal file regardless of its location). Many scientific papers were published about these distributions which are, in fact, merely different outcomes of the old Planck law. Benford's law is used for fraud detection in tax reports [21] . Similarly, it is possible to use Planck-Benford law described in this article for fraud detection in elections and surveys in addition to its predictive value. Moreover, if we have different companies competing in the same market we can see if there is monopolistic power that distorts the equilibrium distribution obtained by Planck-Benford law. If this theory would be applied in economics, each corporate would be able to calculate the equilibrium distribution of salaries among its employees. If the compensation distribution fits equilibrium, probably the employees will accept the salaries' gaps. If the salary distributions fits equilibrium, than β of the company's salaries distribution can be calculated and be compared to other companies as well as to β of the whole country salary distribution. In the present probability theory β replaces the reciprocal value of the average. In a recent paper [22] a network was described as a microcanonical ensemble. In a network, the boxes are the pairs of sites and the balls are the number of links in the pair. It was shown that high number of balls (links) in equilibrium is equivalent to high temperature ensemble. It was found that moving a ball/link from a hot net to a cold one increases entropy. However, moving a site which reduces the number of boxes from a hot net to a cold net decreases entropy [22] [23] . These results explain the tendency of sites/people to emigrate from poor countries to the rich ones as is seen in the immigration waves from Africa to Europe and from Mexico to USA. Similarly this explains the tendency of links/money to flow from rich countries to poor ones as seen in the flow of money from USA to China and from Germany to Greece. These processes are an outcome of the present statistical theory.

A surprising result of this theory is that by adding a property to the balls we obtain a chemical potential. In physics, chemical potential is known as a change in the potential energy due to chemical reaction or phase transition. In fact chemical potential is an outcome of the constraint on the number of boxes. Increasing α means increasing the number of boxes, or decreasing the average number of balls per box. Decreasing the average number of balls brings the system to be more similar to the canonical distribution of Maxwell-Boltzmann that we are so accustomed to. In Figure 4 and Figure 5 we see that in Bose-Einstein distribution the maximum of the functions is at α , which means that the average property of the balls are determined

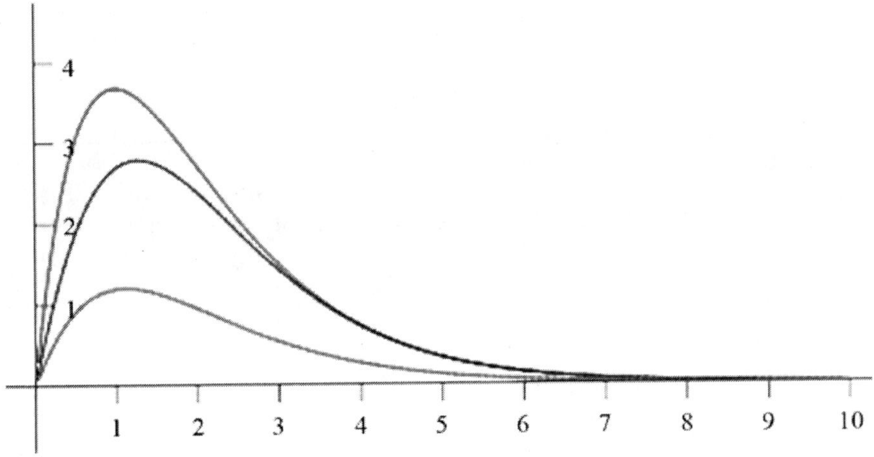

Figure 5. The upper red distribution is the Maxwell-Boltzmann, the blue distribution below is the Fermi-Dirac without chemical potential, and the purple distribution is the Fermi-Dirac with a chemical potential. All these distributions look similar.

by the number of the boxes. In Figure 5 it is seen that in sparse system most of the particles are found in the average which is determine by β. When the property is energy, α is potential energy. However, when the property is money, we have a similar "potential money" that may enable economists to treat financial processes and financial phase transitions with the proven arsenal of the physicists.

It should be mentioned that Planck made historic assumption [11] that the energy of a photon is quantized; namely the energy of a radiation mode is $E(n) = nh\nu$. Here this equation was obtained by assuming that each ball carries an amount of energy h' and the balls are moving at the same velocity c. This paper shows that for small number of ball in a box, which means more energy per ball, according to Zipf's law the boxes have higher frequency. This suggest a statistical explanation to Planck's assumption that the energy of a photon is proportional to its frequency.

It is important to note that in all the derivations presented in this paper we use Stirling formula. The Stirling formula yields inaccurate values for small numbers. In the example of 12 balls distributed randomly in 3 boxes that was presented previously, the exact combinatorial calculation shows that the empty box has the highest probability (14%), the second most probable is a box with one ball (13%), and the least expected is a box with 12 balls (1%). However, using Planck-Benford law we obtain 27% for box with 1 ball, 3% for box with 12 balls and no value for empty boxes. The reason for this discrepancy is that we have not constrained the number of boxes in the entropy maximization process. The Planck-Benford law is independent of the number of boxes. If we use combinatorial calculation for 12 balls in 3 boxes, or in 4 boxes or in 100 boxes we will obtain different values for any number of boxes. Planck-Benford law is universal in this aspect, however we give up the knowledge of the probability of empty boxes and accurate values for small number of boxes and balls.

6. SUMMARY

We present a simple and effective way to calculate many real-life probabilities and distributions such as votes in election polls, market share distributions among competing companies, wealth distributions among populations and many others. By superseding classic probability theory with the maximum entropy principle we are able to calculate these complex distributions easily, and empirical real-life results strongly corroborate this novel approach. In addition, it is shown here that this new approach yields the well-known distributions of Zipf's Law (with Pareto 80:20 law being its outcome), Benford's Law (which is the unequal distribution of digits in random data files), as well as energy distributions among particles in physics. This paper unifies everyday probability calculations done by surveyors, gamblers and economists with the calculations done by physicists, therefore enabling the applications of the methodology of statistical physics in economics and the life sciences.

ACKNOWLEDGEMENTS

We want to thank H. Kafri and Y. Kafri for many comments. Special thanks to Alex E. Kossovsky for many comments and especially for the combinatorial calculations of balls and boxes.

CITE THIS PAPER

Oded Kafri (2016) A Novel Approach to Probability. *Advances in Pure Mathematics*, **06**, 201-211.

REFERENCES

1. 1. Stark, P.B. The Law of Large Numbers. http://stat-www. berkeley .edu/~stark/Java/lln.htm
2. Stein, C. (1972) A Bound for the Error in the Normal Approximation to the Distribution of a Sum of Dependent Random Variables. Proceedings of the Sixth Berkeley Symposium on Mathematical Statistics and Probability, 583-602.
3. See for Example Kafri, O. and Kafri, H. (2013) Entropy—God's Dice Game. Create Space, 208-210. http://www.entropy-book.com/
4. Tribus, M. and Levine, R.D., Eds. (1979) The Maximum Entropy Formalism. MIT Press, Cambridge.
5. Shannon, C.E. (1948) A Mathematical Theory of Communication. Bell System Technical Journal, 27, 379-423, 623-656.
6. Zipf, G.K. (1949) Human Behavior and the Principle of Least-Effort. Addison-Wesley, Boston.
7. Gabarix, X. (1999) Zipf's Law for Cities: An Explanation. The Quarterly Journal of Economics, 114, 739-767.
8. Newcomb, S. (1881) Note on the Frequency of Use of the Different Digits in Natural Numbers. American Journal of Mathematics, 4, 39-401.
9. Benford, F. (1938) The Law of Anomalous Numbers. Proceedings of the American Mathematical Society, 78, 551-572.
10. Kossovsky, A.E. (2015) Benford's Law. World Scientific Publishing, Singapore.
11. Planck, M. (1901) On the Law of Distribution of Energy in the Normal Spectrum. Annalen der Physik, 4, 553-562.
12. Bose, S.N. (1924) Plancks Gesetz und Lichtquantenhypothese. Zeitschrift für Physik, 26, 178.
13. Einstein, A. (1925) Quantentheorie des einatomigen idealen Gases. Sitzungsberichte der Preussischen Akademie der Wissenschaften, 1, 3.
14. Maxwell, J.C. (1860) Illustrations of the Dynamical Theory of Gases. Part I. On the Motions and Collisions of Perfectly Elastic Spheres. Philosophical Magazine Series 4, 19, 19-32.
15. Fermi, E. (1926) Sulla quantizzazione del gas perfetto monoatomico. Rendiconti Lincei, 3, 145-149. (In Italian)
16. Dirac, P.A.M. (1926) On the Theory of Quantum Mechanics. Proceedings of the Royal Society, Series A, 112, 661-677.
17. Ullah, A. (1996) Entropy, Divergence and Distance Measures with Econometric Applications. Journal of Statistical Planning and Inference, 49, 141.
18. http://en.wikipedia.org/wiki/Maximum_entropy_probability_distribution
19. Back, C. (2009) Generalize Information and Entropy Measures in Physics. Contemporary Physics, 50, 495-510.
20. Kafri, O. (2009) The Distributions in Nature and Entropy Principle.
21. Nigrini, M. (1996) A Taxpayer Compliance Application of Benford's Law. The Journal of the American Taxation Association, 18, 72-91.
22. Kafri, O. (2014) Money, Information and Heat in Networks Dynamics. Mathematical Finance Letters, 2014, Article ID: 4.
23. Kafri, O. (2014) Follow the Multitude—A Thermodynamic Approach. Natural Science, 6, 528-531.

CHAPTER 2

Strong Law of Large Numbers under an Upper Probability

Xiaoyan Chen

Graduate Department of Financial Engineering, Ajou University, Suwon, South Korea

ABSTRACT

Strong law of large numbers is a fundamental theory in probability and statistics. When the measure tool is nonadditive, this law is very different from additive case. In 2010 Chen investigated the strong law of large numbers under upper probability V by assuming V is continuous. This assumption is very strong. Upper probabilities may not be continuous. In this paper we prove the strong law of large numbers for an upper probability without the continuity assumption whereby random variables are quasi-continuous and the upper probability is generated by a weakly compact family of probabilities on a complete and separable metric sample space.

Keywords: Strong Law of Large Numbers; Upper Probability; Weakly Compact; Independence; Quasi-Continuous

1. INTRODUCTION

Strong law of large numbers under nonadditive probabilities is a much important theory in uncertainty theories and has more applications in statistics, risk measures, asset pricings and many other fields. In 1999 Marinacci [1] first investigated the strong law of large numbers for sequences of independent and identically distributed (IID for short) random variables $\{X_i\}_{i=1}^{\infty}$ relative to a capacity v which is continuous and totally monotone and proved that under regularity condition the limit inferior and limit superior of $\frac{S_n}{n}$, where $S_n := \sum_{i=1}^{n} X_i$, lie between the two Choquet integrals $\int X_1 \mathrm{d}v$ (submean) and $-\int -X_1 \mathrm{d}v$ (supermean) induced by this capacity with probability 1 under v (that is, quasi surely), and furthermore, if v is null-additive, then that limit inferior attains the submean and the limit superior attains the supermean quasi surely, respectively. This is different from the law under probability measure P whereby under suitable conditions, such as for IID sequences, $\frac{S_n}{n}$ converges to the mathematical expectation of X_1 almost surely relative to P. In 2005 Maccheroni and Marinacci [2] extended the results of Marinacci [1] for a totally monotone capacity on Polish space whereby the bounded variables X_i are continuous or simple, or the capacity v is continuous. But the conditions of these two articles on capacity are too strong and not easy to test. And generally capacities can not uniquely determine the (nonlinear) expectations relative to the capacities. Motivated by robust statistics and limit theories under sublinear expectations given by Peng in 2007, Chen [3] in 2010 investigated the strong law of large numbers for a pair of lower and upper probabilities (v, V) which are induced by a sublinear expectation $\mathbb{E}$ (see Peng (2012) [4]) whereby the sequence $\{X_i\}_{i=1}^{\infty}$ is IID under $\mathbb{E}$ (the independence is different from classical case and the one in Marinacci [1], see Peng (2012) [4]). He proved if

$\mathbb{E}\left[|X_1|^{1+\alpha}\right]<\infty$ for some $\alpha>0$, then those limit inferior and limit superior lie between another submean $-\mathbb{E}[-X_1]$ and supermean $\mathbb{E}[X_1]$ which may not equal the ones given by Choquet integrals (see Chen, Wu and Li (2012) [5] for details). Furthermore, if we futher assume that V is continuous, then

$$V\left(\limsup_{n\to\infty}\frac{S_n}{n}=\mathbb{E}[X_1]\right)$$
$$=V\left(\liminf_{n\to\infty}\frac{S_n}{n}=-\mathbb{E}[-X_1]\right)=1 \quad (1)$$

Hu (2012) [6] extends the results of Chen [3] to the sequence of non-identically distributed random variables for the same independence and continuity assumptions. Chen and Wu (2011) [7] extends Chen [3] to more weaker independence condition without identical distribution assumption, and proves if we further assume that

V is continuous, for any subsequence $\{n_i\}_{i=1}^{\infty}$ of $\mathbb{N}$, $\{S_{n_i}-S_{n_{i-1}}\}_{i=1}^{\infty}$ are pairwise weakly independent under

$\mathcal{V}$, and there exist some constants $\alpha>0,\ n_0\geq 1,$ and $c_0>0$ such that

$\sup_{i\geq 1}\hat{E}\left[|X_i|^{1+\alpha}\right]<\infty, \frac{|S_n|}{n}\leq c_0\ln(n+1), \forall n\geq 1,$ then (1) still holds.

We can see that for the strong law of large numbers (1) under an upper probability, there are two key conditions: well-defined independence and continuity of the upper probability V. The continuity assumption of V is based on the second Borel-Cantelli lemma to get $V(A_n, i.o.)=1$ for certain sequence of measurable events A_n. But as we know in general V is not continuous, since for nonclosed nonincreasing sequence of measurable events G_n, even if $G_n\downarrow G, V(G_n)\downarrow V(G)$ may not hold (see Xu and Zhang (2010) [8] for an example). Hence a natural question is: if V is not continuous, whether does (1) hold? In this paper we will give a confirmative answer. We assume a1) Ω is a complete and separable metric space, F is a σ -algebra of all Borel subsets of Ω, P is a nonempty subset of $\mathcal{M}$ which is a family of all probabilities on (Ω, F), and P is also weakly compact;

a2) For each $i \geq 1,\ X_i$ is quasi-continuous, and

$$\sup_{i\geq 1} \mathbb{E}\left[|X_i|^2\right] < \infty;$$

a3) $\{X_n\}_{n=1}^{\infty}$ is independent sequence of random variables under $\mathbb{E}$ where $\mathbb{E}$ is a sublinear expectation corresponding to P. In this paper we successfully proved the strong law of large numbers under assumptions a1)-a3) without the continuity assumption of V by transforming that an event $A \in F$ occures with probability 1 under V (that is, $V(A)=1$) to the problem of its complementary event A^c, i.e., $v(A^c)=1-V(A)=0$, where v is the conjugate lower probability of V and proving $v(B_n, i.o.)=0$ for some appropriate events B_n with $\limsup_{n\to\infty} B_n \supset A^c$ by using properties of V and v.

This paper is organized as follows. In Section 2 we give some basic concepts and useful lemmas. In Section 3 we mainly prove the strong law of large numbers without continuity assumption of upper probability V for IID and continuous sequences. Section 4 extends results of Section 3 and gets the law for non-identically distributed sequence. Section 5 gives an example.

2. PRELIMINARIES

Let Ω be a separable and complete metric space. F is a σ-algebra of all Borel subsets of Ω. We introduce an upper probability V by

$$V(A) = \sup_{Q\in P} Q(A), \forall A \in F,$$

where P is a family of probabilities on (Ω, F) and weakly compact. Thus its conjugate capacity (see Choquet (1954) [9]), i.e., lower probability is

$$v(A)=\inf_{Q\in P}Q(A)=1-V\left(A^c\right),\forall A\in F,$$

where A^c is the complementary set of A. From Huber and Strassen (1973) [10] V and v also satisfy the following properties.

Proposition 1.

1) $V(\Omega)=v(\Omega)=1, V(\varnothing)=v(\varnothing)=0.$

2) $V(A)\le V(B), v(A)\le v(B)$ if $A\subset B, A, B\in F.$

3) $V\left(\sum_{n=1}^{\infty}A_n\right)\le\sum_{n=1}^{\infty}V(A_n), \forall A_n\in F, n\ge 1.$

4) $V(A)\ge v(A)$ for all $A\in F.$

5) lower-continuity of V for all sets in F: if $A_n\uparrow A$, $A_n\in F$, then $V(A_n)\uparrow V(A).$

6) upper-continuity of V for all closed sets: if $F_n\in F$ closed, $F_n\downarrow F$, then $V(F_n)\downarrow V(F).$

7) lower-continuity of v for all open sets: if $G_n\uparrow G, G_n\in F$ open, then $v(G_n)\uparrow v(G).$

8) upper-continuity of v for all sets: if $B_n\in F, B_n\downarrow B$, then $v(B_n)\downarrow v(B).$

Now we introduce an upper expectation $\mathbb{E}$ by V in the following

$$\mathbb{E}[X]=\sup_{Q\in P}E_Q[X]$$

for all $X\in F$ such that $\sup_{Q\in P}E_Q[X]<\infty$, E_Q is the linear expectation corresponding to $Q\in P$ such that $E_Q[I_A]=Q(A), \forall A\in F.$

Then $\mathbb{E}$ is a sublinear expectation (see Peng [4]) on (Ω, H), where H is a set of all real-valued random variables $X \in F$ such that $\mathbb{E}[X] < \infty$, that is, $\mathbb{E}$ satisfies that for all $X, Y \in H$

1) Monotonicity: $\mathbb{E}[X] \leq \mathbb{E}[Y]$ if $X \leq Y$.

2) Constant preserving: $\mathbb{E}[c] = c, \forall c \in \mathbb{R}$.

3) Sub-additivity: $\mathbb{E}[X+Y] \leq \mathbb{E}[X] + \mathbb{E}[Y]$.

4) Positive homogeneity: $\mathbb{E}[\lambda X] = \lambda \mathbb{E}[X], \forall \lambda \geq 0$.

$(\Omega, H, \mathbb{E})$ is called a sublinear expectation space in contrast with probability space. Given $X = (X_1, X_2, \cdots, X_n)$, we say $X \in H^n$, if $X_i \in H$ for all $i = 1, 2, \cdots, n$. For $X \in H$, $\mathbb{E}[X]$ is called its supermean, whereas $-\mathbb{E}[-X]$ is called its submean. If $\mathbb{E}[X] \neq -\mathbb{E}[-X]$, then X is said to have mean uncertainty.

In the following we introduce some useful concepts (one can refer to Peng (2010) [4] for details).

Definition 2. An n-dimensional random vector $Y \in H^n$ is said to be independent from an m-dimensional random vector $X \in H^m$ under $\mathbb{E}$, if for all bounded Lipschitz continuous functions $\varphi \in C_{b,Lip}(\mathbb{R}^{m+n})$, we have

$$\mathbb{E}[\varphi(X,Y)] = \mathbb{E}\left[\mathbb{E}[\varphi(x,Y)]_{x=X}\right],$$

where $\mathbb{E}[\varphi(x,Y)]_{x=X} \in H^m$.

Remark 3. In general Y being independent of X under $\mathbb{E}$ does not imply X being independent of Y. See Example 3.13 of Chapter I in Peng (2012) [4] as a counterexample.

Definition 4. A sequence $\{X_n\}_{n=1}^{\infty}$ on $(\Omega, H, \mathbb{E})$ is said to be a sequence of independent random variables under $\mathbb{E}$, if for any $i \geq 1,\ X_{i+1}$ is independent of $(X_1, X_2, \cdots, X_i)$ under $\mathbb{E}$

Definition 5. A real random variable $X \in F$ is said to be quasi-continuous (q.c. for short) if for any $\varepsilon > 0,$ there exists an open set O with $V(O) < \varepsilon$ such that X is continuous on O^c.

Lemma 6. (Denis-Hu-Peng (2011) [11] Theorem 2) For any $A \in F,$

$$V(A) = \sup\{V(K) : \text{compact } K \subset A\}.$$

Remark 7. Lemma 6 implies that for any $A \in F,$

$$v(A) = \inf\{v(G) : \text{open } G \supset A\}.$$

The following Borel-Cantelli lemma is obvious (the readers also can refer to Peng [4] or Chen [3]).

Lemma 8 (Borel-Cantelli Lemma). For any sequence of events $\{A_n\}_{n=1}^{\infty}$ in F, if $\sum_{n=1}^{\infty} V(A_n) < \infty,$ then $V(A_n, i.o.) = V\left(\bigcap_{k=1}^{\infty} \bigcup_{n=k}^{\infty} A_n\right) = 0.$

Lemma 9 (Hu [6] Theorem 3.1). Let $\{X_n\}_{n=1}^{\infty}$ be a sequence of independent random variables on

$(\Omega, H, \mathbb{E})$. We assume 1) For any $i \geq 1,$ there exist real constants $\underline{\mu}_i \leq \overline{\mu}_i,$

such that $\mathbb{E}[X_i] = \overline{\mu}_i$ and $-\mathbb{E}[-X_i] = \underline{\mu}_i;$

2) There exist two real constants $\underline{\mu} \leq \overline{\mu}$ such that

$$\lim_{n\to\infty} \frac{1}{n} \sum_{i=1}^{n} |\underline{\mu}_i - \underline{\mu}| = 0, \lim_{n\to\infty} \frac{1}{n} \sum_{i=1}^{n} |\overline{\mu}_i - \overline{\mu}| = 0;$$

3) $\sup_{n\ge1}\mathbb{E}\left[X_n^2\right]<\infty.$

Set $S_n=\sum_{i=1}^{n}X_i, \forall n\ge 1.$ Then for any continuous function φ with linear growth on $\mathbb{R}$, we have

$$\lim_{n\to\infty}\mathbb{E}\left[\varphi\left(\frac{S_n}{n}\right)\right]=\sup_{\underline{\mu}\le y\le\bar{\mu}}\varphi(y).$$

Lemma 10 (Hu [6] Theorem 3.2 (I)). Let $\{X_n\}_{n=1}^{\infty}$ satisfy all the conditions given in Lemma 9, then

$$V\left(\limsup_{n\to\infty}\frac{S_n}{n}>\bar{\mu}\right)=V\left(\liminf_{n\to\infty}\frac{S_n}{n}<\underline{\mu}\right)=0.$$

3. STRONG LAW OF LARGE NUMBERS

Theorem 11. Let $\{X_n\}_{n=1}^{\infty}$ be an independent and continuous sequence under $\mathbb{E}$. We assume there exist two real constants $\underline{\mu}\le\bar{\mu}$ such that $\mathbb{E}[X_i]=\bar{\mu},\ -\mathbb{E}[-X_i]=\underline{\mu}$ for all $i\ge 1,$ and $\sup_{n\ge1}\mathbb{E}\left[|X_n|^2\right]<\infty.$

Set $S_n=\sum_{i=1}^{n}X_i, \forall n\ge 1.$ Then

$$V\left(\limsup_{n\to\infty}\frac{S_n}{n}=\bar{\mu}\right)=1, \tag{2}$$

$$V\left(\liminf_{n\to\infty}\frac{S_n}{n}=\underline{\mu}\right)=1. \tag{3}$$

Proof. It is obvious that we only need to prove one of the Equations (2) and (3), since on Ω

$$\limsup_{n\to\infty}\frac{S_n}{n}=-\liminf_{n\to\infty}\frac{-S_n}{n}.$$

In the following we will prove the Equation (2). It is trivial for $\overline{\mu}=\underline{\mu}$ and this theorem obviously holds true in this case from (I) of Theorem 1.1 of Chen and Wu [7] or Lemma 10. Hence we only need to consider $\overline{\mu}>\underline{\mu}$ By Lemma 10 or (I) of Theorem 1.1 of Chen and Wu [7], we have

$$V\left(\limsup_{n\to\infty}\frac{S_n}{n}>\overline{\mu}\right)=0.$$

Hence we only need to prove

$$V\left(\limsup_{n\to\infty}\frac{S_n}{n}\geq\overline{\mu}\right)=1.$$

For any subsequence $\{n_k\}_{k=1}^{\infty}$ of $\mathbb{N}$, we denote

$$A(a)=\left\{\omega\in\Omega:\limsup_{k\to\infty}\frac{S_{n_k}}{n_k}\geq a\right\},$$

$$A_{n_k}(a)=\left\{\omega\in\Omega:\frac{S_{n_k}}{n_k}\geq a\right\},\forall k\geq 1,a\in\mathbb{R}.$$

Since $\{X_n\}_{n=1}^{\infty}$ is a sequence of continuous random variables, thus $A(a)$ and $A_{n_k}(a)$ are both closed sets in F for all $a\in\mathbb{R}$ and $k\geq 1.$ Thus $A_{n_k}^c(a)$ is an open set in F for any $k\geq 1$ and $a\in\mathbb{R}.$ Then by the upper-continuity of V (see Proposition 1 (6)) for closed sets in F, we only need to prove for any fixed constant $\varepsilon\in\left(0,\overline{\mu}-\underline{\mu}\right),$

$$V\left(\limsup_{n\to\infty}\frac{S_n}{n}\geq\overline{\mu}-\varepsilon\right)=1.$$

Equivalently, we only need to prove for any fixed

$$\varepsilon \in \left(0, \overline{\mu} - \underline{\mu}\right),$$

$$v\left(\limsup_{n\to\infty} \frac{S_n}{n} < \overline{\mu} - \varepsilon\right) = 0.$$

Then it is sufficient to find an increasing subsequence $\{n_k\}_{k=1}^{\infty}$ of $\mathbb{N}$ such that for any fixed $\varepsilon \in \left(0, \overline{\mu} - \underline{\mu}\right)$,

$$v\left(\limsup_{k\to\infty} \frac{S_{n_k}}{n_k} < \overline{\mu} - \varepsilon\right) = 0. \tag{4}$$

Noticing that

$$\begin{aligned}
&v\left(\limsup_{k\to\infty} \frac{S_{n_k}}{n_k} < \overline{\mu} - \varepsilon\right) \\
&\leq v\left(\liminf_{k\to\infty} \frac{S_{n_k}}{n_k} < \overline{\mu} - \varepsilon\right) \\
&= v\left(\bigcap_{m=1}^{\infty} \bigcup_{k=m}^{\infty} A_{n_k}^c\left(\overline{\mu} - \varepsilon\right)\right) \\
&= \lim_{m\to\infty} v\left(\bigcup_{k=m}^{\infty} A_{n_k}^c\left(\overline{\mu} - \varepsilon\right)\right) \\
&= \lim_{m\to\infty}\left[1 - V\left(\bigcap_{k=m}^{\infty} A_{n_k}\left(\overline{\mu} - \varepsilon\right)\right)\right] \\
&= 1 - \lim_{m\to\infty} V\left(\bigcap_{k=m}^{\infty} A_{n_k}\left(\overline{\mu} - \varepsilon\right)\right) \\
&= 1 - \lim_{m\to\infty}\lim_{l\to\infty} V\left(\bigcap_{k=m}^{l} A_{n_k}\left(\overline{\mu} - \varepsilon\right)\right),
\end{aligned} \tag{5}$$

since $A_{n_k}\left(\overline{\mu} - \varepsilon\right)$ are all closed sets and V is uppercontinuous for closed sets.

In addition, for any $m \geq 2, l \geq m$,

$$
\begin{aligned}
&V\left(\bigcap_{k=m}^{l} A_{n_k}(\bar{\mu}-\varepsilon)\right)\\
&\geq V\left(\bigcap_{k=m}^{l}\left[\{Y_k \geq \bar{\mu}-\varepsilon\}\cap\left\{\frac{S_{n_{k-1}}}{n_{k-1}} \geq \bar{\mu}-\varepsilon\right\}\right]\right)\\
&= V\left(\bigcap_{k=m}^{l}\{Y_k \geq \bar{\mu}-\varepsilon\}\cap\left\{\frac{S_{n_{m-1}}}{n_{m-1}} \geq \bar{\mu}-\varepsilon\right\}\right)\\
&\geq \mathbb{E}\left[\prod_{k=m}^{l} g^{\delta}(Y_k)\, g^{\delta}\left(\frac{S_{n_{m-1}}}{n_{m-1}}\right)\right],
\end{aligned}
$$

where $Y_k = \frac{S_{n_k} - S_{n_{k-1}}}{n_k - n_{k-1}}$ and

$$
g^{\delta}(x) = \begin{cases} 1, & x \geq \bar{\mu}-\varepsilon+\delta \\ \frac{1}{\delta}(x-\bar{\mu}+\varepsilon), & \bar{\mu}-\varepsilon < x < \bar{\mu}-\varepsilon+\delta; \\ 0, & x \leq \bar{\mu}-\varepsilon, \end{cases}
$$

where δ is any fixed constant in $(0,\varepsilon)$. It is obvious that for any fixed $\delta \in (0,\varepsilon)$, g^{δ} is a bounded and Lipschitz continuous function on $\mathbb{R}$. Thus by the independence assumption we know that

$$
\begin{aligned}
&V\left(\bigcap_{k=m}^{l} A_{n_k}(\bar{\mu}-\varepsilon)\right)\\
&\geq \prod_{k=m}^{l} \mathbb{E}\left[g^{\delta}\left(\frac{S_{n_k}-S_{n_{k-1}}}{n_k-n_{k-1}}\right)\right]\mathbb{E}\left[g^{\delta}\left(\frac{S_{n_{m-1}}}{n_{m-1}}\right)\right]
\end{aligned}
$$

And then by Lemma 9 for any fixed $m \geq 2$, and $l \geq m$, if we choose a small constant $\tau = \frac{1}{(l+1-m)m}$, then there exists an integer $N = N(\delta,\varepsilon,l,m) \geq 1$, such that for any $n_k - n_{k-1} \geq N$, we have

$$\mathbb{E}\left[g^{\delta}\left(\frac{\bar{S}_{n_k-n_{k-1}}}{n_k-n_{k-1}}\right)\right] \geq \sup_{\mu\leq y\leq\bar{\mu}} g^{\delta}(y)-\tau=1-\tau,$$
$$m\leq k\leq l, \tag{6}$$

where we denote $\bar{S}_{n_k-n_{k-1}} := S_{n_k}-S_{n_{k-1}}$ for all fixed k with $m\leq k\leq l.$

Taking $n_k=N^k$ for any $m\leq k\leq l,$ we can obtain

$$V\left(\bigcap_{k=m}^{l} A_{n_k}(\bar{\mu}-\varepsilon)\right)\geq(1-\tau)^{l+1-m}\,\mathbb{E}\left[g^{\delta}\left(\frac{S_{n_{m-1}}}{n_{m-1}}\right)\right]. \tag{7}$$

Then letting l tend to ∞ and then letting m tend to ∞ on both sides of inequality (7), by Lemma 9 again, we can get

$$\begin{aligned}&\lim_{m\to\infty}\lim_{l\to\infty}V\left(\bigcap_{k=m}^{l} A_{n_k}(\bar{\mu}-\varepsilon)\right)\\&\geq\lim_{m\to\infty}\lim_{\tau\to 0^+}(1-\tau)^{\frac{1}{\tau m}}\,\mathbb{E}\left[g^{\delta}\left(\frac{S_{n_{m-1}}}{n_{m-1}}\right)\right]\\&=\lim_{m\to\infty}\mathbb{E}\left[g^{\delta}\left(\frac{S_{n_{m-1}}}{n_{m-1}}\right)\right]\lim_{m\to\infty}\mathrm{e}^{-\frac{1}{m}}=1\end{aligned} \tag{8}$$

Thus from (5) and (8) we can obtain

$$v\left(\limsup_{k\to\infty}\frac{S_{n_k}}{n_k}<\bar{\mu}-\varepsilon\right)\leq 0.$$

Therefore, (4) holds true. We complete the whole proof. □

Remark 12. If $\underline{\mu}=\bar{\mu}=:\mu,$ then from the Theorem 11 we can see that $V\left(\lim_{n\to\infty}\frac{S_n}{n}=\mu\right)=1.$ This is just a trivial case for sequences without mean uncertainty.

Corollary 13. Let $\{X_n\}_{n=1}^{\infty}$ be a sequence of quasicontinuous random variables and satisfy all other conditions except for the continuity given in Theorem 11, then Theorem 11 still holds.

Proof. Similarly as the arguments in the proof of Theorem 11, we only need to prove

$$V\left(\limsup_{n\to\infty}\frac{S_n}{n}\geq\bar{\mu}\right)=1, \tag{9}$$

when $\bar{\mu}>\underline{\mu}.$

By the assumptions we know that for each $n\geq 1$ and any constant $2^{-n}>0,$ there exists an open subset O_n of Ω with $V(O_n)<2^{-n},$ such that X_n is continuous on O_n^c. Denote $O=\bigcap_{n=1}^{\infty}\bigcup_{k=n}^{\infty}O_k,$ then by Borel-Cantelli lemma (see Lemma 8) we can obtain $v(O)=V(O)=0.$

For any $1>\delta>0,$ there exist an increasing subsequence $\{n_k\}_{k=1}^{\infty}$ of $\mathbb{N}$, an integer $N=N_\delta\geq 1$ and an open set $G^\delta\supset O$ (by Remark 7) satisfying

$$v\left(G^\delta\right)\leq v(O)+\delta=\delta,$$

such that when $n_1\geq N,$ we have

$$G^\delta\supset\bigcup_{k=1}^{\infty}O_{n_k}.$$

Then X_{n_k} is continuous on $\left(G^\delta\right)^c$. By Lemma 6, for any $\varepsilon>0$, we can find a compact set $K^\varepsilon\subset\left(G^\delta\right)^c$

with $a_\varepsilon=V\left(K^\varepsilon\right)\neq 0$ such that

$$a_\varepsilon = V\left(K^\varepsilon\right) \geq V\left(\left(G^\delta\right)^c\right) - \varepsilon \geq 1 - \delta - \varepsilon.$$

Then we have

$$\begin{aligned} & V\left(\limsup_{n\to\infty} \frac{S_n}{n} \geq \bar{\mu}\right) \\ & \geq V\left(\limsup_{k\to\infty} \frac{S_{n_k}}{n_k} I_{\left(G^\delta\right)^c} \geq \bar{\mu}\right) \\ & \geq \left(V \limsup_{k\to\infty} \frac{S_{n_k}}{n_k} I_{K^\varepsilon} \geq \bar{\mu}\right), \forall \varepsilon > 0. \end{aligned} \tag{10}$$

For $\varepsilon > 0,$ we define

$$c_V^\varepsilon\left(A\right) = \frac{1}{a_\varepsilon} V\left(A\right), \forall A \in K^\varepsilon \cap F =: F^\varepsilon. \tag{11}$$

Then it is obvious that c_V^ε is a capacity on K^ε and satisfies all the properties of V given in Proposition 1 where Ω is substituted by K^ε. We also denote by H^ε the set of all random variables $X \in F^\varepsilon$ such that $\mathbb{E}\left[X\right] < \infty$. Thus on $\left(K^\varepsilon, H^\varepsilon, \mathbb{E}\right)$, $\left\{X_{n_k} I_{K^\varepsilon}\right\}_{k=1}^\infty$ is an independent and continuous sequence. Since K^ε is also a complete and separable metric space, by Theorem 11 we have

$$c_V^\varepsilon\left(\limsup_{k\to\infty} \frac{S_{n_k}}{n_k} I_{K^\varepsilon} \geq \bar{\mu}\right) = 1. \tag{12}$$

Then from (10)-(12) we have

$$V\left(\limsup_{n\to\infty} \frac{S_n}{n} \geq \bar{\mu}\right) \geq a_\varepsilon \geq 1 - \delta - \varepsilon. \tag{13}$$

Letting ε and δ tend to 0 in inequality (13) we can derive

$$V\left(\limsup_{n\to\infty}\frac{S_n}{n}\geq\overline{\mu}\right)\geq 1,$$

which implies (9). We complete the whole proof of this corollary. □

4. EXTENSIONS

In Section 3 we get that the submean $\underline{\mu}$ and the supermean $\overline{\mu}$ are the inferior and superior limits of the arithmetic average of the first n random variables $X_1, X_2, \cdots, X_n$ given in Theorem 11, respectively, with probability 1 under the upper probability V. In fact, except the two values, any other value $x\in(\underline{\mu},\overline{\mu})$ is still the limit of some subsequence of $\{X_n\}_{n=1}^{\infty}$, with probability 1 under V. We can see it in the following theorem.

Theorem 14. Under assumptions of Theorem 11, we have for any $x\in[\underline{\mu},\overline{\mu}]$

$$V\left(x\in C\left(\frac{S_n}{n}\right)\right)=1,$$

where $C(y_n)$ is a cluster of limit points of a real sequence $\{y_n\}_{n=1}^{\infty}$.

Proof. For $x=\underline{\mu}$ and $x=\overline{\mu}$, the result has been obtained in Theorem 11. For the trivial case, $\underline{\mu}=\overline{\mu}$, it is obvious. Now we consider $\overline{\mu}>\underline{\mu}$ and any $x\in(\underline{\mu},\overline{\mu})$. We can notice that

$$V\left(x\in C\left(\frac{S_n}{n}\right)\right)=V\left(\liminf_{n\to\infty}\left|\frac{S_n}{n}-x\right|=0\right)$$

$$=\lim_{\varepsilon\to 0}V\left(\liminf_{n\to\infty}\left|\frac{S_n}{n}-x\right|\leq\varepsilon\right),$$

where ε is any constant in $\left(0, \min\left\{\bar{\mu}-x, x-\underline{\mu}\right\}\right)$,

since V is upper-continuous for closed sets. Thus we only need to find an increasing subsequence $\left\{n_k\right\}_{n=1}^{\infty}$ of $\mathbb{N}$ such that for any $\varepsilon \in \left(0, \min\left\{\bar{\mu}-x, x-\underline{\mu}\right\}\right)$, we have

$$v\left(\limsup_{n\to\infty}\left|\frac{S_n}{n}-x\right|>\varepsilon\right)=0, \tag{14}$$

Following the arguments in the proof of Theorem 11 we can obtain

$$\begin{aligned}
&v\left(\limsup_{n\to\infty}\left|\frac{S_n}{n}-x\right|>\varepsilon\right)\\
&\leq 1-\lim_{m\to\infty}\lim_{l\to\infty}V\left(\bigcap_{k=m}^{l}\left\{\left|Y_{k,x}\right|\leq\varepsilon\right\}\cap\left\{\left|Z_{k,x}\right|\leq\varepsilon\right\}\right)\\
&\leq 1-\lim_{m\to\infty}\lim_{l\to\infty}\prod_{k=m}^{l}\mathbb{E}\left[f^{\delta}\left(Y_{k,x}\right)\right]\mathbb{E}\left[f^{\delta}\left(Z_{k,x}\right)\right],
\end{aligned}$$

where $Y_{k,x}=\dfrac{S_{n_k}-S_{n_{k-1}}}{n_k-n_{k-1}}-x$ and $Z_{k,x}=\dfrac{S_{n_{m-1}}}{n_{m-1}}-x$ for any $k\geq 1$ and $x\in\left(\underline{\mu},\bar{\mu}\right)$, δ is any given constant in $\left(0,\varepsilon\right)$ and

$$f^{\delta}\left(y\right)=\begin{cases}0, & y\geq\varepsilon;\\ \dfrac{1}{\delta}\left(\varepsilon-y\right), & \varepsilon-\delta<y<\varepsilon;\\ 1, & -\varepsilon+\delta\leq y\leq\varepsilon-\delta;\\ \dfrac{1}{\delta}\left(y+\varepsilon\right), & -\varepsilon<y<-\varepsilon+\delta;\\ 0, & y\leq-\varepsilon.\end{cases}$$

Then by using the same arguments as in the proof of Theorem 11 we also can prove that (14) holds true. The whole proof is complete. □

The following corollary is obvious.

Corollary 15. Under the conditions of Theorem 11, for any continuous real function φ on $\mathbb{R}$, we have for all $x \in [\underline{\mu}, \overline{\mu}]$,

$$V\left(\varphi(x) \in C\left(\varphi\left(\frac{S_n}{n}\right)\right)\right) = 1.$$

In particular,

$$V\left(\limsup_{n\to\infty} \varphi\left(\frac{S_n}{n}\right) = \sup_{y\in[\underline{\mu},\overline{\mu}]} \varphi(y)\right) = 1,$$

$$V\left(\liminf_{n\to\infty} \varphi\left(\frac{S_n}{n}\right) = \inf_{y\in[\underline{\mu},\overline{\mu}]} \varphi(y)\right) = 1, \forall \varphi \in C(\mathbb{R}).$$

We also can extend Theorem 11, Theorem 14 and Corollary 15 to the sequences with different submeans and supermeans as follows.

Theorem 16. Let $\{X_n\}_{n=1}^{\infty}$ be an independent and continuous sequence under $\mathbb{E}$ and satisfy conditions (1)-(3) of Lemma 9. Set

$$S_n = \sum_{i=1}^{n} X_i, \forall n \geq 1.$$

Then for any $\varphi \in C(\mathbb{R})$ we have

$$V\left(\varphi(x) \in C\left(\varphi\left(\frac{S_n}{n}\right)\right)\right) = 1, \forall x \in [\underline{\mu}, \overline{\mu}].$$

Proof. By Lemma 10 and the proofs of Theorem 11 and Theorem 14 we only need to check whether (6) and (8) hold true under our assumptions of this theorem. In fact, from Lemma 9 they are obviously satisfied. Hence this theorem holds. □

From the proof of Corollary 13 and Theorem 16 we can immediately obtain the following corollary.

Corollary 17. Theorem 16 still holds when continuity assumption is substituted by quasi-continuity condition and condition (2) of Lemma 9 is replaced by the following condition:

(2') there exist real constants $\underline{\mu} \le \overline{\mu}$ such that $\lim_{n\to\infty} \overline{\mu}_n = \overline{\mu}, \lim_{n\to\infty} \underline{\mu}_n = \underline{\mu}.$

5. AN EXAMPLE

Let $\Omega = C[0,T]$ with the supremum norm. Then Ω is a Banach space and compact, thus it is a separable and complete metric space with the distance generated by the norm of the space. Then we can define a G-expectation E_G, a special sublinear expectation (see Peng [4] for details), where $G(a) = \frac{1}{2}(a^+ - \sigma^2 a^-)$, σ is a nonnegative real number less than 1. Then for any bounded and independent sequence $\{X_n\}_{n=1}^{\infty}$ with the same submean $\underline{\mu}$ and supermean $\overline{\mu}$ in $L_G^1(F_T)$ under E_G, by Theorem 11 we have

$$V\left(\limsup_{n\to\infty} \frac{1}{n}\sum_{i=1}^{n} X_i = \overline{\mu}\right) = V\left(\liminf_{n\to\infty} \frac{1}{n}\sum_{i=1}^{n} X_i = \underline{\mu}\right) = 1,$$

where V is generated by E_G, since this sequence is a sequence of quasi-continuous random variables and from Denis, Hu and Peng [11] E_G can be represented as supremum of a family of linear expectations corresponding to a family of probabilities which is weakly compact.

6. ACKNOWLEDGEMENTS

Xiaoyan Chen (World Class University) program through the National Research Foundation of Korea funded by the Ministry of Education, Science and Technology (R31-20007). The author gratefully thanks the referees for their careful reading.

REFERENCES

1. M. Marinacci, "Limit Laws of Non-Additive Probabilities and Their Frequentist Interpretation," Journal of Economic Theory, Vol. 84, No. 2, 1999, pp. 145-195.
2. F. Maccheroni and M. Marinacci, "A Strong Law of Large Numbers for Capacities," Annals of Probability, Vol. 33, No. 3, 2005, pp. 1171-1178.
3. Z. Chen, "Strong Laws of Large Numbers for Capacities," Unpublished, 2010. http://arxiv.org/abs/1006.0749
4. S. Peng, "Nonlinear Expectations and Stochastic Calculus under Uncertainty-with Robust Central Limit Theorem and G-Brownian Motion," Unpublished, 2012. http://arxiv.org/abs/1002.4546 [Citation Time(s):7]
5. Z. Chen, P. Wu and B. Li, "A Strong Law of Large Numbers for Non-additive Probabilities," International Journal of Approximate Reasoning, 2012. http://www.sciencedirect.com/science/article/pii/S0888613X12000783
6. F. Hu, "General Laws of Large Numbers under Sublinear Expectations," 2012. http://arxiv.org/abs/1104.5296
7. Z. Chen and P. Wu, "Strong Laws of Large Numbers for Bernoulli Experiments under Ambiguity," Nonlinear Mathematics for Uncertainty and Its Applications, Advances in Intelligent and Soft Computing, Vol. 100, 2011, pp. 19- 30. http://link.springer.com/chapter/10.1007%2F978-3-642-22833-9_2?LI=true
8. J. Xu and B. Zhang, "Martingale Property and Capacity under G-Framework," Electronic Journal of Probability, Vol. 15, No. 67, 2010, pp. 2041-2068.
9. G. Choquet, "Theory of Capacities," Annales de Institute Fourier, Vol. 5, 1954, pp. 131-295.
10. P. J. Huber and V. Strassen, "Minimax Tests and the Neyman-Pearson Lemma for Capacities," The Annals of Statistics, Vol. 1. No. 2, 1973, pp. 251-263.
11. L. Denis, M. Hu and S. Peng, "Function Spaces and Capacity Related to a Sublinear Expectation: Application to G-Brownian Motion Paths," Potential Analysis, Vol. 34, No. 2, 2011, pp. 139-161.

CHAPTER 3

A Novel Technique for Evaluating Failure Probability of Random Structure

Seifedine Kadry[1], Abdelkhalak El Hami[2]

[1]*American University of the Middle East, Egaila, Kuwait*

[2]*INSA of Rouen Institute, Normandy University, St. Etienne du Rouvray, France*

ABSTRACT

The purpose of this article is to develop a new methodology to evaluate the statistical characteristic of the response of structures subjecting to random excitation, by combining the Finite Element Method (FEM) with the Transforming Density Function (TDF). Uncertainty modeling of structure with random variables encourages the coupling of advanced TDF for reliability analysis to analyze problems of stochastic mechanical systems. The TDF is enthusiastically applicable in the situation where the relationship between input and output of structures is available in explicit analytical form. However, the situation is much more involved when it is necessary to perform the evaluation of implicit expression between input and output of structures through numerical models. For this aim, we propose a new technique that combines the FEM software, and the TDF method to evaluate the most important statistical parameter the Probability Density Function (PDF) of the response where the expression between input and output of structures is implicit. Once the PDF is evaluated, all other statistical parameters are derived easily. This technique is based on the numerical simulations of the FEM and the TDF by making a middleware between Finite Element software and Matlab. Some problems, range from simple to complex, of structures are analyzed using our proposed technique. Its accuracy is validated through Monte-Carlo simulation.

Keywords: Random Parameter, Uncertainty, Transforming Density Function, Finite Element Method, Finite Element Software, Matlab, Structures

1. INTRODUCTION

Many applications in mechanics require consideration of stochastic properties of materials, geometry or loads. The problem of reliability analysis of stochastic mechanical systems is of central importance in the safety assessment of structures. In a stochastic system, a large number of random variables influence the performance of the system, e.g. young's modulus, external loads... The performance of the system is evaluated by a best-esti- mate code. Consider a performance criterion Y of the system depending on the input variables $X_1, X_2, \cdots, X_n$. the function $Y = g(X_1, X_2, \cdots, X_n)$ is a random variable to be determined. In order to get the information about the uncertainty of Y, a number of FE runs have to be performed. For each of these runs, all identified uncertain parameters are varied simultaneously. According to the exploitation of the result of these studies, the uncertainty on the response can be evaluated either in the form of an uncertainty range, or in the form of a probability density function.

Reliability methods focus on the calculation of the probability of failure associated with a limit-state function. First-order and second-order reliability approximations methods (FORM and SORM) and various simulation methods are commonly used in reliability analysis [1] . Because of the typically high level of reliability of civil structures, the failure probability is usually small (in the order of 102 - 106). To account for the spatial variability of uncertain quantities (for instance, that of a material property), a characterization in terms of a random field is usually employed. Through a process of discretization, it is possible to represent the random field by a vector of random variables. One of the methods mentioned above may then be used to carry out second-moment or reliability analysis. The spectral stochastic finite element method (SSFEM) proposed by Ghanem and Spanos [2] - [5] is an approach well suited to analysis involving random fields. It is based on two discretization of the system of (stochastic) partial differential equations governing the problem under consideration, one in the spatial domain and one in the probabilistic domain. The response (e.g. the random vector of nodal displacements) is cast as a series expansion in standard normal random variables. This can be interpreted as an "intrinsic" representation of the random response, from which quantities such as statistical moments can be computed by post-processing, either analytically or by simulation.

In this paper, a proposed technique: the Finite Element Method (FEM) software coupled with the Transforming Density Function (TDF) program is applied in order to evaluate numerically the probabilistic and statistical characteristics of the response of stochastic mechanical system. To demonstrate the efficiency of the proposed technique, we have carried out different applications to cover several structural problems

2. RELIABILITY ANALYSIS

Reliability methods aim at evaluating the probability of failure of a system whose modeling takes into account randomness. The design of structures and the prediction of their good functioning lead to the verification of a certain number of rules resulting from the knowledge of physical and mechanical experience of designers and constructors. These rules traduce the necessity to limit the loading effects such as stresses and displacements. Each rule represents an elementary event and the occurrence of several events leads to a failure scenario. The objective is then to evaluate the failure probability corresponding to the occurrence of critical failure modes. In addition to the vector of deterministic variables x to be used in the system control and optimization, the uncertainties are modeled by a vector of stochastic physical variables Y affecting the failure scenario. The knowledge of these variables is not, at best, more than statistical information and we admit a representation in the form of random variables. For a given design rule, the basic random variables are defined by their probability distribution associated with some expected parameters; the vector of random variables is noted herein Y whose realizations are written y. The safety is the state where the structure is able to fulfill all the functioning requirements: mechanical and serviceability, for which it is designed. To evaluate the failure probability with respect to a chosen failure scenario, a limit state function $G(x,y)$ is defined by the condition of good functioning of the structure. The limit between the state of failure $G(x,y) \leq 0$ and the state of safety $G(x,y) > 0$ is known as the limit state surface $G(x,y) = 0$ (Figure 1). The failure probability is then calculated by [1] :

$$P_f = \Pr\left[G(x,y) \leq 0\right] = \int_{G(x,y)\leq 0} f_Y(y)\,\mathrm{d}y_1 \cdots \mathrm{d}y_n \tag{1}$$

where P_f is the failure probability, $f_Y(y)$ is the joint density function of the random variables Y and Pr [6] is the probability operator. The evaluation of integral (1) is not easy, because it represents a very small quantity and all the necessary information for the joint density function are not available. For these reasons, the First and the Second Order Reliability Methods FORM/SORM have been developed. They are based on the reliability index

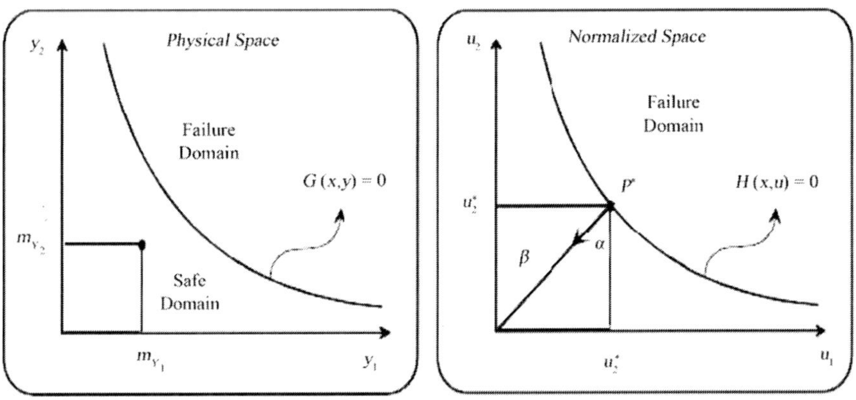

Figure 1. Physical and normalized spaces.

concept, followed by an estimation of the failure probability. The invariant reliability index β was introduced by working in the space of standard independent Gaussian variables instead of the space of physical variables. The transformation from the physical variables y to the normalized variables u is given by:

$$u = T(x,y) \quad \text{and} \quad y = T^{-1}(x,u) \tag{2}$$

This transformation T(.) is called the probabilistic transformation. In this standard space, the limit state function takes the form:

$$H(x,u) \equiv G(x,y) = 0 \tag{3}$$

For practical engineering, Equation (3) gives sufficiently accurate estimation of the failure probability.

2.1. Approximate Reliability Methods (FORM & SORM)

The First Order Reliability Method (FORM) uses the closest point on the limit state function H(u) = 0 to the origin in the standard normal space as a measure of the reliability. This point is called design point u^* and its absolute beta value $\beta = \|u^*\|$ specifies the reliability or complementary the failure probability P_f,

$P_f \approx \Phi(-\beta) = \Phi\left(-\|u^*\|\right)$, where $\Phi(.)$ denotes the standard normal cumulative distribution function. FORM

leads to exact results in case a single linear limit state function H(u) = 0 is considered.

Second Order Reliability Methods (SORM) approximate the limit state function by an incomplete second order polynomial, which assumes in a rotated space the simple form

$$H(u) \approx \tilde{H}(u) = \beta - u_1 + \frac{1}{2}\sum_{i=2}^{n} k_i u_i^2 \tag{4}$$

The form is incomplete since the terms $\{(\beta - u_1)u_i\}_{i=2}^{n}$, leading to a complete second order approximation

around the design point, are missing. An exact result in form of a one dimensional integral has been derived in [7] for the incomplete representation and an asymptotic result, sufficiently accurate for large β value, has been developed [8] .

2.2. Importance Sampling

Importance Sampling has been one of the most prevalent approaches in the context of Simulation based methods for the estimation of structural reliability [9] . The underlying concept is to draw samples of the vector of random parameters y from a distribution $f(y)$ which is concentrated in the "important region" of the random parameter space.

2.3. Line Sampling

An alternative, quite suitable for large dimension n and efficient in context with FEA, is the following approach denoted as "line sampling". It requires an important direction a as starting point, defining the direction along the limit state G(x, y) will be determined. Each point (x, y) in the standard point is decomposed into the one dimensional space [10] .

2.4. Some Factors of Comparison

Procedures for estimating probabilities of failure developed over the last twenty years like FORM/SORM, Importance Sampling, and all its variants built on it, lack robustness or computational efficiency as the number of random variables (dimension n) tends to infinity. However, the robust and simple straight forward Line sampling procedures is able to overcome most difficulties encountered in traditional procedures. Summarizing the arguments, the following can be concluded:

- FORM provides a point estimate, subject to linearization errors, without confidence. Moreover, it requires the evaluation of the design point, which becomes difficult in high dimension for nonlinear limit state functions in the standard normal space. The efforts to compute the design point grows proportional with the dimension n.
- SORM requires in addition to the design point $n > 1$ main curvature with cannot be obtained in a feasible manner for high dimensions n. The procedure implies that the domain close to the design point is the important domain which is not the case for high dimensions.
- Importance Sampling is more robust and accurate as FORM or SORM, but not competitive to Line Sampling, because it is generally impossible to sample according to the optimal sampling density. The approach has difficulties to deal with multiple failure domains if they are not well separated.

Line Sampling is capable to take advantage of simple flat limit states in standard normal space and samples in the most important domain, without assuming a linear or quadratic limit state surface or requiring a design point computation.

3. FINITE ELEMENT METHOD AND TDF

3.1. Finite Element Method

The Finite Element Method is the standard tool for certain classes of partial differential equations arising in various fields of engineering and in particular for those arising in solid mechanics. For linear systems enforcing global static or dynamic equilibrium the FE Analysis leads to a system of linear equations, respectively

$$KU = F \tag{5}$$

$$M\ddot{U}(t) + C\dot{U}(t) + KU(t) = F(t) \tag{6}$$

where the matrices K and M are the global stiffness and mass matrices, respectively, obtained by adding the contributions of all element matrices

$$K = \sum_e K^e = \sum_e \int_{\Omega^e} D^e B^e \mathrm{d}\Omega^e \tag{7}$$

$$M = \sum_e M^e = \sum_e \int_{\Omega^e} \rho^e H^{e\mathrm{T}} H^e \mathrm{d}\Omega^e \tag{8}$$

where B^e is the matrix relating element displacements and strains, D^e is the elasticity matrix relating stresses and strains, ρ^e is the mass density, H^e is the shape functions and e is the spatial domain of the element. The global damping matrix C is typically formulated in terms of M and K.

3.2. Transforming Density Function TDF

The Transforming Density Function is based on the following theorem

Theorem: Suppose that X is a continuous random variable with PDF, f(x) and $A \subset \Re$ is the one dimensional space where $f(X) \succ 0$, is differentiable and monotonic. Consider the random variable $Y = u(X)$, where $y = u(x)$ defines a one-to-one transformation that maps the set A onto a set $B \subset \Re$ so that the equation $y = u(x)$ can be uniquely solved for x in terms of y, say $x = u^{-1}(y)$. Then, the PDF of Y is

$$f_Y(Y) = f_X\left[u^{-1}(y)\right] * |J| \tag{9}$$

where $J = \frac{dx}{dy} = \frac{du^{-1}(y)}{dy}$ is the transformation Jacobean, which must be continuous for all points $y \in B$. The

TDF is based on one-to-one mapping between the random output(s) and input(s) where the transformation Jacobean J can be computed. The PDF of the output(s) is then computed through the known joint PDF of the inputs multiplied by the determinant of transformation Jacobean matrix. The idea of TDF is based on the following formula [11]

$$f_u(u) = |J| \cdot f_z(z) = \left|\frac{dz}{du}\right| \cdot f_z(z) \tag{10}$$

$f_u(u)$ is the probability density function of the variable u. $f_z(z)$ is the probability density function of the variable z.

The general steps in the application of the TDF program: 1) the random variable input is generating and the stochastic equation of equilibrium is solved first using FEM software. 2) This solution is used to compute numerically the function between the input and the output by the Cubic Spline Interpolation, which is then inverted for the calculation of the determinant of the transformation Jacobean. Finally, the PDF of the response at any point in the domain can be deduced by using the formula (7). This approach has the advantage of giving a closed-form of the density function of the response, which is very helpful for reliability analysis of mechanical systems [3] [5] . To overcome the drawback of TDF, in analysis of complicated structures, we propose in this paper an interface between the FEM software and the TDF program.

4. FINITE ELEMENT METHOD COUPLED WITH TRANSFORMING DENSITY FUNCTION: FEMCTDF

Before you begin to format your paper, first write and save the content as a separate text file. Keep your text and graphic files separate until after the text has been formatted and styled. Do not use hard tabs, and limit use of hard returns to only one return at the end of a paragraph. Do not add any kind of pagination anywhere in the paper. Do not number text heads—the template will do that for you.

Finally, complete content and organizational editing before formatting. Please take note of the following items when proofreading spelling and grammar:

The FEM software is used to perform the structural analysis to obtain the structure weight, maximal displacement, and maximal stress, corresponding to a set of given design variables. These analysis results are sent to the TDF program to conduct the Probabilistic Density Function PDF, and the probability of failure and generate new random variables. The newly generated variables are then used to update the input file. The FEM software is then invoked again to perform the structural analysis with the new input parameters. This process is repeated until satisfactory results are obtained.

4.1. Finite Element Analysis

Finite element codes are less complicated than many of the word processing and spreadsheet packages found on modern microcomputers. Nevertheless, they are complex enough that most users do not find it effective to program their own code. A number of prewritten commercial codes are available, representing a broad price range and compatible with machines from microcomputers to supercomputers 1. However, users with specialized needs should not necessarily shy away from code development, and may find the code sources available in such texts as that by Zienkiewicz 2 to be a useful starting point. Most finite element software is written in Fortran, but some newer codes such as felt are in C or other more modern programming languages. In practice, a finite element analysis usually consists of three principal steps:

1) Preprocessing: The user constructs a model of the part to be analyzed in which the geometry is divided into a number of discrete sub regions, or "elements," connected at discrete points called "nodes." Certain of these nodes will have fixed displacements, and others will have prescribed loads. These models can be extremely time consuming to prepare, and commercial codes vie with one another to have the most user-friendly graphical "preprocessor" to assist in this rather tedious chore. Some of these preprocessors can overlay a mesh on a preexisting CAD file, so that finite element analysis can be done conveniently as part of the computerized drafting-and-design process.

2) Analysis: The dataset prepared by the preprocessor is used as input to the finite element code itself, which constructs and solves a system of linear or nonlinear algebraic equations $K_{ij}u_j = f_i$where u and f are the displacements and externally applied forces at the nodal points. The formation of the K matrix is dependent on the type of problem being attacked, and this module will outline the approach for truss and linear elastic stress analyses. Commercial codes may have very large element libraries, with elements appropriate to a wide range of problem types. One of FEA's principal advantages is that many problem types can be addressed with the same code, merely by specifying the appropriate element types from the library.

3) Postprocessing: In the earlier days of finite element analysis, the user would pore through reams of numbers generated by the code, listing displacements and stresses at discrete positions within the model. It is easy to miss important trends and hot spots this way, and modern codes use graphical displays to assist in visualizing the results. A typical postprocessor display overlays colored contours representing stress levels on the model, showing a full-field picture.

4.2. Algorithm of Method FEMCTDF

We will implement our method in Matlab. The outline of the proposed technique FEMCTDF is as follows:

1) Generate the input random variables;

2) Calculate the value of output variables by FE software, for each value of input the correspondent value of output is estimating using FEM and it is stocking in solution file;

3) Approximate the function between input and output variables using Cubic Spline Interpolation;

4) Calculate the determinant of Jacobean of input and output variables (Equation (10));

5) Apply the basic relation of TDF (Equation (10));

6) Evaluate the graphic of PDF of output variable in function of this output variable (in our case the PDF of displacement in function of displacement);

7) Approximate the Probability of failure P_f.

We start with the creation of file variables. Out that serves to stock the random variables generating by TDF program in Matlab in order to use them by the executable file under FEM software. Once the FEM software receives these variables it introduces them in the modelling of the considered structure, and by the FEM and Static analysis done by the FEM software we get the results (nodal displacement, strains, and so on). These results are stocked in the file response. out. This later file is communicated to the TDF program in order to calculate the PDF of the output variable of the considered structure (in our case the vertical displacement u_y) and the probability of failure of this structure.

5. APPLICATION

Three case studies presented in this section all have implicit function of output variable in terms of the input random variable. The proposed technique use the FEM for modelling the structure in every iteration and calculate the value of displacement (output variable) corresponding at each input variable value, and in other hand it would be necessary to use the TDF program for estimating the value of P_f (probability of failure) of the considered structure. In order to validate the FEMCTDF technique and to compare the results, the Monte Carlo Simulation is also employed as reference method for comparison.

5.1. Three Bar Truss

In this example, the three-bar truss (Figure 2 and Figure 3) is analysed using our proposed technique. For that purpose, statistic models must be defined for each random variable involved. Hence, uniform distribution in the range [5] [10] is assumed for the Young's modulus E and the other parameters are considered as deterministic variables. These parameters

Figure 2. Real application of three bar truss.

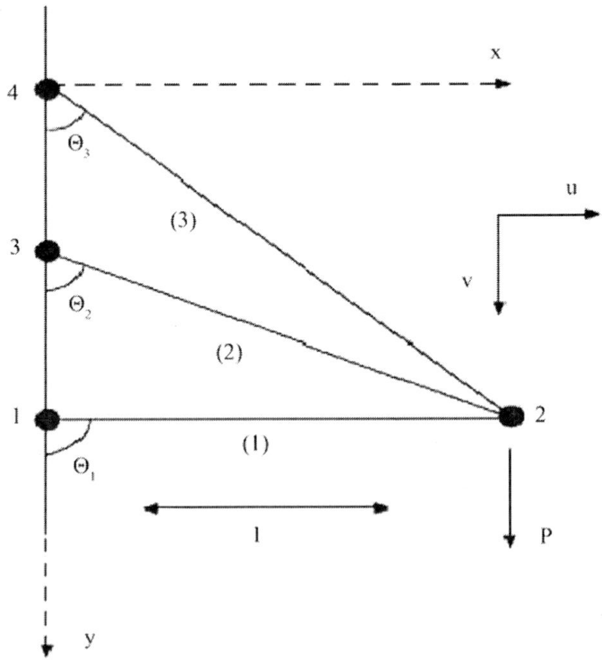

Figure 3. Three bar truss.

(Geometrical and material proprieties) are: the section of each beam A = 1 m^2, the length l = 3 m, the load W = 1.2 N, the mass density r = 7800 kg/m^3.

From Figure 4, the PDF's of the normalized vertical displacement u_y of the 3bar structure calculated by our proposed technique FEMCTDF are independent and uniformly distributed in the range [1.2505; 2.5008].

Also in this case the obtained results are accurate as shown by favourable comparison with classical Monte Carlo simulations. For Reliability Analysis, let us suppose that the limit displacement is $u_{\text{limit}} = 2\ \text{mm}$, It is required to find the failure probability $P_f = P\left(u \le u_{\text{limit}}\right)$.

The numerical values of probabilistic characteristics of the displacement of this 3bar structure are listed in this Table 1.

5.2. Spatial Truss

This application treats structural analysis on one hand of the arrow of a crane of construction that one assimilates to a spatial truss. It is constituted by identical beams (Figure 5 and Figure 6). To the extremity of this truss is applied a load M = 5 t. The bars forming the structure are in steel of which Young's modulus E is uniformly distributed in the range [100 GPa, 300 GPa] and the Poisson coefficient σ = 0.29. The bars have circular sections of which the big one is worth A1 = 22 $mm^2$and the small one A2 = 11 mm^2. The weight of each of the bars of the truss is not negligible in front of the load M. The goal of this analysis is to determine the maximum displacement generated by the applied load to his extremity.

From Figure 7, the PDF's of the normalized vertical displacement u_y of the spatial truss calculated by our proposed technique FEMCTDF are independent and uniformly distributed in the range [0.00704, 0.02027]. Also the obtained results are accurate as shown by favourable comparison with classical Monte Carlo simulations. For Reliability Analysis, let us suppose that the limit displacement is u_{limit} =0.015 mm, It is required to find the failure probability $P_f = P(u \le u_{\text{limit}})$. The numerical values of probabilistic characteristics of the displacement of this spatial truss are listed in this Table 2.

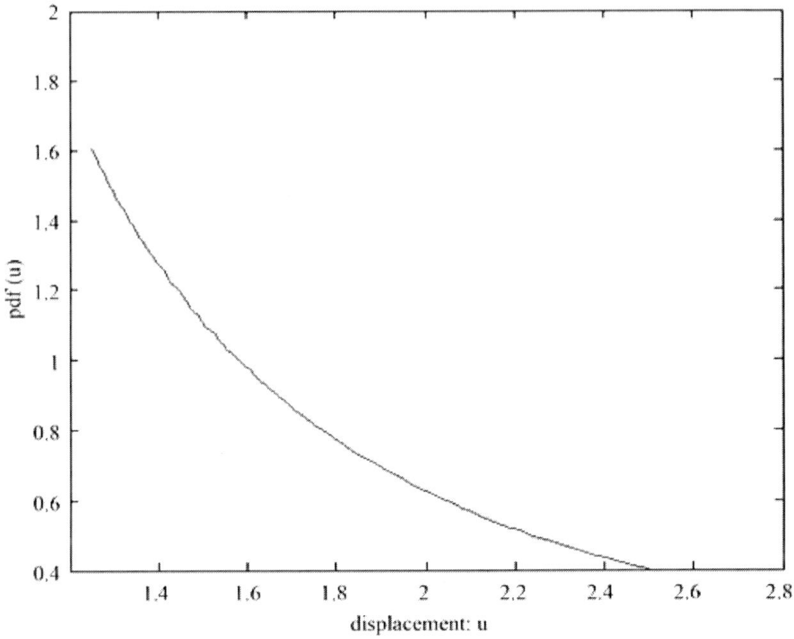

Figure 4. PDF (u) when E is uniformly distributed.

Figure 5. Real application of spatial truss.

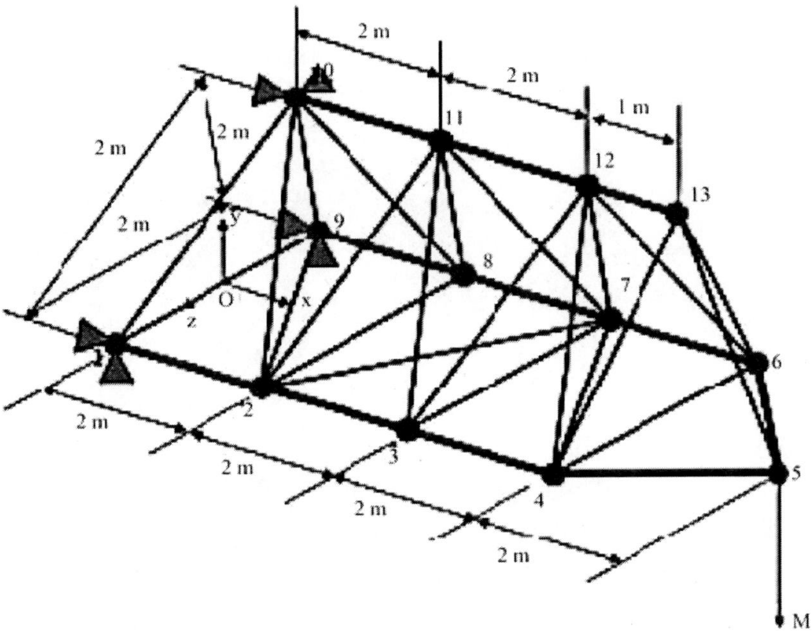

Figure 6. The spatial truss.

Table 1. FEMCTDF versus Monte Carlo simulation.

	FEMCTDF	Monte Carlo simulation
U_{max}	2.5009	2.5008
U_{min}	1.2505	1.2505
U_{mean}	1.7349	1.7334
Stdev	0.3556	0.3496
P_f	0.2536	0.2504

Table 2. FEMCTDF versus Monte Carlo simulation.

	FEMCTDF	Monte Carlo simulation
U_{max}	2.027e−2	2.11e−2
U_{min}	7.04e−3	7.039e−3
U_{mean}	1.120e−2	1.156e−2
Stdev	3.6e−3	3.75e−3
P_f	0.2031	0.203092

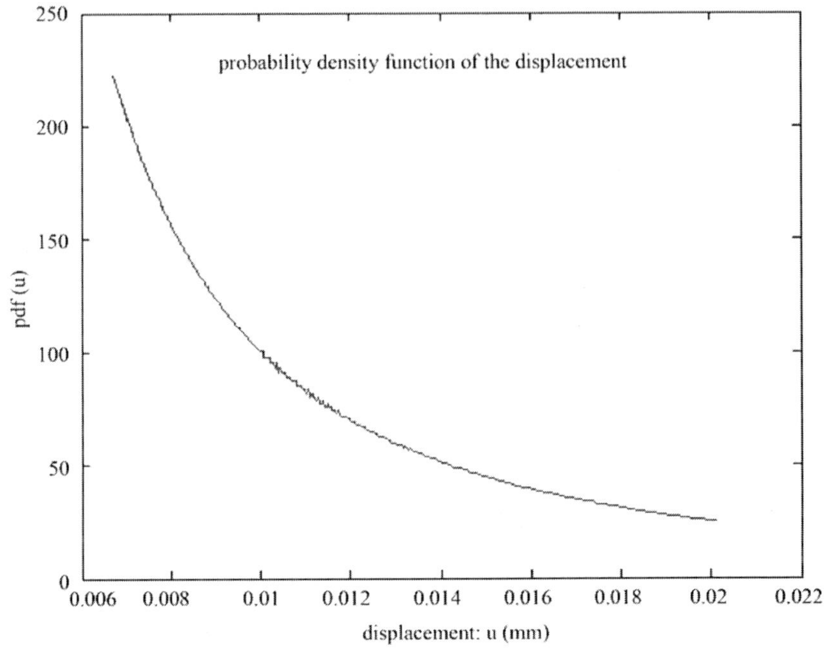

Figure 7. PDF (u) when E is uniformly distributed.

5.3. Pylon of a Line of Transportation of Electricity

We are going to analyze the reliability of a more complex structure, the pylon of a line of transportation of electricity. Two identical loads F of 1.8 KN are applied to the two superior extremities of the following pylon an angle of $\theta = 15$. The bars forming the pylon are in steel of which Young's modulus is uniformly distributed E = [100 GPa, 300 GPa] and the Poisson coefficient $\sigma = 0.29$. The section of every bar is worth A = 27.90 cm^2. The hypothesis for this problem is that the weight of each bar of the pylon is negligible in front of the applied efforts (see Figure 5). Using the proposed technique FEMCTDF, we obtain the following graph (see Figure 8 and Figure 9):

The PDFs of the normalized vertical displacement u_y are plotted in Figure 10 assuming that the variable u_y is independent and uniformly distributed in the range [2.9489e04, 8.8466e04]. Also in this case the results are accurate as shown by favorable comparison with classical Monte Carlo simulation. Let us suppose the limit displacement is u = 6.635e04 mm. It is required to find the failure probability $P_f = P(u \leq u_{limit})$. Mean Squared Error

(MSE), and Absolute Percentage Error (APE) are calculated by using the following equation:

Figure 8. Real application of pylon

$$\mathrm{MSE} = \left(u_1 - u_{mc}\right)^2, \ \mathrm{APE} = \left|\frac{\left(u_1 - u_{mc}\right)}{u_{mc}} \times 100\right|$$

where u_1 is displacement obtained with FEMCTDF and u_{mc} the Monte Carlo simulation one.

Table 3 reports the results (U_{max}, U_{min}, U_{mean} are in mm) obtained by our technique and the Monte Carlo simulation (10,000 iterations). This table also illustrates the efficiency of the FEMCTDF, since a number of 800 (less than 1000) iterations suffice to obtain results close to those obtained by Monte Carlo simulation. To compare the results, the MSE and APE are calculated. The values of MSE and APE are very small, which shows the accuracy and efficiency of FEMCTDF.

.

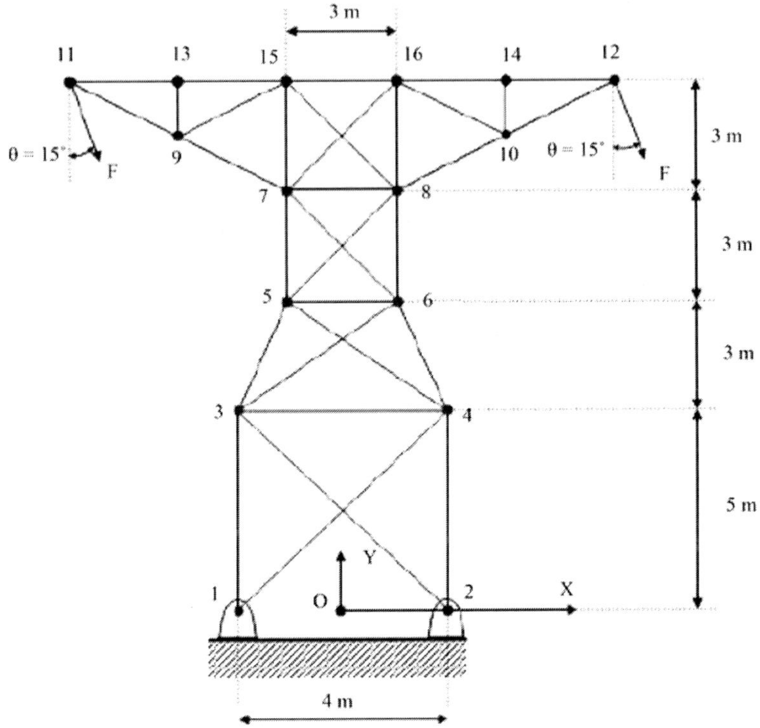

Figure 9. The pylon.

Table 3. FEMCTDF versus Monte Carlo simulation.

	FEMCTDF	M-C	MSE	APE (%)
u_{min}	2.94e−04	2.94e−04	1.64e−16	0.0044
u_{max}	8.84e−04	8.846e−04	2.53e−15	0.0057
u_{mean}	4.86e−04	4.85e−04	1.03e−14	0.0209
Stdev	1.57e−04	1.57e−04	9.10e−14	0.1919
P_f	1.88e−01	1.66e−01	4.78e−04	13.1181

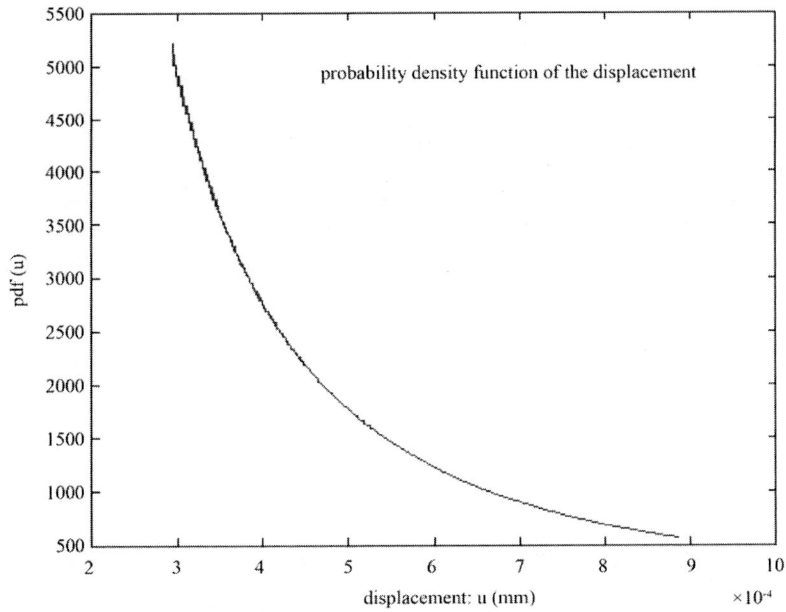

Figure 10. PDF (u) when E is uniformly distributed.

6. CONCLUSION

In this paper, an efficient, robust technique is proposed to solve the reliability problems that required the evaluation of implicit function between the input variable of the structure and his response variable. This technique has been developed with the motivation to facilitate the application of structural reliability analysis methods to numerical models of mechanical structures. It integrates the treatment by the FEM software and the TDF program. In fact, the FEM is used to estimate the structural response function, once the implicit response function is numerically found that the TDF program can easily be applied to solve the structural reliability problem.

CITE THIS PAPER

Seifedine Kadry, Abdelkhalak El Hami (2015) A Novel Technique for Evaluating Failure Probability of Random Structure. *World Journal of Mechanics*,**05**,139-150.

REFERENCES

1. Kadry, S. (2007) A Proposed Method for Reliability Analysis in Higher Dimension. Engineering Letters, 15, 7-12.
2. Liu, W.-K., Belytschko, T. and Mani, A. (1986) Random Field Finite Elements. International Journal for Numerical Methods in Engineering, 23, 1831-1845.
3. Deodatis, G. (1991) The Weighted Integral Method, I: Stochastic Stiffness Matrix. Journal of Engineering Mechanics, 117, 1851-1864.
4. Ditlevsen, O. and Madsen, H. (1996) Structural Reliability Methods. Wiley, Chichester.
5. Ghanem, R.-G. and Spanos, P.-D. (1991) Stochastic Finite Elements—A Spectral Approach. Springer, Berlin.
6. Kadry, S., Chateauneuf, A. and El-Tawil, K. (2006) One Dimensional Transformation Methods in Reliability Analysis. CST, Madrid.
7. Tvedt, L. (1990) Distribution of Quadratic Forms in Normal Space-Application to Structural Reliability. Journal of Engineering Mechanics, 116, 1183-1197.
8. Breitung, K. (1994) Asymtotic Approximations for Probability Integrals. Lecture Notes in Mathematics, 1592. Springer, Berlin.
9. Schueller, G. (1998) Structural Reliability-Recent Advances-Freudenthal Lecture. In: Shiraishi, N., Shinozuka, M. and Wen, Y., Eds., Proceedings of the 7th International Conference on Structural Safety and Reliability (ICOSSAR'97), A.A. Balkema Publications, Rotterdam, 3-35.
10. Koutsourelakis, P., Pradlwarter, H. and Schuëller, G. (2003) Reliability of Structures in High Dimensions. Proceedings in Applied Mathematics and Mechanics, 3, 495-496.
11. Papoulis, A. and Pillai, S.U. (2002) Probability, Random Variables and Stochastic Processes. 4th Edition, McGraw-Hill, Boston.

CHAPTER 4

Stochastic Approximation Method for Fixed Point Problems

Ya. I. Alber[1], C. E. Chidume[2], Jinlu Li[3]

[1]*Department of Mathematics, The Technion-Israel Institute of Technology, Haifa, Israel*

[2]*The Abdus Salam International Centre for Theoretical Physics, Trieste, Italy*

[3]*Department of Mathematical Sciences, Shawnee State University, Portsmouth, USA*

ABSTRACT

We study iterative processes of stochastic approximation for finding fixed points of weakly contractive and nonexpansive operators in Hilbert spaces under the condition that operators are given with random errors. We prove mean square convergence and convergence almost sure (a.s.) of iterative approximations and establish both asymptotic and nonasymptotic estimates of the convergence rate in degenerate and non-degenerate cases. Previously the stochastic approximation algorithms were studied mainly for optimization problems.

Keywords: Hilbert Spaces; Stochastic Approximation Algorithm; Weakly Contractive Operators; Nonexpansive Operators; Fixed Points; Convergence in Mean Square; Convergence Almost Sure (a.s.); Nonasymptotic Estimates of Convergence Rate

1. INTRODUCTION

In this paper the following problem is solved: To find a fixed point x^* of the operator $T:G\to H$, in other words, to find a solution $x^*\in G$ of the equation

$$x=Tx, \tag{1.1}$$

where T is a Lipschitz continuous mapping, H is a Hilbert space, $G\subseteq H$ is a closed convex subset. We suppose that x^* exists, i.e., the fixed point set N of T is nonempty. Note in different particular cases of the Equation (1.1), for example, when $T:G\to G,$ the solution existence and solution uniqueness can be proved under some additional assumptions.

We separately consider two classes of mappings T: the class of weakly contractive maps and more general class of nonexpansive ones. Let us recall their definitions.

Definition 1.1. A mapping $T:G\to H$ is said to be weakly contractive of class $C_{\phi(t)}$ on a closed convex subset $G\subset H$ if there exists a continuous and increasing function $\phi(t)$ defined on IR^+ such that ϕ is positive on $IR^+/\{0\},\phi(0)=0,\lim_{t\to+\infty}\phi(t)=+\infty,$ and for all

$$x,y\in G,\|Tx-Ty\|\le\|x-y\|-\phi(\|x-y\|). \tag{1.2}$$

Remark 1.2. It follows from (1.2) that $\psi(t)\le t$ and in real problems an argument t of the function $\psi(t)$ doesn't necessary approaches to ∞ obeying the condition $T:G\to H$ (see the example in Remark 3.4).

Definition 1.3. A mapping $T:G\to H$ is said to be nonexpansive on the closed convex subset $G\subset H$ if for all $x,y\in G$

$$\|Tx - Ty\| \le \|x - y\|.$$

It is obvious that the class of weakly contractive maps is contained in the class of nonexpansive maps because the right-hand side of (1.2) is estimated as

$$0 \le \|x - y\| - \phi(\|x - y\|) \le \|x - y\|, \tag{1.3}$$

and it contains the class of strongly contractive maps because $\phi(t) = (1-q)t$ with $0 < q < 1$ gives us

$$\|Tx - Ty\| \le q\|x - y\|. \tag{1.4}$$

We study the following algorithm of stochastic approximation:

$$x_{n+1} = Pr_G\left(x_n - \alpha_n S_n x_n\right), n = 1, 2, \cdots, x_1 \in G, \tag{1.5}$$

where Pr_G is the metric projection operator from H onto G and deterministic step-parameters α_n satisfy the standard conditions:

$$\sum_{n=1}^{\infty} \alpha_n = \infty \text{ and } \sum_{n=1}^{\infty} \alpha_n^2 < \infty. \tag{1.6}$$

The factor $S_n x_n$ in (1.5) is an infinite-dimensional vector of random observations of the clearance operator $F = I - T$ at random points $x_n \in G$ given for all $n \ge 1$ on the same probability space (Ω, A, P). We set

$$S_n x_n = \zeta_n + \xi_n, \tag{1.7}$$

where $\zeta_n = x_n - Tx_n$ and ξ_n is a sequence of independent random vectors with the conditions

$$E[\xi_n] = 0 \text{ and } E\left[\|\xi_n\|^2\right] \le C_1, 0 < C_1 < \infty. \tag{1.8}$$

Here E is a symbol of the mathematical expectation. In order to calculate conditional mathematical expectations of different random variables we define the σ- subalgebra $A_n := \sigma(x_1, x_2, \cdots, x_n)$ on (Ω, A, P). And then $E[\xi_n / A_n]$ means A_n-measurable function with the following property: for any $B \in A_n$ $\int_B \xi_n dP(\omega) = \int_B E[\xi_n / A_n] dP(\omega).$

We also assume in the sequel that ζ_n is A_n-measurable for all $n \geq 1.$

Let us recall the mean square convergence and almost sure (a.s.) convergence.

We say that the sequence $\{\xi_n\}$ of random variables $\xi_n(\omega)$ converges in mean square to ξ if ξ exists and $\lim_{n\to\infty} E\left[\|\xi_n - \xi\|^2\right] = 0.$

The sequence ξ_n converges to ξ almost surely or with probability 1 if

$$P\left(\lim_{n\to\infty} \xi_n(\omega) = \xi(\omega)\right) = 1.$$

Almost sure convergence and convergence in mean square imply convergence in the sense of probability: The sequence $\{\xi_n\}$ of random variables $\xi_n(\omega)$ converges in the sense of probability to $\xi(\omega)$ if for all $\varepsilon > 0$ $\lim_{n\to\infty} P\left(\|\xi_n(\omega) - \xi(\omega)\| \geq \varepsilon\right) = 0.$

So, we consider iterative processes of stochastic approximation in the form (1.5) for finding fixed points of weakly contractive (Definition 1.1) and nonexpansive (Definition 1.3) mappings in Hilbert spaces under the conditions (1.8). We prove mean square convergence and convergence almost sure of iterative approximations and establish both asymptotic and nonasymptotic estimates of the convergence rate. Perhaps, we present here the first results of this sort for fixed point problems. Formerly the stochastic approximation methods were studied mainly to find minimal and maximal points in optimization problems (see, for example, [1-6] and references within).

2. AUXILIARY RECURRENT INEQUALITIES

Lemma 2.1. [3,4] Let $\{\mu_k\}, \{\rho_k\}$ and $\{\gamma_k\}$ be sequences of nonnegative real numbers satisfying the recurrent inequality.

$$\mu_{k+1} \le (1+\rho_k)\mu_k + \gamma_k, k = 1, 2, \cdots \tag{2.1}$$

Assume that $\sum_{k=1}^{\infty} \rho_k < \infty$ and $\sum_{k=1}^{\infty} \gamma_k < \infty.$ Then $\{\mu_k\}$ is bounded and converges to some limit.

Lemma 2.2. [3,4] Let $\{\mu_k\}, \{\zeta_k\}, \{\alpha_k\}$ and $\{\gamma_k\}$ be sequences of nonnegative real numbers satisfying the recurrent inequality.

$$\mu_{k+1} \le (1+\rho_k)\mu_k - \alpha_k \psi(\mu_k) + \gamma_k, k = 1, 2, \cdots, \tag{2.2}$$

where $\sum_{k=1}^{\infty} \alpha_k = \infty, \sum_{k=1}^{\infty} \rho_k < \infty$ and either $\sum_{k=1}^{\infty} \gamma_k < \infty$ or

$$\lim_{k \to \infty} \frac{\gamma_k}{\alpha_k} = 0. \tag{2.3}$$

Assume that $\psi(t)$ is continuous and increasing function defined on IR^1 such that ψ is positive on $IR^1 / \{0\}$, $\psi(0) = 0.$ Then $\lim_{k \to \infty} \mu_k = 0.$ There exists an infinite subsequence $\{k_l\}, l = 1, 2, \cdots,$ such that

$$\mu_{k_l} \le C_0 \psi^{-1} \left(\frac{1}{\sum_{k=1}^{k_l} \alpha_k} + \frac{\gamma_{k_l}}{\alpha_{k_l}} \right),$$

where $C_0 \ge \prod_{k=1}^{\infty} (1 + \rho_k).$

In the following two lemmas we want to present nonasymptotic estimates for the whole sequence $\mu_k, k \ge 1.$ For this the stronger requirements are made of parameters α_k and function $\psi(t)$ in the recurrent inequality.

Suppose that $\alpha(t)$ such that $\alpha(k)=\alpha_k$, $F(t)$ and $\Phi(t)$ are antiderivatives from $\alpha(t)$ and $\frac{1}{\psi(t)}$, respectively, with arbitrary constants C (without loss of generality, one can put $C=0$), i.e.

$$F(t)=\int\alpha(t)\mathrm{d}t,\Phi(t)=\int\frac{\mathrm{d}t}{\psi(t)}.$$

Observe that $F(t)$ has the following properties:

i) $F'(t)=\alpha(t);$

ii) $F(t)$ is strictly increasing on $[1,\infty)$ and $F(t)\to\infty$ as $t\to\infty;$

iii) The function $g(t)=\frac{\alpha(t)}{F(t)}=\frac{F'(t)}{F(t)}$ is decreasing;

iv) $G(t)=\ln F(t)\to\infty$ as $t\to\infty.$

Introduce the following denotations:

1) $\psi^{-1}(z)$ and $\Phi^{-1}(z)$ are the inverse functions to $\psi(t)$ and $\Phi(t)$, respectively;

2) $v(s)=\psi^{-1}\left[\frac{c_0\gamma(s-1)}{\alpha(s-1)}\right]+\gamma(s-1),s\geq 2,c_0>1$ is a fixed control parameter;

3) $u(s,C)=\Phi^{-1}\left[\Phi(C)-a(F(s)-F(2))\right],$

$C\geq 0,a=\frac{c_0-1}{c_0}<1;$

4) $w(s,v(r))=\Phi^{-1}\left[\Phi(v(r))-a(F(s)-F(2))\right],$

$2\leq r<\infty;$

5) $Q(\overline{c})=\Phi^{-1}\left[\Phi(\overline{c})-a(F(2)-F(1))\right],$ where

$\bar{c} > 0$ is an arbitrary fixed number;

We present now the based condition (P): The graphs of the scalar functions $v(s)$ and $w(s, v(r))$ with any fixed $r \in [2, \infty)$ are intersected on the interval $[2, \infty)$ not more than at two points s_1 and s_2 (we do not consider contact points as intersection ones excepting $s = 2$ if any).

For example, the graphs of the functions $v(s)$ and

$w(s, v(r))$ calculated for $\alpha(s) = \frac{b}{s^{\kappa}}, b > 0, 0 < \kappa \le 1,$

and $\psi(t) = t^{\nu}, \nu \ge 1,$ satisfy the condition (P).

Lemma 2.3. [3,4] Assume that 1) the property (P) is carried out for the function $w(s, v(2)) = u(s, v(2))$ and

$v(s)$; 2) $u(s, v(2)) \ge v(s)$ as $s \to \infty$; 3) the control parameter c_0 is chosen such that

$$u(s, v(2)) \ge v(s) \text{ as } s \to 2. \tag{2.4}$$

Then for the sequence $\{\mu_k\}$ generated by the inequality

$$\mu_{k+1} \le \mu_k - \alpha_k \psi(\mu_k) + \gamma_k, k = 1, 2, \cdots, \tag{2.5}$$

it follows: $\lim_{k \to \infty} \mu_k = 0$ and for all $k \ge 1$

$$\mu_k \le u(k, C), C = \max\{Q(\mu_1), v(2)\}. \tag{2.6}$$

Lemma 2.4. [3,4] Assume that 1) the property (P) is carried out for all the function $w(s, v(2r))$ and $v(s)$; 2) $u(s, v(2)) \le v(s)$ as $s \to \infty$. Then for the sequence $\{\mu_k\}$ generated by the inequality (2.5) $\lim_{k \to \infty} \mu_k = 0$. In additiona) if $Q(\mu_1) \le v(2)$ and the control parameter c_0 is chosen such that $u(s, v(2)) \le v(s)$ as $s \to 2$, then for all $k \ge 2$

$$\mu_k \le v(k); \tag{2.7}$$

b) in all remaining cases

$$\mu_k \le u(k,C), C = \max\{Q(\mu_1), v(2)\},$$
$$1 \le k \le \bar{s}, \tag{2.8}$$

$$\mu_k \le v(k), k > \bar{s}, \tag{2.9}$$

where $\bar{s}$ is a unique root of the equation

$$u(s,C) = v(s) \tag{2.10}$$

on the interval $[2,\infty)$.

The following lemmas deal with another sort of recurrent inequalities:

Lemma 2.5. [7,8] Let $\{\mu_k\}, \{\alpha_k\}, \{\beta_k\}$ and $\{\gamma_k\}$ be sequences of non-negative real numbers satisfying the recurrence inequality.

$$\mu_{k+1} \le \mu_k - \alpha_k \beta_k + \gamma_k, k = 1,2,\cdots. \tag{2.11}$$

Assume that

$$\sum_{k=1}^{\infty} \alpha_k = \infty \text{ and } \sum_{k=1}^{\infty} \gamma_k < \infty.$$

Then:

i) There exists an infinite subsequence $\{\beta_{\ell_k}\} \subset \{\beta_k\}$ such that

$$\beta_{\ell_k} \le \frac{1}{\sum_{j=1}^{\ell_k} \alpha_j}, \tag{2.12}$$

and, consequently, $\lim_{k\to\infty} \beta_{\ell_k} = 0$;

ii) if $\lim_{k\to\infty} \alpha_k = 0$ and there exists $\kappa > 0$ such that

$$|\beta_{k+1} - \beta_k| \le \kappa \alpha_k \tag{2.13}$$

for all $k \geq 1$, then $\lim_{k\to\infty}\beta_k = 0.$

Lemma 2.6. [7,8] Let $\{\mu_k\}, \{\alpha_k\}, \{\beta_k\}$ and $\{\gamma_k\}$ be sequences of non-negative real numbers satisfying the recurrence inequality (2.11). Assume that $\sum_{k=1}^{\infty}\alpha_k = \infty$ and (2.3) is satisfied. Then there exists an infinite subsequence $\{\beta_{\ell_k}\} \subset \{\beta_k\}$ such that $\lim_{k\to\infty}\beta_{\ell_k} = 0.$

3. MEAN SQUARE CONVERGENCE OF STOCHASTIC APPROXIMATIONS

Theorem 3.1. Assume that $T: G \to H$ is a weakly contractive mapping of the class $C_{\phi(\tau)}$, $\psi_1(\tau) = \tau\phi(\tau)$ is a convex function with respect to $t = \tau^2$ and $E\left[\left\|x_1 - x^*\right\|^2\right] < \infty.$ Then the sequence $\{x_n\}$ generated by (1.5)-(1.7) converges in mean square to a unique fixed point x^* of T. There exists an infinite subsequence $\{n_l\}, l = 1, 2, \cdots,$ such that

$$E\left[\left\|x_{n_l} - x^*\right\|^2\right] \leq C_0\psi^{-1}\left(\frac{1}{2\sum_{n=1}^{n_l}\alpha_n} + C_1\alpha_{n_l}\right), \tag{3.1}$$

where $\psi(\tau^2) = \psi_1(\tau)$ and some positive constant C_0 satisfies the inequality

$$\prod_{n=1}^{\infty}\left(1 + 8\alpha_n^2\right) \leq C_0. \tag{3.2}$$

Remark 3.2. $1 < C_0 < \infty$ exists in view of the second condition in (1.6).

Proof. First of all, we note that the method (1.5)-(1.7) guarantees inclusion $x_n \in G$ for all $n \geq 1$. Since the metric projection operator Pr_G is nonexpansive in a Hilbert space and $x^* \in G$ exists, we can write

$$\begin{aligned} &\left\|x_{n+1} - x^*\right\|^2 \\ &= \left\|Pr_G\left(x_n - \alpha_n\left(x_n - Tx_n + \xi_n\right)\right) - Pr_G x^*\right\|^2 \\ &\leq \left\|x_n - x^*\right\|^2 - 2\alpha_n\left(\zeta_n + \xi_n, x_{n+1} - x^*\right) \\ &= \left\|x_n - x^*\right\|^2 - 2\alpha_n\left(\zeta_n, x_n - x^*\right) \\ &-2\alpha_n\left(\xi_n, x_n - x^*\right) + \left\|x_{n+1} - x_n\right\|^2. \end{aligned} \tag{3.3}$$

Let us evaluate the first scalar product in (3.3). We have

$$\begin{aligned} &\left(x_n - Tx_n, x_n - x^*\right) \\ &= \left(x_n - x^* - Tx_n + Tx^*, x_n - x^*\right) \\ &= \left\|x_n - x^*\right\|^2 - \left(Tx_n - Tx^*, x_n - x^*\right) \\ &\geq \left\|x_n - x^*\right\|^2 - \left\|Tx_n - Tx^*\right\|\left\|x_n - x^*\right\| \\ &\geq \left\|x_n - x^*\right\|^2 - \left(\left\|x_n - x^*\right\| - \phi\left(\left\|x_n - x^*\right\|\right)\right)\left\|x_n - x^*\right\| \\ &\geq \phi\left(\left\|x_n - x^*\right\|\right)\left\|x_n - x^*\right\| = \psi_1\left(\left\|x_n - x^*\right\|\right) \end{aligned} \tag{3.4}$$

We remember that $\psi\left(\tau^2\right) = \psi_1\left(\tau\right)$. Then the inequalities (3.3) and (3.4) yield

$$\begin{aligned} &\left\|x_{n+1} - x^*\right\|^2 \leq \left\|x_n - x^*\right\|^2 - 2\alpha_n\psi\left(\left\|x_n - x^*\right\|^2\right) \\ &-2\alpha_n\left(\xi_n, x_n - x^*\right) + \alpha_n^2\left\|\zeta_n + \xi_n\right\|^2. \end{aligned} \tag{3.5}$$

Applying the conditional expectation with respect to A_n to the both sides of (3.5) we obtain

$$E\left[\left\|x_{n+1}-x^*\right\|^2/A_n\right]$$
$$\leq\left\|x_n-x^*\right\|^2-2\alpha_n\psi\left(\left\|x_n-x^*\right\|^2\right)$$
$$-2\alpha_n E\left[\left(\xi_n,x_n-x^*\right)/A_n\right]+\alpha_n^2 E\left[\left\|\zeta_n+\xi_n\right\|^2/A_n\right]. \tag{3.6}$$

It is easy to see that

$$E\left[\left\|\zeta_n+\xi_n^{\,2}\right\|/A_n\right]$$
$$\leq 2E\left[\left\|\zeta_n\right\|^2/A_n\right]+2E\left[\left\|\xi_n\right\|^2/A_n\right]$$
$$=2\left\|\zeta_n\right\|^2+2E\left[\left\|\xi_n\right\|^2/A_n\right]. \tag{3.7}$$

Since T is weakly contractive and therefore nonexpansive, one gets

$$\left\|x_n-Tx_n\right\|^2\leq 4\left\|x_n-x^*\right\|^2.$$

Taking into account (3.7), the inequality (3.9) is estimated as follows:

$$E\left[\left\|x_{n+1}-x^*\right\|^2/A_n\right]$$
$$\leq\left(1+8\alpha_n^2\right)\left\|x_n-x^*\right\|^2-2\alpha_n\psi\left(\left\|x_n-x^*\right\|^2\right)$$
$$-2\alpha_n E\left[\left(\xi_n,x_n-x^*\right)/A_n\right]+2\alpha_n^2 E\left[\left\|\xi_n\right\|^2/A_n\right]. \tag{3.8}$$

Now the unconditional expectation implies

$$E\left[\left\|x_{n+1}-x^*\right\|^2\right]$$
$$\leq\left(1+8\alpha_n^2\right)E\left[\left\|x_n-x^*\right\|^2\right]-2\alpha_n E\left[\psi\left(\left\|x_n-x^*\right\|^2\right)\right]$$
$$-2\alpha_n E\left[\left(\xi_n,x_n-x^*\right)\right]+2\alpha_n^2 E\left[\left\|\xi_n\right\|^2\right] \tag{3.9}$$

Next we need the Jensen inequality for a convex function $\psi\left(\left\|x_n-x^*\right\|^2\right)$:

$$E\left[\psi\left(\|x_n - x^*\|^2\right)\right] \le \psi\left(E\left[\|x_n - x^*\|^2\right]\right)$$

(see [9,10]). This allows us to rewrite (3.9) in the form

$$\begin{aligned} &E\left[\|x_{n+1} - x^*\|^2\right] \\ &\le \left(1+8\alpha_n^2\right)E\left[\|x_n - x^*\|^2\right] \\ &-2\alpha_n\psi\left(E\left[\|x_n - x^*\|^2\right]\right) + 2\alpha_n^2 E\left[\|\xi_n\|^2\right] \end{aligned} \tag{3.10}$$

because of

$$E\left[\left(\xi_n, x_n - x^*\right)\right] = 0.$$

Denoting $\lambda_n = E\left[\|x_n - x^*\|^2\right]$ we have

$$\lambda_{n+1} \le \left(1+8\alpha_n^2\right)\lambda_n - 2\alpha_n\psi\left(\lambda_n\right) + 2C_1\alpha_n^2, \tag{3.11}$$

where in view of Definition 1.1, $\psi(\lambda)$ is a continuous and increasing function with $\psi(0) = 0.$ Due to (6), from Lemma 2.2 it follows

$$\lim_{n\to\infty} E\left[\|x_n - x^*\|^2\right] = 0$$

and the estimate (3.1) holds too. The theorem is proved. □

Remark 3.3. If a fixed point of weakly contractive mapping $T : G \to H$ exists, then it is unique [11].

Remark 3.4. The following example was presented in [11]: Let $Tx = \sin x,$ $G = [0,1]$ and $T : G \to G.$ It has been shown in [11] that

$$|\sin x - \sin y| \le |x-y| - \frac{1}{8}|x-y|^3$$

for all $0 \le x \le y \le 1.$ Then

$$\phi(\tau)=\frac{1}{8}\tau^3, \psi_1(\tau)=\frac{1}{8}\tau^4 \text{ and } \psi(t)=\frac{1}{8}t^2.$$

Definition 3.5. Let a nonexpansive mapping

$T:G \to H$ have a unique fixed point x^*. T is said to be weakly sub-contractive on the closed convex subset $G \subset H$ if there exists continuous and increasing function $\psi(t)$ defined on IR^+ such that ψ is positive on $IR^+/\{0\}$, $\psi(0)=0$, $\lim_{t\to+\infty}\phi(t)=+\infty,$ such that for all $x \in G$.

$$\langle x-Tx, x-x^* \rangle \geq \psi\left(\left\|x-x^*\right\|^2\right). \tag{3.12}$$

Theorem 3.6. Assume that a mapping $T:G \to H$ is weakly sub-contractive and the function $\psi(t)$ in (3.12) is convex on IR^+ Then the results of Theorem 3.1 holds for the sequence $\{x_n\}$ generated by (1.5)-(1.7).

The second inequality in (1.6) can be omitted if we assume not less than linear growth of $\psi(\lambda)$ "on infinity" and put $\alpha_n \to 0$ as $n \to \infty$

Theorem 3.7. Assume that a mapping $T:G \to H$ is weakly sub-contractive and the function $\psi(t)$ in (3.12) is convex on IR^+ Suppose that instead of (1.6) the conditions

$$\lim_{n\to\infty}\alpha_n = 0 \text{ and } \sum_{n=1}^{\infty}\alpha_n = \infty. \tag{3.13}$$

hold. In addition, let $\alpha_n \leq \bar{\alpha} < 0.5$ and

$$\lim_{\lambda\to\infty}\frac{\psi(\lambda)}{\lambda} \geq 4\bar{\alpha}. \tag{3.14}$$

Then the sequence $\{x_n\}$ generated by (1.5), (1.7) and (3.13) converges in mean square to x^*. There exists an infinite subsequence $\{n_l\}, l = 1, 2, \cdots,$ such that

$$E\left[\left\|x_{n_l} - x^*\right\|^2\right] \le \psi^{-1}\left(\frac{1}{2\sum_{n=1}^{n_l}\alpha_n} + \frac{C_3\alpha_{n_l}}{2}\right).$$

where

$$C_3 = 2C_1 + 8\max\left\{\left\|x^* - x_1\right\|^2, C_2\right\},$$
$$C_2 = \lambda_* + 2C_1\bar{\alpha}^2 + 8\bar{\alpha}^2\lambda_*,$$
$$\lambda_* = \psi^{-1}\left(C_1\bar{\alpha} + 4\bar{\alpha}\lambda_*\right). \tag{3.15}$$

Proof. Consider the inequality (11) in the form

$$\lambda_{n+1} \le \lambda_n - 2\alpha_n\psi\left(\lambda_n\right) + 2C_1\alpha_n^2 + 8\alpha_n^2\lambda_n, \tag{3.16}$$

where $\lambda_n = E\left[\left\|x_n - x^*\right\|^2\right].$ Observe that it is derived by making use of (3.4) and the nonexpansivity property of T. We shall show that λ_n are bounded for all $n \in I = \{1, 2, \cdots\}$. Indeed, since $\psi(\lambda)$ is a convex increasing continuous function, we conclude that

$\psi_1(\lambda) = \frac{\psi(\lambda)}{\lambda}$ is nondecreasing and since (3.14) holdsthe inequality $\psi(\lambda) \le C_1\bar{\alpha} + 4\bar{\alpha}\lambda$ has a solution $\lambda \le \lambda_*$, where λ_* is the unique root of the scalar equ0 ation $\psi(\lambda) = C_1\bar{\alpha} + 4\bar{\alpha}\lambda.$ Together with this, (3.4) and (3.14) are co-ordinated by the parameter $\bar{\alpha} < 0.5.$

Only one alternative can happen for each $n \in I:$ either

$$H_1 := -2\alpha_n \psi(\lambda_n) + 2C_1\alpha_n^2 + 8\alpha_n^2\lambda_n \geq 0$$

or

$$H_2 := -2\alpha_n \psi(\lambda_n) + 2C_1\alpha_n^2 + 8\alpha_n^2\lambda_n < 0.$$

Denote $I_1 = \{n \in I | H_1 \text{ is true}\}$ and $I_2 = \{n \in I | H_2 \text{ is true}\}$. It is clear that $I_1 \cup I_2 = I$. From the hypothesis H_1, it arises $\psi(\lambda_n) \leq C_1\bar{\alpha} + 4\bar{\alpha}\lambda_n$,

and then $\lambda_n \leq \lambda_*$ for all $n \in I_1$. From the hypothesis H_2, we have: $\lambda_{n+1} < \lambda_n$ for all $n \in I_2$. Consider all the possible cases:

1) $I_2 = \varnothing$. Then $\lambda_n \leq \lambda_*$ for all $n \in I$.

2) $I_1 = \varnothing$. Then $\lambda_n \leq \lambda_1$ for all $n \in I$.

3) Let $I_1 = \{1, 2, \cdots, N_0\}$ and $I_2 = \{N_0 + 1, N_0 + 2, \cdots\}$. Then $\lambda_n \leq \lambda_*$ for $n = 1, 2, \cdots, N_0$. By (3.16), $\lambda_{N_0+1} \leq C_2$. It is obvious that $\lambda_n \leq C_2$ for $N_0 + 2, N_0 + 3, \cdots$. Therefore, $\lambda_n \leq C_2$ for all $n \in I$.

4) Let $I_2 = \{1, 2, \cdots, N_0\}$ and $I_1 = \{N_0 + 1, N_0 + 2, \cdots\}$. Then $\lambda_n \leq \lambda_1$ for $n = 1, 2, \cdots, N_0$. and $\lambda_n \leq \lambda_*$ for $N_0 + 1, N_0 + 2, \cdots$ Thus, $\lambda_n \leq \lambda_*$ for all $n \in I$.

5) Let I_1 and I_2 be unbounded sets. Consider an arbitrary interval

$I_s = [n_s + 1, n_{s+1} - 1] \subset I_2,$

where $n_s, n_{s+1} \in I_1, s = 1, 3, 5, \cdots$. It is easy to be sure that $\lambda_{n_s} \leq \lambda_*$ and $\lambda_n \leq C_2$ for all $n \in I_s$.

6) The other situations of bounded and unbounded sets I_1 and I_2 are covered by the items 1)-5). Consequently, we have the final result: $\lambda_n \leq \max\{\lambda_1, C_2\}$ for all $n \in I$.

Thus, we obtain the inequality

$$\lambda_{n+1} \leq \lambda_n - 2\alpha_n \psi(\lambda_n) + C_3 \alpha_n^2, \tag{3.17}$$

where C_3 is defined by (3.15). Now Lemma 2.2 with the condition (2.3) implies the result. □

Remark 3.8. For a linear function $\psi(\tau) = c\tau$ which is convex and concave at the same time we suppose $2 \geq c > 4\bar{\alpha}$.

Remark 3.9. If G is bounded or more generally $\{x_n\}$ is bounded, then the inequality (3.17) (with some different constant C_3) immediately follows from (3.16).

4. ESTIMATES OF THE MEAN SQUARE CONVERGENCE RATE

Using Lemmas 2.3 and 2.4 we are able to give two general theorems on the nonasymptotic estimates of the mean square convergence rate for sequence $\{x_n\}$ generated by the stochastic approximation algorithm (1.5)-(1.7).

Again we introduce denotations 1)-5) from Section 2 induced now by the recurrent inequality (3.11):

1) $\psi^{-1}(z)$ and $\Phi^{-1}(z)$ are the inverse functions to $\psi(t)$ and $\Phi(t)$, respectively;

2) $v(s) = \psi^{-1}\left[C_1 c_0 \alpha(s-1)\right] + 2C_1\alpha^2(s-1), c_0 > 1$ is a fixed control parameter;

3) $u(s, C_2) = \Phi^{-1}\left[\Phi(C_2) - a(F(s) - F(2))\right], C_2 > 0,$ $a = \dfrac{c_0 - 1}{c_0} < 1$

4) $w(s,v(r)) = \Phi^{-1}\left[\Phi(v(r)) - a(F(s) - F(2))\right],\ 2 \le s, r < \infty;$

5) $Q = \Phi^{-1}\left[\Phi\left(\left\|x_1 - x^*\right\|^2\right) - a(F(2) - F(1))\right].$

Introduce also the basic condition (P).

Theorem 4.1. Assume that all the conditions of Theorem 3.1 are fulfiled and i) the condition (P) holds for the functions

$w(s,v(2)) = u(s,v(2))$ and $v(s)$;

ii) $u(s,v(2)) \ge v(s)$ as $s \to \infty$;

iii) $c_0 > 1$ is chosen such that $u(s,v(2)) \ge v(s)$ as $s \to 2$.

Then the sequence $\{x_n\}$ generated by (1.5)-(1.7) converges in average to a unique fixed point x^* of T and for all $n \ge 1$

$$E\left[\left\|x_n - x^*\right\|^2\right] \le C_0 u(n, C_2),$$
$$C_2 = \max\{Q, v(2)\}. \tag{4.1}$$

Theorem 4.2. Assume that all the conditions of Theorem 3.1 are fulfiled and i) the condition (P) holds for the functions $v(s)$

and $w(s,v(z))$ with any fixed $z \in [2,\infty)$;

ii) $u(s,v(2)) \le v(s)$ as $s \to \infty$;

iii) If $Q \le v(2)$ and $c_0 > 1$ is chosen such that $u(s,v(2)) \le v(s)$ as $s \to 2$, then the sequence $\{x_n\}$

generated by (1.5)-(1.7) converges in average to a unique fixed point x^* of T and for all $n \ge 1$

$$E\left[\left\|x_n - x^*\right\|^2\right] \le C_0 v(n); \tag{4.2}$$

iv) In all the remaining cases, (4.1) holds for $1 \le n \le \overline{s}$ and (4.2) for $n \ge \overline{s}$, where $\overline{s}$ is a unique root of the equation $u(s, C_2) = v(s)$ on the interval $[2, \infty)$.

Let us provide the examples of functions $\psi(\tau)$ and $\Phi(\lambda)$ suitable for Theorems 4.1 and 4.2 (see [12,13]).

1) Below in Corollaries 4.3-4.6 we use the functions $\psi(\tau) = \tau^{\nu}$ with $\nu \ge 1$. For them

$$\Phi(\lambda) = \begin{cases} \ln \lambda, \text{ if } \nu = 1, \\ \dfrac{\lambda^{1-\nu}}{1-\nu}, \text{ if } \nu \ne 1, \end{cases} \tag{4.3}$$

and

$$\Phi^{-1}(z) = \begin{cases} \exp z, \text{ if } \nu = 1, \\ \left[(1-\nu)z\right]^{\frac{1}{1-\nu}}, \text{ if } \nu \ne 1, \end{cases} \tag{4.4}$$

2) If $\psi(\tau) = \exp \tau - 1,\ \tau > 0,$ then

$$\Phi(\tau) = \ln(1 - \exp \tau) \text{ and } \Phi^{-1}(z) = -\ln(1 - \exp z).$$

3) If $\psi(\tau) = \dfrac{\tau}{\tau + 1},\ \tau > 0,$ then

$$\Phi(\tau) = \ln \tau + \tau \text{ and } \Phi^{-1}(z) \le \ln(1 + \exp z).$$

4) If $\psi(\tau) = \dfrac{\tau^2}{\tau + 1},\ \tau > 0,$ then

$$\Phi(\tau) = \ln \tau - \frac{1}{\tau}.$$

In this example we are unable to define $\Phi^{-1}(z)$ in analitical form, therefore suggest to calculate it numerically by computer.

We next present very important corollaries from Theorems 4.1 and 4.2, where their assumptions automatically guarantee accomplishment of the condition (P) (see [4]). The functions $\psi(\tau)$ coincide with the point 1) above.

Corollary 4.3. Assume that $T:G\to H$ is a strongly contractive mapping, that is, (1.4) is satisfied with

$0<q<1$. Let in (1.5) $\alpha_n=\frac{b}{n}, b>0.$ Then

$$F(t)=b\ln t, \Phi(\tau)=\frac{\ln\tau}{q_1}, \Phi^{-1}(z)=e^{q_1 z}, q_1=1-q,$$

$$v(s)=\left(\frac{2c_0C_1}{b}+2C_1\right)\frac{1}{s-1}, u(s,C)=C\left(\frac{2}{s}\right)^{abq_1},$$

$$Q=\left\|x_1-x^*\right\|^2\left(\frac{1}{2}\right)^{abq_1}.$$

I. Suppose that $b>\frac{1}{q_1}$ and $c_0\ge\frac{bq_1}{bq_1-1}.$ Then

$\lim_{n\to\infty}E\left[\left\|x_n-x^*\right\|^2\right]=0$ and 1) If $Q\le v(2)$ and $c_0\ge\frac{bq_1}{bq_1-2}$, we have for all $n>1$

$$E\left[\left\|x_n-x^*\right\|^2\right]\le v(n); \tag{4.5}$$

2) In all the remain cases

$$E\left[\left\|x_n-x^*\right\|^2\right]\le u(n,C),$$
$$C=\max\{Q,v(2)\}, 1<n\le\bar{s}, \tag{4.6}$$

and

$$E\left[\left\|x_n - x^*\right\|^2\right] \le v(s),\ n \ge \bar{s}, \tag{4.7}$$

where $\bar{s}$ is a unique root of the equation

$u(s,C) = v(s)$ on the interval $[2,\infty)$.

II. Suppose that $\frac{1}{2q_1} < b < \frac{1}{q_1}$ and $c_0 \ge \frac{bq_1}{1-bq_1} > 1.$

Then $\lim_{n\to\infty} E\left[\left\|x_n - x^*\right\|^2\right] = 0$ and the estimate (4.6) holds for all $n \ge 1.$

Corollary 4.4. Assume that $T : G \to H$ is a strongly contractive mapping, that is, in (1.4) is satisfied with

$0 < q < 1.$ Let in (1.5) $\alpha_n = \frac{b}{n^\kappa}, b > 0, 0 < \kappa < 1.$ Then

$$F(t) = \frac{b}{1-\kappa} t^{1-\kappa}, \Phi(\tau) = \frac{\ln \tau}{q_1},$$

$$\Phi^{-1}(z) = \left(\frac{1-\kappa}{b} z\right)^{\frac{1}{1-\kappa}}, q_1 = 1 - q,$$

$$v(s) = \left(\frac{2c_0 C_1}{b} + 2C_1\right) \frac{1}{s-1},$$

$$u(s,C) = C \exp\left\{-\frac{abq_1}{1-\kappa}\left(s^{1-\kappa} - 2^{1-\kappa}\right)\right\},$$

$$Q = \left\|x_1 - x^*\right\|^2 \exp\left\{-\frac{abq_1}{1-\kappa}\left(2^{1-\kappa} - 1\right)\right\}.$$

Suppose that $c_0 \ge \frac{bq}{bq-1}.$ Then $\lim_{n\to\infty} E\left[\left\|x_n - x^*\right\|^2\right] = 0$

and 1) If $Q \le v(2)$ and $c_0 \ge \frac{bq_1}{bq_1 - 2},$ we have for all

$$n > 1$$

$$E\left[\left\|x_n - x^*\right\|^2\right] \le v(n); \tag{4.8}$$

2) In all the remain cases the estimates (4.6) and (4.7) hold.

Corollary 4.5. Assume that $T:G \to H$ is a weakly contractive mapping of the class $C_{\phi(t)} = t^{2\nu-1}, \nu > 1,$ that is, in Theorem 3.1 $\psi(\tau) = \tau^{\nu}$. Let in (1.5)

$\alpha_n = \frac{b}{n}, b > 0.$ Then

$$F(t) = b\ln t, \Phi(\tau) = \frac{\tau^{1-\nu}}{1-\nu}, \Phi^{-1}(z) = \left[(1-\nu)z\right]^{\frac{1}{1-\nu}},$$

$$u(s,C) = C\left[1+(\nu-1)C^{\nu-1}ab\ln\frac{s}{2}\right]^{-\frac{1}{\nu-1}},$$

$$Q = \left\|x_1 - x^*\right\|^2\left[1+(\nu-1)\left\|x_1 - x^*\right\|^{2(\nu-1)}ab\ln 2\right]^{-\frac{1}{\nu-1}}.$$

If $c_0 > 1$ is chosen from the condition

$$\left(\frac{2c_0C_1}{b} + 2C_1\right)^{\nu-1} ab \le \nu^{-1},$$

then $\lim_{n\to\infty} E\left[\left\|x_n - x^*\right\|^2\right] = 0$ and for all $n > 1$

$$E\left[\left\|x_n - x^*\right\|^2\right] \le u(n,C),$$

$$C = \max\left\{Q, \left(\frac{2C_1c_0}{b}\right)^{\frac{1}{\nu}} + 2C_1\right\}. \tag{4.9}$$

Corollary 4.6. Assume that $T:G \to H$ is a weakly contractive mapping of the class $C_{\phi(t)} = t^{2\nu-1}, \nu > 1,$ that is, in Theorem 3.1 $\psi(\tau) = \tau^{\nu}$. Let in (1.5)

$\alpha_n = \frac{b}{n^{\kappa}}, b > 0, 0 < \kappa < 1.$ Then

$$F(t) = \frac{b}{1-\kappa} t^{1-\kappa}, \Phi(\tau) = \frac{\tau^{1-\nu}}{1-\nu}, \Phi^{-1}(z) = [(1-\nu)z]^{\frac{1}{1-\nu}},$$

$$v(s) = \left[\left(\frac{2c_0 C_1}{b} \right)^{\frac{1}{\nu}} + 2C_1 \right] \left(\frac{1}{s-1} \right)^{\frac{\kappa}{\nu}},$$

$$u(s, C) = C \left[1 + \frac{\nu - 1}{1-\kappa} C^{\nu-1} ab \left(s^{1-\kappa} - 2^{1-\kappa} \right) \right]^{-\frac{1}{\nu-1}},$$

$$Q = \left\| x_1 - x^* \right\|^2 \left[1 + \frac{\nu - 1}{1-\kappa} \left\| x_1 - x^* \right\|^{2(\nu-1)} ab \left(2^{1-\kappa} - 1 \right) \right]^{-\frac{1}{\nu-1}}.$$

I. Suppose that

$$\frac{\kappa}{\nu} < \frac{1-\kappa}{\nu - 1}.$$

1) If $Q \le v(2)$ and $c_0 > 1$ is chosen from the condition

$$\left[\left(\frac{2c_0 C_1}{b} \right)^{\frac{1}{\nu}} + 2C_1 \right]^{\nu-1} ab \ge \frac{2^{\kappa} \kappa}{\nu},$$

then $\lim_{n \to \infty} E\left[\left\| x_n - x^* \right\|^2 \right] = 0$ and for all $n > 1$

$$E\left[\left\| x_n - x^* \right\|^2 \right] \le v(n). \tag{4.10}$$

2) In all the remain cases the estimates (4.6) and (4.7) hold.

II. Suppose that

$$\frac{\kappa}{\nu} \ge \frac{1-\kappa}{\nu - 1}.$$

If $c_0 > 1$ is chosen from the condition

$$\left[\left(\frac{2c_0C_1}{b}\right)^{\frac{1}{\nu}}+2C_1\right]^{\nu-1}ab\leq\frac{\kappa}{\nu},$$

then $\lim_{n\to\infty}E\left[\left\|x_n-x^*\right\|^2\right]=0$ and for all $n\geq 1$

$$E\left[\left\|x_n-x^*\right\|^2\right]\leq u(n,C),C=\max\{Q,v(2)\}. \tag{4.11}$$

In addition to the examples presented in this section, we produce the functions $\psi(\tau)$ and $\Phi(\lambda)$ which have $\tau=0$ as a tangency point of the infinite degree multiplicity and given logarithmic estimates of the convergence rate.

We define the function $\psi(\tau)$ by the following way:

$$\psi(\tau)=-\exp(-f(\tau))[f'(\tau)]^{-1},\tau>0,$$

where $f(\tau)$ is differentiable and decreasing function, $\lim_{\tau\to 0}f(\tau)=\infty,$ and

$$\psi^{(l)}(0)=\lim_{\tau\to 0+}\psi^{(l)}(\tau),l=0,1,2,\cdots,$$

where (l) denote the derivative degrees of the function

$\psi(\tau),\psi^{(0)}(\tau)=\psi(\tau).$ It is easy to see that $\Phi(\tau)=-\exp f(\tau)$

And $\Phi^{-1}(z)=f^{-1}(\ln(-z)).$

In particulari) $f(\tau)=\frac{1}{\tau},\psi(\tau)=\tau^2\exp\left(-\frac{1}{\tau}\right).$ We have $\Phi(\tau)=-\exp\frac{1}{\tau}$

and $\Phi^{-1}(z)=\frac{1}{\ln(-z)}.$ We have to verify that $\psi(\tau)$ is convex. In fact, it is true because

$$\frac{d^2\psi(\tau)}{d\tau}=\left(2+\frac{2\tau+1}{\tau^2}\right)\exp\left(-\frac{1}{\tau}\right)>0,\forall\tau>0.$$

Beside this, it is easy to see that $\psi(\tau) \le \tau,$ at least, on the interval $[0,1]$. In the next examples we leave to readers to check these properties.

ii) $f(\tau) = \frac{1}{\tau^s},\ s \ge 1,\ \psi(\tau) = \exp\left(-\frac{1}{\tau^s}\right)\tau^{s+1}.$

We have $\Phi(\tau) = -\exp\frac{1}{\tau^s}$ and $\Phi^{-1}(z) = (\ln(-z))^{-\frac{1}{s}}.$

iii) $f(\tau) = \exp\frac{1}{\tau^s}, \psi(\tau) = \exp\left(-\left[\frac{1}{\tau^s} + \exp\left(\frac{1}{\tau^s}\right)\right]\right)\tau^{s+1}$

We have $\Phi(\tau) = -\exp\left(\exp\frac{1}{\tau^s}\right)$ and $\Phi^{-1}(z) = (\ln\ln(-z))^{-\frac{1}{s}}.$

5. ALMOST SURE CONVERGENCE OF STOCHASTIC APPROXIMATIONS FOR NONEXPANSIVE MAPPINGS

Consider next the almost surely convergence of stochastic approximations. First of all, we need the stochastic analogy of Lemma 2.5:

Lemma 5.1. Let $\{\alpha_k\}$ be sequences of non-negative real numbers and $\{\beta_k\}$ be sequence of random A_k measurable variables, a.s. nonnegative for all $k \ge 1.$ Assume that

$$\sum_{k=1}^{\infty} \alpha_k = \infty \text{ and } \sum_{k=1}^{\infty} \alpha_k \beta_k < \infty.$$

If $\lim_{k\to\infty} \alpha_k = 0$ and there exists $c > 0$ such that for all $k \ge 1$

$$\left|\beta_{k+1} - E[\beta_k / A_n]\right| \le c\alpha_k \text{ a.s.,} \tag{5.1}$$

then $\lim_{k\to\infty} \beta_k = 0$ a.s.

The proof can be provided by the scheme of nonstochastic case (see Proposition 2 in [8]) or as it was done in [5].

We need also the following lemma from [14] as applied to our case of Hilbert spaces (the concepts of modulus of convexity $\delta_B(\varepsilon)$ of Banach spaces B or Hilbert spaces H can be found in [15] and [16]).

Lemma 5.2. If $F = I - T$ with a nonexpansive mapping $T : G \to H$, then for all $x, y \in D(T)$,

$$(Fx - Fy, x - y) \geq R_1^2 \delta_H\left(\frac{\|Fx - Fy\|}{2R_1}\right),$$

where

$$R_1 = \sqrt{2^{-1}\left(\|x - y\|^2 + \|Tx - Ty\|^2\right)} \leq \|x - y\|.$$

If $\|x\| \leq R$ and $\|y\| \leq R$ with $x, y \in D(T)$, then $R_1 \leq 2R$ and

$$(Fx - Fy, x - y) \geq L^{-1} R^2 \delta_H\left(\frac{\|Fx - Fy\|}{4R}\right).$$

Theorem 5.3. Assume that a mapping $T : G \to H$ is nonexpansive and its fixed point set N is nonempty. If

(1.8) holds and $E\left[\|\xi_n\| / A_n\right] \leq C_0$, then the sequence

$\{x_n\}$ generated by (1.5)-(1.7) weakly almost surely converges to some $\tilde{x} \in N$.

Proof. Let $x^* \in N$. We next use Lemma 5.2 and the estimate (see [17], p. 49) $\delta_H(\varepsilon) \geq \frac{\varepsilon^2}{8}$ to get $\left(x - Tx, x - x^*\right) \geq R_1^2 \delta_H\left(\frac{\|x - Tx\|}{2R_1}\right) \geq \frac{\|x - Tx\|^2}{32}$.

In this case the inequality (3.3) implies

$$\|x_{n+1}-x^*\|^2 \le \left(1+8\alpha_n^2\right)\|x_n-x^*\|^2-(32)^{-1}\|\alpha_n x_n-Tx_n\|^2 -2\alpha_n\left(\xi_n,x_n-x^*\right)+2\alpha_n^2\|\xi_n\|^2. \tag{5.2}$$

Similarly to (3.10), we have

$$E\left[\|x_{n+1}-x^*\|^2/A_n\right] \le \left(1+8\alpha_n^2\right)\|x_n-x^*\|^2-2\alpha_n E\left[\|x_n-Tx_n\|^2/A_n\right] -2\alpha_n E\left[\left(\xi_n,x_n-x^*\right)/A_n\right]+2\alpha_n^2 E\left[\|\xi_n\|^2/A_n\right]. \tag{5.3}$$

Denote $\lambda_n=E\left[\|x_n-x^*\|^2\right]$ and $\beta_n=E\left[\|x_n-Tx_n\|^2\right]$ and apply the unconditional expectation to both sides of (5.3). Then

$$\lambda_{n+1}\le\left(1+8\alpha_n^2\right)\lambda_n-2\alpha_n\beta_n+2C_1\alpha_n^2. \tag{5.4}$$

It follows from this that

$$\lambda_{n+1}\le\left(1+8\alpha_n^2\right)\lambda_n+2C_1\alpha_n^2.$$

Since $\sum_1^\infty\alpha_n^2<\infty$ and due to Lemma 2.1, we conclude that $\{\lambda_n\}$ is bounded. Consequently, $\{x_n\}$ is bounded a.s. that follows from the theory of convergent quasimartingales (see [5,18]).

We now need Lemma 5.1. It is not difficult to see that

$$\sum_1^\infty\alpha_n\beta_n=\sum_1^\infty\alpha_n E\left[\|x_n-Tx_n\|^2\right]<\infty. \tag{5.5}$$

The last gives us

$$\sum_1^\infty\alpha_n\|x_n-Tx_n\|^2<\infty\text{ a.s.}$$

Next we evaluate the following difference:

$$\begin{aligned}&\|x_{n+1}-Tx_{n+1}\|^2-\|x_n-Tx_n\|^2\\ &\le\left(\|x_{n+1}-Tx_{n+1}\|+\|x_n-Tx_n\|\right)\\ &\times\left(\|x_{n+1}-Tx_{n+1}\|-\|x_n-Tx_n\|\right)\end{aligned}$$

It is easy to see that $\|x_n-Tx_n\|$ is bounded a.s. Indeed, since $\|x_n\|\le C_1$ a.s., there exists a constant $C_2>0$ such that

$$\|x_n-Tx_n\|=2\|x_n-x^*\|\le C_2 \text{ a.s.}$$

Therefore

$$\|x_{n+1}-Tx_{n+1}\|+\|x_n-Tx_n\|\le 2C_2 \text{ a.s.}$$

It is obviously that

$$\begin{aligned}&\|x_{n+1}-Tx_{n+1}\|-\|x_n-Tx_n\|\\ &\le\|(x_{n+1}-x_n)+(Tx_n-Tx_{n+1})\|\le 2\|x_{n+1}-x_n\|\\ &=2\|\Pr_G x_n-\alpha_n(x_n-Tx_n+\xi_n)-\Pr_G x_n\|\\ &\le 2\alpha_n\|x_n-Tx_n\|+2\alpha_n\|\xi_n\|.\end{aligned}$$

Thus,

$$\begin{aligned}&E\left[\|x_{n+1}-Tx_{n+1}\|^2/A_n\right]-\|x_n-Tx_n\|^2\\ &\le 4\alpha_n C_2^2+4\alpha_n C_2 E\left[\|\xi_n\|/A_n\right]\\ &\le 4C_2(C_2+C_0)\alpha_n \text{ a.s..}\end{aligned}$$

By Lemma 5.1, $\|x_n-Tx_n\|\to 0$ a.s. as $n\to\infty.$

Since $\{x_n\}$ is bounded a.s., there is a subsequence $\{x_{n_k}\}$ weakly convergent to some point $\tilde{x}.$ Since G is convex and closed, consequently, weakly closed, we assert that $\tilde{x}\in G.$ It is known that a nonexpansive mapping T is weakly demiclosed, therefore, $\tilde{x}\in N$ a.s. Weak almost surely convergence of whole sequence $\{x_n\}$ is shown by the standard way [8]. □

Corollary 5.4. Assume that $T : G \to H$ is a weakly contractive mapping of the class $C_{\phi(t)}$. If $\sum_1^\infty \alpha_n = \infty$, $\sum_1^\infty \alpha_n^2 < \infty$ and $E\left[\|\xi_n\| / A_n\right] \le C_0$, then the sequence

$\{x_n\}$ generated by (1.5)-(1.7) strongly almost surely converges to unique fixed point x^* of T.

Proof. We have from (3.4)

$$\|x_n - Tx_n\|\|x_n - x^*\| \ge \psi\left(\|x_n - x^*\|^2\right). \tag{5.6}$$

Since $\{x_n\}$ is bounded a.s. and $\|x_n - Tx_n\| \to 0$ a.s.

as $n \to \infty$, we conclude that $\psi\left(\|x_n - x^*\|^2\right) \to 0$ a.s.

The proof follows due to the properties of the function ψ. □

Remark 5.5. It is clear that all the results remain still valid for self-mappings $T : G \to G$. However, in this case, unlike any deterministic situation, the algorithm (1.5)-(1.7) must use the projection operator Pr_G because the vector $v_n = x_n - \alpha_n S_n x_n$ not always belongs to G. If $T_n x_n = Tx_n + \xi_n \in G$ for all $n \ge 1$ and $0 < \alpha_n < 1$, then (1.5) can be replaced by Ya. I. Alber, C. E. Chidume, Jinlu Li $x_{n+1} = (1 - \alpha_n)x_n + \alpha_n T_n x_n, n = 1, 2, \cdots, x_1 \in G.$

REFERENCES

1. M. T. Wasan, "Stochastic Approximation," Cambridge University Press, Cambridge, 1969.
2. M. B. Nevelson and R. Z. Hasminsky, "Stochastic Approximation and Recursive Estimation," AMS Providence, Rhode Island, 1973.
3. Ya. Alber and S. Shilman, "Nonasymptotic Estimates of the Convergence Rate of Stochastic Iterative Algorithms," Automation and Remote Control, Vol. 42, 1981, pp. 32-41.

4. Ya. Alber and S. Shilman, "General Nonasymptotic Estimates of the Convergence Rate of Iterative Stochastic Algorithms," USSR Computational Mathematics and Mathematical Physics, Vol. 25, No. 2, 1985, pp. 13-20.
5. K. Barty, J.-S. Roy and C. Strugarek, "Hilbert-Valued Perturbed Subgradient Algorithms," Mathematics of Operation Research, Vol. 32, No. 3, 2007, pp. 551-562.
6. X. Chen and H. White, "Asymptotic Properties of Some Projection-Based Robbins-Monro Procedures in a Hilbert Space," Studies in Nonlinear Dynamics & Econometrics, Vol. 6, No. 1, 2002, pp. 1-53.
7. Ya. I. Alber and A. N. Iusem, "Extension of Subgradient Techniques for Nonsmooth Optimization in Banach Spaces," Set-Valued Analysis, Vol. 9, No. 4, 2001, pp. 315-335.
8. Ya. Alber, A. Iusem and M. Solodov, "On the Projection Subgradient Method for Nonsmooth Convex Optimization in a Hilbert Space," Mathematical Programming, Vol. 81, No. 1, 1998, pp. 23-35.
9. A. P. Dempster, N. M. Laird and D. B. Rubin, "Maximal Likelihood from Incomplete Data via the EM Algorithm," Journal of the Royal Statistical Society, Vol. 39, No. 1, 1977, pp. 185-197.
10. W. Rudin, "Real and Complex Analysis," McGraw-Hill, New York, 1978.
11. Ya. I. Alber and S. Guerre-Delabriere, "Principle of Weakly Contractive Maps in Hilbert Spaces," Operator Theory, Advances and Applications, Vol. 98, 1997, pp. 7-22.
12. Ya. I. Alber, "Recurrence Relations and Variational Inequalities," Soviet Mathematics Doklady, Vol. 27, 1983, pp. 511-517.
13. Ya. Alber, S. Guerre-Delabriere and L. Zelenko, "The Principle of Weakly Contractive Maps in Metric Spaces," Communications on Applied Nonlinear Analysis, Vol. 5, No. 1, 1998, pp. 45-68.
14. Ya. I. Alber, "New Results in Fixed Point Theory," Preprint, Haifa Technion, 2000.
15. J. Diestel, "The Geometry of Banach Spaces," Lecture Notes in Mathematics, No. 485, Springer, Berlin, 1975.
16. T. Figiel, "On the Moduli of Convexity and Smoothness," Studia Mathematica, Vol. 56, No. 2, 1976, pp. 121-155.
17. Ya. Alber and I. Ryazantseva, "Nonlinear Ill-Posed Problems of Monotone Type," Springer, Berlin, 2006.
18. M. Métivier, "Semimartingales," De Gruyter, Berlin, 1982.

CHAPTER 5

Stochastic Process Optimization Technique

Hiroaki Yoshida[1], Katsuhito Yamaguchi[2], Yoshio Ishikawa[1]

[1]*College of Science and Technology, Nihon University, Chiba, Japan*

[2]*Junior College, Nihon University, Chiba, Japan*

ABSTRACT

The conventional optimization methods were generally based on a deterministic approach, since their purpose is to find out an accurate solution. However, when the solution space is extremely narrowed as a result of setting many inequality constraints, an ingenious scheme based on experience may be needed. Similarly, parameters must be adjusted with solution search algorithms when nonlinearity of the problem is strong, because the risk of falling into local solution is high. Thus, we here propose a new method in which the optimization problem is replaced with stochastic process based on path integral techniques used in quantum mechanics and an approximate value of optimal solution is calculated as an expected value instead of accurate value. It was checked through some optimization problems that this method using stochastic process is effective. We call this new optimization method "stochastic process optimization technique (SPOT)". It is expected that this method will enable efficient optimization by avoiding the above difficulties. In this report, a new optimization method based on a stochastic process is formulated, and several calculation examples are shown to prove its effectiveness as a method to obtain approximate solution for optimization problems.

Keywords: Optimization, Stochastic Process, Path Integral, Expected Value, Quantum Mechanics

1. INTRODUCTION

Optimization methods are useful to system design, and especially the applications to engineering problems are expected. In order to obtain the accurate optimal solution, generally the conventional optimization methods were based on a deterministic approach. However, when design variables have many inequality restriction conditions, in order to search for a solution efficiently, it is necessary to adjust a parameter with trial and error. Moreover, if the problem is complex, the risk of falling into a local solution is also high.

Our method obtains an approximate value of an optimal solution using stochastic process. The optimization methods using stochastic process like simulated annealing (SA) [1] are found. However, in these methods, stochastic process is used as a means of searching for an accurate solution efficiently. Our method is based on probability theory, and essentially differs from the concepts of the conventional optimization methods. The purpose of the conventional optimization is to obtain the accurate optimal solution, but that of our method is to obtain the approximate value of the optimal solution.

We showed that the approximate value of the optimal solution could be obtained using stochastic process [2] [3] . And even if it was an approximate value of the optimal solution obtained by this method, it was shown that it is useful in engineering [4] . Furthermore, it was shown that this method can apply to general engineering optimization problems containing static design variables and dynamic design variables [5] .

As mentioned above, it was shown that the approximate value of the optimal solution can be obtained by this method using stochastic process. The aim of this paper is not to show the results of the applications of our method, but to bring our original idea to a wide reading audience.

Thus, we propose a new method in which an optimization problem is replaced with a stochastic process based on path integral techniques used in quantum mechanics and an approximate value of an optimal solution is calculated as an expected value instead of an accurate value. We call this new optimization method the "stochastic process optimization technique (SPOT)". It is expected that this method will enable efficient optimization by avoiding the above difficulties.

Since expected values are stochastic average values, a stochastic approach is not appropriate to obtain an accurate solution as a deterministic approach. However, its characteristics are simple algorithms and the fewer number of manual parameters. The resultant solution is approximate enough to a global solution of engineering problems whose solution spaces are expected to be relatively smooth. The approximate solution can be an initial value to obtain an accurate solution with a deterministic process.

In this paper, a new optimization method is formulated based on a stochastic process. Some calculation examples are introduced to show how effective the approximate solutions are for optimization problems. The examples are a simple benchmark test, a classic line of swiftest descent problem, and an optimized glide problem of a hang glider as an introductory engineering optimization problem.

2. PATH INTEGRAL AND OPTIMIZATION

According to the principle of least action, Newtonian motion of a mass point makes the action integral minimum:

$$I = \int_{t_1}^{t_2} L\left(x_i, \mathrm{d}x_i/\mathrm{d}t, t\right)\mathrm{d}t, \tag{1}$$

where $L\left(x_i, \mathrm{d}x_i/\mathrm{d}t, t\right)$ is the Lagrange function. Also, the Euler-Lagrange equation obtained with variation of I is equal to an equation of Newtonian motion. On the other hand, the motion of a particle is determined stochastically according to the quantum mechanics. A motion on the classical mechanics, which keeps I minimum, occurs at the "highest possibility". However, other motions can occur at certain probabilities. This results in some particular phenomena, such as the tunnel effect, which never occur on the classical mechanics.

The well-known simulated annealing (SA) method adopts such a quantum mechanical approach to the solution of optimization problems. SA introduces a probability distribution that reaches its maximum at an optimal value. This method numerically searches the optimal value (peak of the probability distribution) with a temperature parameter using the probability distribution that reaches its maximum of an optimal value instead of directly obtaining a variational problem of a performance index I This method searches solution spaces in the large, which include solutions that never obtained with the variational method. Therefore, one advantage is a low probability of reaching local solutions. On the other hand, there are some disadvantages. For example, the operation of a cooling schedule, which is a temperature parameter, is difficult. In addition, it may take much time in calculation to converge a solution.

R. Feynman constructed the "path integral" theory in 1948 and showed that it is equivalent to the quantum mechanics. The theory provides a countless number of paths that meet given boundary conditions, and the prob- ability (probability amplitude) of each particle taking its path is expressed in the form of multiple integral [6] . A stochastic behavior is the nature of quantum mechanics. It is therefore impossible to predict a measured physical quantity deterministically under a certain condition. Expected values only can be predicted. Thus, the expected value of a physical quantity is expressed in the form of multiple integral in the path integral method. Furthermore, the multiple integral calculation of an expected value can fit into numerical calculation with the Monte Carlo method.

In this paper, we propose a new numerical calculation method for optimization problems based on the quantum mechanics rather than SA. It means that we show a method of obtaining an approximate solution of an optimal solution as an expected value. An expected value is predicted to be located close to an optimal solution in a continuous probability distribution. This method can thus be applied to various engineering optimal problems.

The introduced probability distribution P here is basically the same as that in SA, and to obtain the minimum of performance function I, the following equation is set:

$$P = e^{-I/h} . \quad (2)$$

In SA, the peak of P is searched with $h \to 0$. On the other hand, we automatically generate a state variable of various time histories at probability P as finite value h. According to the generated frequency, the stochastic average (expected value) of the quantity to be obtained is calculated.

3. FORMULATION OF SPOT

According to the path integral method, the path of motion for a naturally stochastic particle such as photon and electron is obtained from an expected value as a path that keeps the action integral I minimum. We replace the action integral I with a performance index to recognize the motion of a general object as a stochastic process. Thus, we formulate the path integral method for a naturally stochastic particle in physics as an optimal method of obtaining a function that keeps the performance function I minimum. Note that since this performance function is stochastic, an optimal solution is obtained as an expected value, which results in an approximate solution instead of an accurate solution.

According to the path integral method, the action integral $I[x(t)]$ described in the above section is replaced with a performance function to define probability distribution P as shown in Equation (3). Here, $x(t)$ does not always show a path, but shows a general variable to be optimized. Also, A is the normalization constant.

$$P = \frac{1}{A}\exp(-I/h), \tag{3}$$

where h is the Plank constant in quantum mechanics. A stochastic mechanics system matches a classical mechanics system as $h \to 0$. However, in this situation, h is an arbitrary parameter that gives a fluctuation of solution. It means that h is a parameter to define a distribution of P. The larger h is, the wider the distribution is. Also, A is the normalization constant to make the summation of the probability 1. To define the probability distribution as shown in Equation (3), the expected value is obtained from the following:

$$\langle x \rangle = \int_{-\infty}^{\infty} [x(t) \cdot P] Dx(t), \tag{4}$$

where $Dx(t) = \mathrm{d}x_0 \mathrm{d}x_1 \cdots \mathrm{d}x_N$ shows the summation for the paths.

Here, the summation for the paths is the integral for all the countless paths (multiple integral), which for example connect each point x_i with lines at each time point t_i $(0 \le i \le N)$ divided by N in one-dimensional motion. Figure 1 shows some paths that are generated in a probability distribution of Equation (3) when a function defined in Equation (1) is a performance index in one-dimensional upcast motion.

The horizontal axis and vertical axis show time and vertical positions, respectively. The better performance values the paths have, in other words, the closer to the accurate solution (red line) the values are, the higher the density of the paths is. Also, there are some paths that never exist with a deterministic method. The optimal solution is the expected value of all the paths. In this case, that solution is an approximate solution of an accurate solution. The expected solution that is thus obtained shows a path of the highest probability for move of a stochastic behavioral particle, such as an electron, between two points. When those paths are replaced with arbitrary functions, approximate solutions that make the performance function values minimum are obtained.

This idea can be extended to obtain an expected value of a m-dimensional functional that consists of $m \times N$ functions of time in the following way. First, independent variable t is divided by N, and $x_k^{(i)}$ values at time t_i $(0 \leq i \leq N)$ are randomly chosen. Next, a path that connects these points is set to the time history of variable x_k (Figure 2).

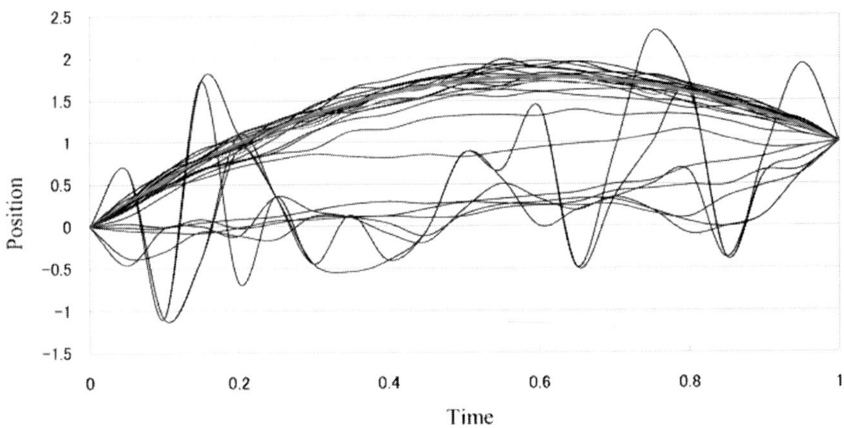

Figure 1. Paths generated by a probability distribution.

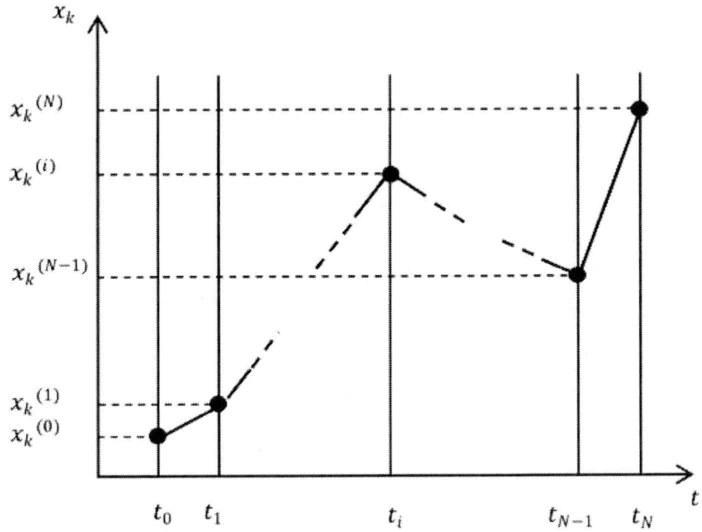

Figure 2. Time history of a variable.

Note that the subscript k shows the k-th variable $(1 \le k \le m)$. From Equation (4), the expected value $\langle x_k^{(i)} \rangle$ of x_k at t_i $(0 \le i \le N)$ is expressed as follows:

$$x_k^{(i)} = \int_{-\infty}^{\infty} \left[x_k^{(i)} \cdot P \right] Dx^{(0)}(t)\, Dx^{(1)}(t) \cdots Dx^{(N)}(t),$$
$$Dx^{(i)} = \mathrm{d}x_1^{(i)} \cdot \mathrm{d}x_2^{(i)} \cdots \mathrm{d}x_m^{(i)}. \quad (5)$$

The integral range is from $-\infty$ to ∞ in Equation (5) according to the form of the path integral. The range in actual calculation is between the upper and lower limits of each variable. Also, Equation (5) is formulated to set the values according to the probability distribution for variable $x_k^{(i)}$ at time 0 and N. However, a variable may be fixed in some problems. The end points should be set in accordance with the boundary conditions.

Note that since Equation (5) shows the multiple integral, the Monte Carlo method can be applied to actual numerical calculations.

4. CALCULATION ALGORITHM FOR SPOT

First, a performance index is determined, and if necessary, variables are discretized for numerical calculation. Next, the values of all variables are randomly chosen to generate solutions. Generation probabilities of variables should agree with the probability distribution of Equation (3). Thus, a solution with a fine performance value is generated at a high frequency, and a solution with a poor performance index value is generated at a low frequency. Also, the algorithm is based on the Markov process so that future solutions will have no dependency on the former solutions. Lastly, expected values are calculated using all the solutions generated with a probability distribution to obtain an approximate solution.

The procedure of calculation is shown in Figure 3.

1) The values of all the variables are generated randomly to obtain initial solutions.

2) Time point t_i is chosen according to a random number, and function value x_i at t_i is randomly rewritten to generate a different solution.

3) The performance index value of the newly generated solution is calculated.

4) The adoption or rejection of this solution is decided to generate solutions depending on the probability distribution of Equation (3).

The procedure from Steps 2) to 4) is a pure explanation of the Metropolis method [7] to merely conduct the multiple integral. This method has an advantage of no necessity to obtain the normalization constant: A in Equation (3). A prime characteristic of this method is to calculate the expected values in Step 5).

5) The expected value is calculated.

6) If the final condition is met, the calculation ends. If not, it returns to the generation of solutions.

The final condition can be a certain range of variation of an expected value. However, in this paper, the final condition is the number of iterations, $K_{\max}$. This is because the generation of solutions during the calculation is a Markov process, and therefore, the still more iterations of calculation can result in fine performance index values if the convergence of expected values is not enough.

Equation (3) for probability distribution in this paper is the same as that in SA. SA is a deterministic approach to search only optimal solution at $h \to 0$.

On the other hand, h is finite and an optimal solution is obtained as an expected value (stochastic average value) in this paper. It means that the two approaches are based on fundamentally different concepts. In the concrete, SA starts a search with an initial solution for candidate solutions according to the probability distribution. It adopts a solution of the best performance function value as an optimal solution. On the other hand, our method determines an approximate solution to an optimal solution as an expected value with all the solutions generated according to the probability distribution. In this method, there is no process of the so-called cooling schedule which is in SA.

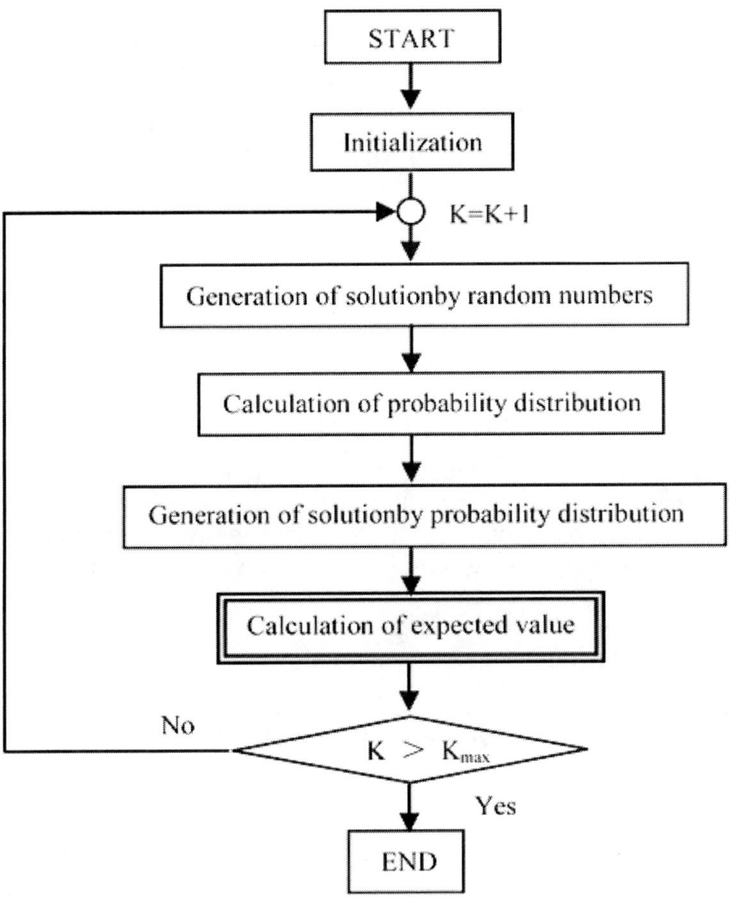

Figure 3. Flow chart of calculation.

5. CALCULATION EXAMPLE

This section describes examples of numerical calculation with SPOT to demonstrate the validity. First, SPOT is applied to a problem where the minimum value of a multi-peaked function is obtained without staying at other minimal values and the minimum value is always only one with no dependency on the initial value. Next, we apply SPOT to a problem of obtaining the minimum value of a functional to show the validity in a physical optimization problem. Lastly, we apply SPOT to an engineering optimal problem on the glide of a hang glider to show the high applicability of an approximate solution by SPOT.

We already applied SPOT to some engineering problems [8] [9] .

5.1. Minimization Problem of Bivariate Function

First, we apply SPOT to a problem of obtaining the minimum value of a bivariate function expressed in Equation (6). Here, performance function I is the function value $f(x,y)$ as follows:

$$f(x,y)=\{\cos(2\pi x)+\cos(2.5\pi x)-2.1\}\times\{2.1-\cos(3\pi y)-\cos(3.5\pi y)\},\quad 0.0\le x\le 3.0,\ 0.0\le y\le 1.5 \tag{6}$$

Figure 4(a) shows the solution space of Equation (6). The set of $f(x,y)$ that minimizes this function is (0.439, 0.306). This function has multiple minimal values and thus is a multi-peaked function. The contours are shown in the bottom plane, and Figure 4(b) shows the top view.

This example is a problem for obtaining the minimum value of a bivariate function. Therefore, the variables are not discretized, but the values of two variables are generated according to the probability distribution (3) defined by the performance values to obtain an expected value. The closed dots in Figure 4(a) show multiple initial values chosen to check the dependency on the initial values. Also, the cross marks in Figure 4(b) show the minimum values obtained with this method.

If calculation starts with any initial value marked in the closed dot, the resultant values do not stay at any of the many minimal values in the solution space, but all the values concentrate near the minimum value. Since every solution generated in each calculation is obtained stochastically with this method, the calculation results are not influenced by the initial value in principle. However in reality, the number of calculations is finite. Therefore, the expected value as the minimum value may have a variation.

The minimum values in Figure 4(b) have a little variation though it may be invisible. This variation is caused by the limited number of calculations.

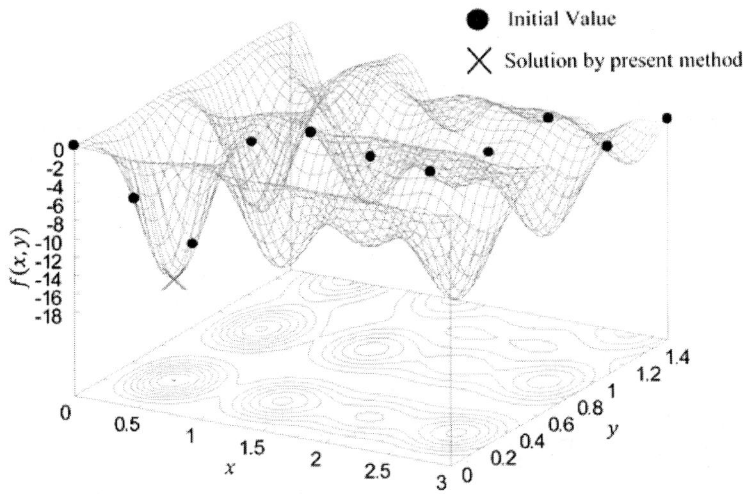

(a)

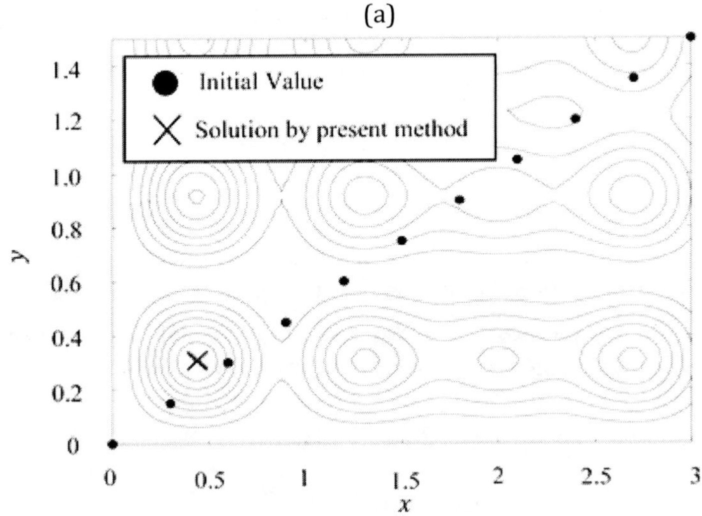

(b)

Figure 4. (a) Function space; (b) Contour lines of solution space.

Note that it took about 18 seconds to perform one million times of calculations on a 1.4 GHz Pentium 4 machine.

5.2. LINE OF SWIFTEST DESCENT PROBLEM

Next, we applied SPOT to a problem on brachistochrone as one of the classical optimization problems to obtain a function. This problem is demonstrated to obtain a mass point path that shows the brachistochrone between a fixed start point and an end point at different altitudes under the uniform gravitation field. Here, performance function I is time t until the mass point reaches the end point as expressed in the following equation:

$$t = \frac{1}{\sqrt{2g}} \int_0^{y_f} \frac{\sqrt{1+(dx/dy)^2}}{\sqrt{y}} dy \quad (7)$$

The boundary conditions are as follows:

Initial condition: $y(0) = x(0) = 0$, final condition: $y(t_f) = 1.0$, $x(t_f) = \pi/2$.

In this problem, we discretized the vertical axis y at an even division, and generate the values of the horizontal axis x corresponding to y according to the probability distribution of Equation (3) defined by the performance index of Equation (7) to obtain an expected value.

Figure 5 shows the paths obtained with SPOT in the closed dots and the analytic solution in the solid line. It took about 438 seconds on a Pentium 4 machine to perform 20 million times of iterations.

SPOT is to obtain an approximate solution of an optimal solution. However, the approximate solutions agree well with the analytic solutions. Thus, this method can be applied to general optimization problems where performance functions are expressed in functional.

5.3. Optimal Glide Problem of Hang Glider

Lastly, we apply SPOT to a problem of a hang glider that starts the fall and glide at an altitude of 12 m toward the ground to obtain its flight operation (time history of angle of attack) for the maximum down range (as shown in Figure 6).

This problem was firstly solved by Suzuki et al. [10] and their results were used to check the validity of the results of SPOT.

This problem is based on the former problem of [10] and the specifications of the hang glider used here are based on those of the optimal glider obtained in [10] . The shape of the hang glider is shown in Figure 7.

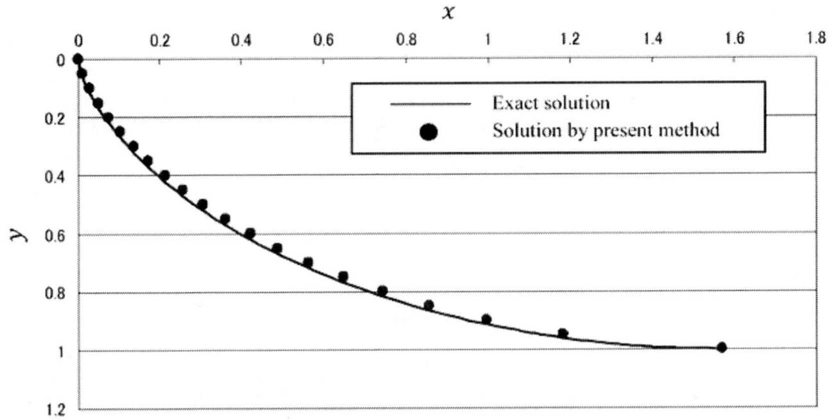

Figure 5. Line of swiftest descent.

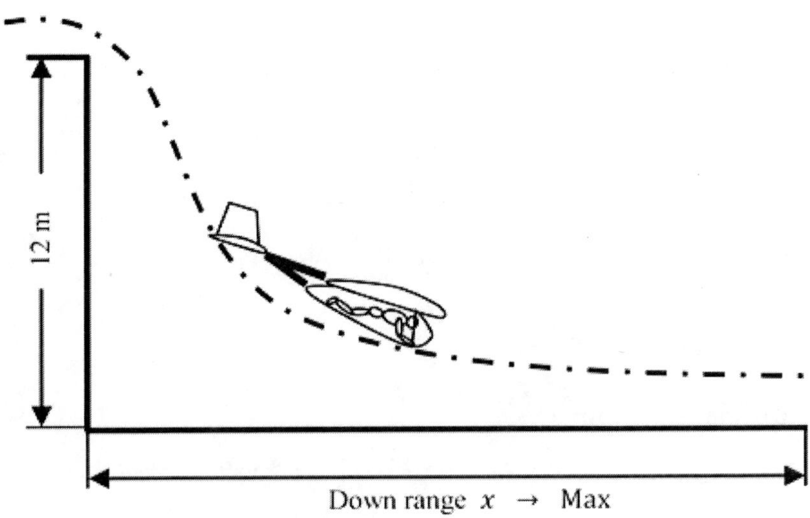

Figure 6. Flight trajectory optimization.

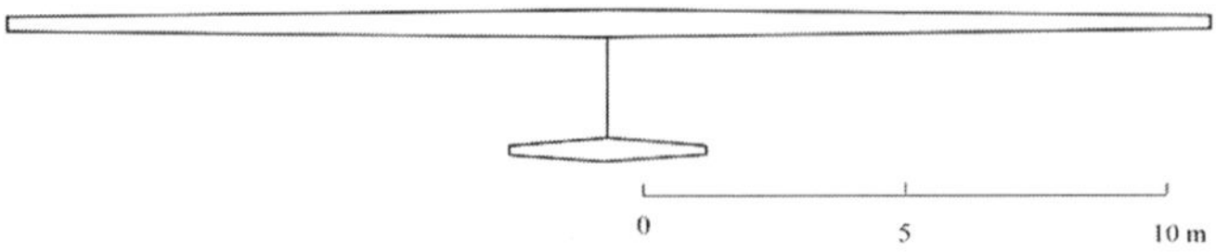

Figure 7. Top view of the hang glider.

Note that the performance index is the reciprocal number of the flight distance to set this problem to a minimal value problem. Also, the maximum value of the load factor is set to the restriction condition. If this condition exceeds the limitation during the flight, a penalty is added to the performance index value.

The equation of motion is as follows:

$$\dot{V} = -(1/2m)\rho V^2 S_1 C_D^{\bullet} - g\sin\gamma \,, \tag{8}$$

$$\dot{\gamma} = (1/2m)\rho V S_1 C_L^{\bullet} - (g/V)\cos\gamma \,, \tag{9}$$

$$\dot{x} = V\cos\gamma \tag{10}$$

$$\dot{z} = V\sin\gamma \tag{11}$$

The main conditions are shown as follows:

Performance index:

$$I = 1/x_f + \sum_{n_i > n_{\max}} n_i/x_f \tag{12}$$

Initial condition: $z_0 = 12.0$ m, $V_0 = 5.0$ m/s, $\gamma_0 = -3.5$ deg.

Final condition: $z_f = 2.0$ m.

Restriction condition: $n_{\max} = 3.0$.

Specifications of hang glider: $S_1 = 14.6$ m², $A_1 = 36.0$, $b_1 = 22.9$ m, $\lambda_1 = 0.5$, $S_2 = 2.3$ m², $A_2 = 6.0$, $\lambda_2 = 0.5$, $m_0 = 35.0$ kg, $m_p = 60.0$ kg.

The performance index value is the reciprocal number of the down range at the point where the hung glider reaches at an altitude of 2 m. First, the control variable, which is the angle of attack in this case, is discretized on the axis of time. Next, values of angle of attack at each time are specified with random numbers. The equations of motion are numerically calculated according to the time history of angle of attack to obtain the flight path. Then, the performance index value is determined as a reciprocal number of the down range. Therefore, the performance index values are obtained when the series of calculations ends with the final conditions being met.

Figure 8(a) and Figure 8(b) show the relation between the down range and the altitude, and that between the down range and the angle of attack, respectively.

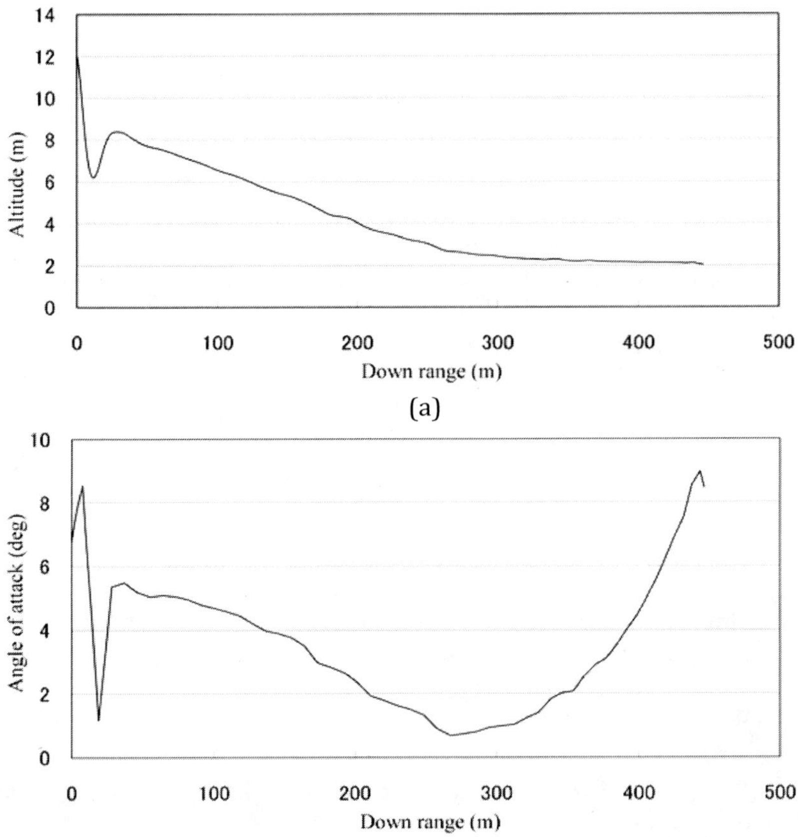

(b)

Figure 8. (a) Altitude vs. down range; (b) Angle of attack vs. down range.

This result may be an approximate solution close to the former optimization [9] . Note that it took about 15 hours on a Pentium 4 (1.4 GHz) machine to obtain the calculation result in this example that set the number of iterations to 3 millions.

6. DISCUSSION

We propose a new optimization method with a stochastic process called SPOT, and then demonstrate numerical calculations.

Generally, when proposing a new method, comparison of performance between the conventional method and a new method should be performed. However, since SPOT is the only method of obtaining the approximate value of the optimal solution, there is no sense in comparing them. Also fair comparison of computation time could not be performed.

As a result, SPOT has the following characteristics:

1) Advantages of Stochastic Process

SPOT results in an expected value from all the solutions generated by calculations. One of the advantages is to avoid influences from the initial value (a start point of the calculation) to the final expected value. The reason is that the generation of solutions during the calculation is a simple stochastic process, and the initial value is only the first-generated solution. As shown in Section 6.1, the calculation results all do not stay at local minimal values even though the calculation starts with any initial value, but the resultant values concentrate near the minimum value. Another advantage is to update an expected value from optimal solutions separately obtained under the same condition. This enables a parallel calculation and the recycling of the former calculation results. This characteristic is unique to SPOT and advantageous to other numerical calculation methods. For example, the calculation of a bivariate function, as shown in Section 6.1, gives the results starting with the multiple initial points as different minimum values. (The cross marks actually show almost the same point.) On the other hand, it is possible to obtain an expected value with all these expected values.

Also, SPOT seems to use asymptotic calculation in the algorithm. However, the former solutions are not used to generate the next solution. This means that the series of calculation is independent and thus the calculation time can easily be estimated with the number of calculations. Therefore, it is easy to estimate the calculation cost.

2) Independency on Experience

Another characteristic is that SPOT has only fluctuation parameter, h, set by operators. There may be a case of extremely low convergence of a solution because of a large fluctuation during calculation with an conventional optimization method using an inclination of the solution curved surface if the features of the solution space are not well known. On the other hand, in principle, the scale of the fluctuation parameter in our method has almost no influence on the process of obtaining an expected value. Although variations of convergence times of calculations due to the scale of the parameter are observed in actual calculations with the limited number of calculations, any large influence on the obtained expected values has not been known. Therefore, this method can be applied to unknown problems, including the surroundings, on solution spaces.

Also, the smaller the fluctuation parameter during calculation, the shorter the calculation time is, though this characteristic is not mentioned in this paper.

3) Restriction Conditions

In the conventional calculation approaches that start with initial solutions to search better performance index values, it may be difficult to generate an initial solution itself and thus to start the numerical calculation if the solution space is narrowed with restriction conditions. However, since the generation of solutions during calculation in our method is a stochastic process, it is possible to perform the calculation regardless of the presence of solutions.

Also, if there are restriction conditions on parameters, which are not the fluctuation parameter: h in (b) but are the load factor: n and the angle of attack: α etc. in Section 6.3 for example, it is necessary to keep parameters within the restriction conditions with general methods. However, with our method, such a device is not necessary. Furthermore, the restriction conditions are rather advantageous to numerical calculations. For example, if there are restrictions on parameter values and the range is specified, solutions can be generated only with values within the restrictions in the process where the series of solutions are generated. This is largely advantageous to perform numerical calculations. The harder the restriction conditions are and the narrower the range of the resultant parameter values is, the more densely the solutions are generated within the restriction conditions and the more easily the calculations are.

Next, in case of restriction conditions on state variables, when a solution generated during calculation is trapped on the restriction, the solution may be abandoned. However, expected values may approach but do not reach the restriction surfaces using only parameter values within constraint ranges and abandoning solutions that break the restrictions. Parameters and solutions that break the restrictions can be applied to a calculation using a penalty method to approach the expected value toward the restriction surface more closely. However, if results with this method are used as initial solutions of deterministic approaches, there is no necessity to make a device.

4) Scope of SPOT

The characteristics of SPOT are to obtain expected values of solution spaces with multiple integral. Thus, there is no problem if the solution spaces have restrictions as mentioned above. However, SPOT is not appropriate to be used for problems that have no restriction on the infinite solution spaces. Also, if a problem has a large solution space and a sharp peak on the performance index value, this method can be applied in principle. However, it may be difficult to obtain an approximate solution with high accuracy in the finite number of calculations. However, it is rare to have infinite range of a variable in an engineering problem, and the upper and lower limits can generally be set. Also, the fluctuation of a performance index value to that of a variable is expected to be smooth. Furthermore, the accuracy required for a variable is finite. Therefore, it is possible to obtain an approximate value for practical use with the finite number of calculations.

7. CONCLUSIONS

We propose a new optimization method based on the path integral called SPOT, and thus perform a numerical calculation on an engineering optimization problem. As a result, we obtain an approximate value of an optimal solution with high accuracy. In addition, this stochastic method has advantages on numerical calculation to other methods.

Hiroaki Yoshida, Katsuhito Yamaguchi, Yoshio Ishikawa Also, SPOT has more effective features of optimization than the conventional methods on optimization problems of an unknown system or narrow solution space that leads to difficulty finding a feasible solution.

REFERENCES

1. Kirkpatrick, S., Gelatt Jr., C.D. and Vecchi, M.P. (1983) Optimization by Simulated Annealing. Science, 220, 671-680.
2. Terasaki, M., Kondo, M., Yoshida, H., Yamaguchi, K. and Ishikawa, Y. (2004) Integrated Optimization of Airplane Design and Flight Trajectory by New Optimization Method Using a Stochastic Process. In: CONPUTATIONAL MECHANICS WCCM VI in Conjunction with APCOM'04, Tsinghua University Press & Springer-Verlag, Beijing, 328.

3. Yoshida, H., Konno, T., Yamaguchi, K. and Ishikawa, Y. (2005) New Optimization Method Using Stochastic Process Based on Path Integral. Journal of the Japan Society for Computational Methods in Engineering, 5, 145-150. (in Japanese)
4. Yoshida, H., Yamaguchi, K. and Ishikawa, Y. (2006) Application of a New Optimization Method for System Design. Joint 3rd International Conference on Soft Computing and Intelligent Systems (SCIS) and 7th International Symposium on Advanced Intelligent Systems (ISIS), Tokyo, 20-24 September 2006, 1542-1547.
5. Yoshida, H., Yamaguchi, K. and Ishikawa, Y. (2007) Design Tool Using a New Optimization Method Based on a Stochastic Process. Transactions of the Japan Society for Aeronautical and Space Sciences, 49, 220-230.
6. Feynman, R.P. and Hibbs, A.R. (1965) Quantum Mechanics and Path Integrals. McGraw-Hill, New York.
7. Metropolis, N., Rosenbluth, A.W., Rosenbluth, M.N., Teller, A.H. and Teller, E. (1953) Equation of State Calculations by Fast Computing Machines. Journal of Chemical Physics, 21, 1087-1092.
8. Nakane, M., Kobayashi, D., Yoshida, H. and Ishikawa, Y. (2009) Feasibility Study on Single Stage to Orbit Space Plane with RBCC Engine. 16th AIAA/DLR/DGLR International Space Planes and hypersonic Systems and Technologies Conference, Bremen, AIAA 2009-7331.
9. Ishimori, Y., Nakane, M., Ishikawa, Y., Yoshida, H. and Yamaguchi, K. (2011) Feasibility Study for Spaceplane's Concepts in Ascent Phase Using the Optimization Method. Journal of the Japan Society for Aeronautical and Space Sciences, 59, 291-297. (in Japanese)
10. Suzuki, S. and Kawamura, N. (1996) Simultaneous Optimization of Sailplane Design and Its Flight Trajectory. Journal of Aircraft, 33, 567-571.

NOMENCLATURE

$A_1 =$ wing aspect ratio;

$A_1^* =$ wing aspect ratio including the ground effect $\left\{33(z/b_1)^{3/2}+1\right\}/33(z/b_1)^{3/2}$;

$b_1 =$ wing span;

$C_L =$ lift coefficient;

$C_{D0} =$ parasite drag coefficient $C_{DW0}+C_{Df}\,S_f/S_1+C_{DrW0}\,S_2/S_1$;

$C_{Df} =$ fuselage drag coefficient;

$C_{DW0} =$ wing minimum drag coefficient;

$C_{DrW0} =$ horizontal stabilizer minimum drag coefficient;

$C_D^* =$ drag coefficient including the ground effect $C_{D0} + C_{Di}^*$;

$C_L^* =$ lift coefficient including the ground effect $C_L A_1^* / A_1$;

$C_{Di}^* =$ induced drag coefficient including the ground effect $C_L^{*2} / \pi e A_1^*$;

$e =$ Oswald's airplane efficiency;

$g =$ gravitational constant;

$m =$ total mass $m_0 + m_p$;

$m_0 =$ mass of the hung glider;

$m_p =$ mass of the pilot;

$n_i =$ load factor at i -th discrete time;

$S_1 =$ wing area;

$S_2 =$ horizontal stabilizer area;

$S_f =$ fuselage representative area;

$t =$ time;

$t_f =$ final time;

$V =$ speed;

$x =$ position, down range;

$x_f =$ down range at the end point;

$y =$ position;

$z =$ altitude;

$\gamma =$ path angle;

$\rho =$ air density.

CHAPTER 6

On the Stability of Stochastic Jump Kinetics

Stefan Engblom

Division of Scientific Computing, Department of Information Technology, Uppsala University, Uppsala, Sweden

ABSTRACT

Motivated by the lack of a suitable constructive framework for analyzing popular stochastic models of Systems Biology, we devise conditions for existence and uniqueness of solutions to certain jump stochastic differential equations (SDEs). Working from simple examples we find reasonable and explicit assumptions on the driving coefficients for the SDE representation to make sense. By "reasonable" we mean that stronger assumptions generally do not hold for systems of practical interest. In particular, we argue against the traditional use of global Lipschitz conditions and certain common growth restrictions. By "explicit", finally, we like to highlight the fact that the various constants occurring among our assumptions all can be determined once the model is fixed. We show how basic long time estimates and some limit results for perturbations can be derived in this setting such that these can be contrasted with the corresponding estimates from deterministic dynamics. The main complication is that the natural path-wise representation is generated by a counting measure with an intensity that depends nonlinearly on the state.

Keywords: Nonlinear Stability, Perturbation, Continuous-Time Markov Chain, Jump Process, Uncertainty, Rate Equation

1. INTRODUCTION

The observation that detailed modeling of biochemical processes inside living cells is a close to hopeless task is a strong argument in favor of stochastic models. Such models are often thought to be more accurate than conventional rate-diffusion laws, yet remain more manageable than, say, descriptions formed at the level of individual molecules. Indeed, several studies [1] -[3] have showed that noisy models have the ability to capture relevant phenomena and to explain actual, observed dynamics.

In this work we shall consider some "flow" properties of a stochastic dynamical system in the form of a quite general continuous-time Markov chain. Since the pioneering work of Gillespie [4] [5] , in the Systems Biology context this type of model is traditionally described in terms of a (chemical) master equation (CME). This is the forward Kolmogorov equation of a certain jump stochastic differential equation (jump SDE for brevity), driven by independent point processes with state-dependent intensities. Despite the popularity of the master equation approach, little analysis on a per trajectory-basis of actual models has been attempted.

In the general literature, when discussing existence/uniqueness and various types of perturbation results, different choices of assumptions with different trade-offs have been made. One finds that the treatment often falls into one of two categories taking either a "mathematical" or a "physical" viewpoint. Both of the conditions are highly general but with subsequently less transparent proofs and resulting in more abstract bounds. Or the conditions are formed out of convenience, say, involving global Lipschitz constants, and classical arguments carry through with only minor modifications.

Protter ([6] Chap. V) offers a nice discussion from the mathematical point of view and in ascending order of generality, including the arguably highly unrestrictive assumption of locally Lipschitz continuous coefficients. Other authors ([7] , Chap. 6, [8] , Chap. 3-5) also treat the evolution of general jump-diffusion SDEs in conti- nuous state spaces.

A study of the flow properties of jump SDEs is found in [9] , where the setting is scalar and the state is continuous. In [10] jump stochastic partial differential equations are treated, and existence/uniqueness results as well as ergodic results for the case of a multiplicative noise, are found in [11] [12] . Numerical aspects in a similar setting are discussed in [13] .

In a more applied context, stability is often thought of as implied from physical premises and the solution is tactically assumed to be confined inside some bounded region ([14] , Chap. V). The fundamental issue here is that for open systems in a stochastic setting, there is a non-zero probability of reaching any finite state and global assumptions must be formed with great care. The analysis of open networks under an a priori assumption of boundedness is therefore quite difficult to interpret other than in a qualitative sense. Notable examples in this setting include time discretization strategies [15] [16] , time-parallel simulation techniques [17] , and parameter perturbations [18] .

Evidently, essentially no systems of interest satisfy global Lipschitz assumptions since the fundamental inte- raction almost always takes the form of a quadratic term. Interestingly, for ordinary differential equations, it has been shown [19] that Lipschitz continuous coefficients imply a computationally polynomial-space complete solution; thus providing a kind of explanation for the convenience with this weak feedback assumption. It is also known [20] , that with SDEs, superlinearly growing coefficients may in fact cause the forward Euler method to diverge.

1.1. Agenda

Besides its expository material, the purpose of this paper is to devise simple conditions that imply stability for finite and, in certain cases, infinite times, and that, when applied to systems of practical interest, yield explicit expressions for the associated stability estimates. As a result the framework developed herein applies in a constructive way to any chemical network, of arbitrary size and topology, formed by any combination of the elementary reactions (2.3) to be presented in Section 2. Additionally, it will be clear how to encompass also other types of nonlinear reactions that typically result from adiabatic simplifications.

As an argument in favor of this bottom-up approach one can note that, for evolutionary reasons, biochemical systems tend to operate close to critical points in phase-space where the efficiency is the highest. Clearly, for such dynamical systems, an analysis by analogy might be highly misleading.

We also like to argue that our results are of interest from the modeling point of view. Due to the type of phenomenological arguments often involved, judging the relative effect of the (non-probabilistic) epistemic uncer- tainty is a fundamental issue which has so far not rendered a consistent analysis.

1.2. Outline

The expository material in Section 2 is devoted to formulating the type of processes we are interested in. We state the master equation as well as the corresponding jump SDE and we also look at some simple, yet informative actual examples. Since it is expected that the properties of the stochastic dynamics are somehow similar to those of the deterministic version, we search for a set of minimal assumptions in the latter setting in Section 3. Techniques for finding explicit values of the constants occurring among our assumptions are also devised. The main results of the paper are found in Section 4 where we put our theory together and prove existence and uniqueness, as well as long time estimates and limit results for perturbations. A concluding discussion is found in Section 5.

2. STOCHASTIC JUMP KINETICS

In this section we start with the physicist's traditional viewpoint of pure jump processes and write down the governing master equations. These are evolution equations for the probability densities of continuous-time Markov chains over a discrete state space. Although the application considered here is mesoscopic chemical kinetics, identical or very similar stochastic models are also used in Epidemiology [21] , Genetics [22] and Sociodynamics [23] , to name just a few.

We then proceed with discussing a path-wise representation in terms of a stochastic jump differential equation. The reason the sample path representation is interesting is the possibility to reason about flow properties and thus compare functionals of single trajectories. This is generally not possible with the master equation approach.

For later use we conclude the section by looking at some prototypical models. A simple analysis shows, somewhat surprisingly, that an innocent-looking example produces second moments that grow indefinitely.

2.1. Reaction Networks and the Master Equation

We consider a chemical network consisting of D different chemical species interacting according to R prescribed reaction pathways. At any given time t, the state of the system is an integer vector $X(t) \in \mathbf{Z}_+^D = \{0,1,2,\ldots\}^D$ counting the number of individual molecules of each species. A reaction law is a prescribed change of state with an intensity defined by a reaction propensity, $w_r : \mathbf{Z}_+^D \to \mathbf{R}_+$. This is the transition probability per unit of time for moving from the state x to $x - \mathrm{N}_r$:

$$\mathbf{P}\left[X(t+\mathrm{d}t) = x - \mathrm{N}_r \,\middle|\, X(t) = x\right] = w_r(x)\,\mathrm{d}t + o(\mathrm{d}t). \tag{2.1}$$

where $\mathrm{N}_r \in \mathbf{Z}^D$ is the transition step and is the rth column in the stoichiometric matrix $\mathrm{N} \in \mathbf{Z}^{D \times R}$. Informally, for states $x(t) \in \mathbf{R}_+^D$, we can picture (2.1) as a stochastic version of the time-homogeneous ordinary differential equation

$$x'(t) = -\sum_{r=1}^{R} \mathrm{N}_r w_r(x) = -\mathrm{N}w(x) =: F(x), \tag{2.2}$$

where $w(x) \equiv [w_1(x), \mathrm{L}\ , w_R(x)]^{\mathrm{T}}$ is the column vector of reaction propensities.

The physical premises leading to a description in the form of discrete transition laws (2.1) often imply the existence of a system size V (e.g. physical volume or total number of individuals). For instance, in a given volume V the elementary chemical reactions can be written using the state vector $x = [a,b]^{\mathrm{T}}$,

$$\begin{aligned}
&\varnothing \xrightarrow{k_1 V} A,\ \text{¥}_1 = [-1,0]^{\mathrm{T}},\\
&A \xrightarrow{k_2 a} \varnothing,\ \text{¥}_2 = [1,0]^{\mathrm{T}},\\
&A + A \xrightarrow{k_3 a(a-1)/V} \varnothing,\ \text{¥}_3 = [2,0]^{\mathrm{T}},\\
&A + B \xrightarrow{k_4 ab/V} \varnothing,\ \text{¥}_4 = [1,1]^{\mathrm{T}},
\end{aligned} \tag{2.3}$$

with the names of the species in capitals. These propensities are generally scaled such that $w_r(x) = Vu_r(x/V)$ for some dimensionless function u_r. Intensities of this form are called density dependent and arise naturally in a number of situations ([24] , Chap. 11). For the rest of this paper, we conveniently take $V = 1$ and defer system's size analysis to another occasion.

The models we consider here all have states in the positive integer lattice $\mathbf{Z}_+^D$ and the assumption that no transition can yield a state outside is therefore natural. We make this formal as follows ([25] , Chap. 8.2.2, Definition 2.4):

Assumption 2.1 (Conservation and stability). For all propensities, $w_r(x)=0$ for any $x\in\mathbf{Z}_+^D$ such that $x-\mathrm{N}_r\notin\mathbf{Z}_+^D$, and we also restrict initial data to $\mathbf{Z}_+^D$. Furthermore, $w_r:\mathbf{Z}_+^D\to\mathbf{R}_+$ such that $w_r(x)$ is finite for all finite arguments x.

To state the chemical master equation (CME), let for brevity $p(x,t)=\mathbf{P}(X(t)=x|X(0)=x_0)$ be the probability that a certain number x of molecules is present at time t conditioned upon an initial state x_0. The CME is then given by ([14] , Chap. V)

$$\frac{\partial p(x,t)}{\partial t}=\sum_{r=1}^{R}w_r(x+\mathrm{N}_r)p(x+\mathrm{N}_r,t)-w_r(x)p(x,t)=:\mathrm{M}^{\mathrm{T}}p(x,t). \tag{2.4}$$

The convention of the transpose of the operator to the right of (2.4) is the standard mathematical formulation of Kolmogorov's forward differential system ([25] , Chap. 8.3) in terms of which $\mathbf{M}$ is the infinitesimal generator of the associated Markov process. This is also the adjoint of the master operator $\mathbf{M}^{\mathrm{T}}$ in the sense that $(\mathrm{M}^{\mathrm{T}}p,q)=(p,\mathrm{M}q)$ in the Euclidean inner product over the state space. An explicit representation is

$$\mathrm{M}q(x)=\sum_{r=1}^{R}w_r(x)\left[q(x-\mathrm{N}_r)-q(x)\right], \tag{2.5}$$

such that the propensities in (2.1) can be retrieved,

$$\mathrm{M}(x,x-\mathrm{N}_r)=w_r(x). \tag{2.6}$$

Under assumptions to be prescribed in Section 4.1 it holds that the dynamics of the expected value of some time-independent unknown function f, conditioned upon the initial state x_0, can be written

$$\begin{aligned}\frac{\mathrm{d}}{\mathrm{d}t}E^{x_0}\left[f(X_t)\right]&=\sum_{x\in Z_+^D}\frac{\partial p(x,t)}{\partial t}f(x)=\left(\mathrm{M}^{\mathrm{T}}p,f\right)=(p,\mathrm{M}f)\\&=\sum_{r=1}^{R}E^{x_0}\left[w_r(X_t)\left(f(X_t-\mathrm{N}_r)-f(X_t)\right)\right].\end{aligned} \tag{2.7}$$

We now consider a path-wise representation for the stochastic process X_t.

2.2. The Sample Path Representation

In the present context of analyzing models in stochastic chemical kinetics, the path-wise jump SDE representation seems to have been first put to use in [26] , and it was later further detailed in [16] . It should be noted, however, that an equivalent representation was used much earlier by Kurtz (see the monograph [24]).

We thus assume the existence of a probability space $(\Omega, \mathrm{F}, \mathbf{P})$ with the filtration $\mathrm{F}_{t\geq 0}$ containing R- dimensional Poisson processes. The state of the system $X(t) \in \mathbf{Z}_+^D$ will be constructed from a stochastic integral with respect to suitably chosen Poisson random measures.

The transition probability (2.1) defines a counting process $\pi_r(t)$ counting at time t the number of reactions of type r that has occurred since $t=0$. It follows that these processes fully determine the state $X(t)$,

$$X_t = X_0 - \sum_{r=1}^{R} ¥_r \pi_r(t). \tag{2.8}$$

The counting processes are obtained from the transition intensities (cf. (2.1))

$$\mathbf{P}\left[\pi_r(t+\mathrm{d}t) - \pi_r(t) = 1 \middle| \mathrm{F}_t\right] = w_r(X_{t-})\mathrm{d}t + o(\mathrm{d}t), \tag{2.9}$$

where by $X(t-)$ we mean the value of the process prior to any transitions occurring at time t, and where the little-o notation is understood uniformly with respect to the state variable. Alternatively, using Kurtz's random time change representation ([24] , Chap. 6.2), we can produce the counting process from a standard unit-rate Poisson process Π_r,

$$\pi_r(t) = \Pi_r\left(\int_0^t w_r(X_{s-})\mathrm{d}s\right). \tag{2.10}$$

The marked counting measure ([27] , Chap. VIII) $\mu_r(\mathrm{d}t\times\mathrm{d}z;\omega)$ with $\omega\in\Omega$ defines an increasing sequence of arrival times $\tau_i\in\mathbf{R}_+$ with corresponding "marks" $z_i\in I:=[0,1]$ according to some probability distribution which we will take to be uniform. The intensity $m_r(\mathrm{d}t\times\mathrm{d}z)$ of $\mu_r(\mathrm{d}t\times\mathrm{d}z)$ is the Lebesgue measure scaled by the corresponding propensity, $m_r(\mathrm{d}t\times\mathrm{d}z)=w_r(X_{t-})\mathrm{d}t\times\mathrm{d}z$. Using this formalism, (2.8) and (2.10) can be written in the jump SDE form

$$\mathrm{d}X_t=-\int_I \mathrm{N}\mu(\mathrm{d}t\times\mathrm{d}z), \tag{2.11}$$

where $\mu=[\mu_1,\ldots,\mu_R]^{\mathrm{T}}$. Here, the time $\tau-t$ to the arrival of the next reaction of type r is exponentially distributed with intensity $w_r(X_{t-})$. Note that, by virtue of the nature of the propensities, the intensities of the counting processes therefore depend nonlinearly on the state ([27] , Chap. II.3).

Using that the point processes are independent and therefore have no common jump times ([25] , Chap. 8.1.3), we can obtain a sometimes more transparent notation in terms of a scalar counting measure. Define for this purpose and for any state x the cumulative intensities

$$W_r(x)=\sum_{s=1}^{r}w_s(x), \tag{2.12}$$

such that the total intensity is given by $W(x)\equiv W_R(x)$. Let the marks z_i be uniformly distributed on I. Then the frequency of each reaction can be controlled through a set of indicator functions $\hat{w}_r:\mathbf{Z}_+^D\times I\to\{0,1\}$ defined according to

$$\hat{w}_r(x;z)=\begin{cases}1 & \text{if } W_{r-1}(x)<zW(x)\le W_r(x),\\ 0 & \text{otherwise.}\end{cases} \tag{2.13}$$

Put $\hat{w}(x)\equiv[\hat{w}_1(x;z),\ldots,\hat{w}_R(x;z)]^{\mathrm{T}}$ and define also for later use the indicator form

$$\hat{F}(x;z) = -\mathrm{N}\hat{w}(x;z), \tag{2.14}$$

such that

$$F(x) = \int_I \hat{F}(x;z) W(x)\,\mathrm{d}z, \tag{2.15}$$

where $F(x)$ is defined in (2.2).

The jump SDE (2.11) can now be written in terms of a scalar counting random measure μ through a state- dependent thinning procedure ([28] , Chap. 7.5),

$$\mathrm{d}X_t = -\int_I \mathrm{N}\hat{w}(X_{t-};z)\,\mu(\mathrm{d}t \times \mathrm{d}z). \tag{2.16}$$

Equation (2.16) expresses exponentially distributed reaction times that arrive according to a point process of intensity $m(\mathrm{d}t \times \mathrm{d}z) = W(X_{t-})\,\mathrm{d}t \times \mathrm{d}z$ carrying a mark which is uniformly distributed in I. This mark implies the ignition of one of the reaction channels according to the acceptance-rejection rule (2.13).

One frequently decomposes (2.16) into its "drift" and "jump" parts,

$$\mathrm{d}X_t = -\mathrm{N}w(X_t)\,\mathrm{d}t - \int_I \mathrm{N}\hat{w}(X_{t-};z)(\mu - m)(\mathrm{d}t \times \mathrm{d}z). \tag{2.17}$$

The second term in (2.17) is driven by the compensated measure $(\mu - m)$ and is a local martingale provided in essence that the path is absolutely integrable (see [27] , Chap. VIII.1, Corollary C4 for details).

2.2.1. Localization: Itô's and Dynkin's Formulas

In analytic work it is often necessary to "tame" the process by deriving results under a stopping time $\tau_P := \inf_{t \geq 0}\{\|X_t\| > P\}$ in some norm. Results for the stopped process $X_{t \wedge \tau_P}$ can then be transferred to the original process by letting $P \to \infty$ under suitable conditions.

Although there are many general versions of Itô's change of variables formula available in the setting of semi- martingales (see for example [8] , Chap. 2.7 and [6] , Chap. II.7), we shall get around with the following simple version ([7] , Chap. 4.4.2). By the properties of the semi-martingale pure jump process we have for $\hat{t} = t \wedge \tau_P$

$$f(X_{\hat{t}}) - f(X_0) = \sum_{0<s\le\hat{t}} f(X_s) - f(X_{s-}) = \int_0^{\hat{t}}\int_I f(X_s) - f(X_{s-})\mu(\mathrm{d}s \times \mathrm{d}z), \tag{2.18}$$

where the sum is over jump times $s \in (0, \hat{t}]$. Using that $X_s = X_{s-} - ¥\hat{w}(X_{s-}; z)$ we can write this in differential form as

$$\mathrm{d}f(X_t) = \int_I f(X_{t-} - \mathrm{N}\hat{w}(X_{t-}; z)) - f(X_{t-})\mu(\mathrm{d}t \times \mathrm{d}z). \tag{2.19}$$

Alternatively, decomposing (2.18) into drift- and jump parts and taking expectation values we get, since the compensated measure is a local martingale,

$$\begin{aligned} Ef(X_{\hat{t}}) - Ef(X_0) &= E\int_0^{\hat{t}}\int_I f(X_{s-} - \mathrm{N}\hat{w}(X_{s-}; z)) - f(X_{s-})m(ds \times \mathrm{d}z) \\ &= E\int_0^{\hat{t}}\int_I \left[f(X_{s-} - \mathrm{N}\hat{w}(X_{s-}; z)) - f(X_{s-})\right]W(X_{s-})\mathrm{d}s \times \mathrm{d}z \\ &= E\int_0^{\hat{t}}\sum_{r=1}^{R}\left[(f(X_s - \mathrm{N}_r) - f(X_s))w_r(X_s)\right]\mathrm{d}s. \end{aligned} \tag{18}$$

This is Dynkin's formula ([25] , Chap. 9.2.2) for the stopped process and we note that (2.7) is just a differential version.

2.2.2. Coupled Processes

When considering stability properties we will need to compare different trajectories with respect to the same noise. The details of this coupling are not defined in either (2.11) or (2.16) and must in fact be chosen explicitly. Since this equality is easy to inspect for a unit-rate Poisson process, the viewpoint of local time expressed in (2.10) provides an answer; two processes X_t and Y_t may be regarded as coupled if and only if they are evolved using identical Poisson processes Π_r, $r = 1, \mathrm{L}, R$ in (2.8) and (2.10). This approach was first used by Kurtz [29] in the context of the random time change representation. Algorithmically it implies the Common Reaction Path (CRP) method for simulating coupled processes [30] (see also [17]).

A refinement of this construction was devised, also by Kurtz, in ([31] , see Equations (2.2), (2.3)). In turn, this approach implies the Coupled Finite Difference method [18] (but see also [16] [26]), and is more amenable to analysis. This is also the construction formalized below under our current framework.

To obtain such a coupled version of (2.16) we will have to make the thinning dependent on both trajectories. This is achieved by firstly replacing the cumulative intensities in (2.12) with the base (or minimal) intensities

$$W_r^{(0)}(x,y) = \sum_{s=1}^{r} w_s(x) \wedge w_s(y), \tag{2.21}$$

and use the new total base intensity $W^{(0)}(x,y) \equiv W_R^{(0)}(x,y)$ as the intensity of the counting measure μ_0; $m_0(\mathrm{d}t \times \mathrm{d}z) = W^{(0)}(X_{t-}, Y_{t-})\mathrm{d}t \times \mathrm{d}z$. We also modify (2.13) accordingly,

$$\hat{w}_r^0(x,y;z) = \begin{cases} 1 & \text{if } W_{r-1}^{(0)}(x,y) < zW^{(0)}(x,y) \le W_r^{(0)}(x,y), \\ 0 & \text{otherwise.} \end{cases} \tag{2.22}$$

Secondly, we also define the remainder intensity,

$$W_r^{(\delta)}(x,y) = \sum_{s=1}^{r} w_s(x) \vee w_s(y) - W_r^{(0)}(x,y) = \sum_{s=1}^{r} \left| w_s(x) - w_s(y) \right|. \tag{2.23}$$

In analogy with the previous construction we have the associated total intensity $W^{(\delta)}(x,y) \equiv W_R^{(\delta)}(x,y)$ and counting measure μ_δ; $m_\delta(\mathrm{d}t \times \mathrm{d}z) = W^{(\delta)}(X_{t-}, Y_{t-})\mathrm{d}t \times \mathrm{d}z$. This time the thinning procedure is non-symmetric in its two first arguments,

$$\hat{w}_r^\delta(x,y;z) = \begin{cases} 1 & \text{if } W_{r-1}^{(\delta)}(x,y) < zW^{(\delta)}(x,y) \le W_{r-1}^{(\delta)}(x,y) + d(x,y), \\ 0 & \text{otherwise,} \end{cases} \tag{2.24}$$

with the non-symmetricity due to

$$d(x,y) = w_r(x) - w_r(x) \wedge w_r(y). \tag{2.25}$$

As a concrete example of how this comparative thinning might be used, consider the following variant of (2.19),

$$\mathrm{d}f(X_t - Y_t) = \int_I f\left(X_{t-} - \mathrm{N}\hat{w}^\delta(X_{t-}, Y_{t-}; z) - Y_{t-} + \mathrm{N}\hat{w}^\delta(Y_{t-}, X_{t-}; z)\right) - f(X_{t-} - Y_{t-})\mu_\delta(\mathrm{d}t \times \mathrm{d}z). \tag{2.26}$$

For this specific example, the terms governed by the base counting measure μ_0 cancel out altogether.

We mention also that an equivalent construction, but one that leads to different algorithms, can be obtained via a thinning of a single measure [16] [26] . Defining instead

$$W_r^{(+)}(x,y) = \sum_{s=1}^{r} w_s(x) \vee w_s(y), \tag{2.27}$$

implying the total intensity $W^{(+)}(x,y) \equiv W_R^{(+)}(x,y)$ and associated counting measure μ_+; $m_+(\mathrm{d}t \times \mathrm{d}z) = W^{(+)}(X_{t-}, Y_{t-})\mathrm{d}t \times \mathrm{d}z$. By construction the indicator functions are now non-symmetric in their first two arguments,

$$\hat{w}_r^+(x,y;z) = \begin{cases} 1 & \text{if } W_{r-1}^{(+)}(x,y) < zW^{(+)}(x,y) \leq W_{r-1}^{(+)}(x,y) + w_r(x), \\ 0 & \text{otherwise.} \end{cases} \tag{2.28}$$

In analogy to (2.26) we get

$$\mathrm{d}f(X_t - Y_t) = \int_r f(X_{t-} - \mathbb{N}\hat{w}^+(X_{t-}, Y_{t-}; z) - Y_{t-} + \mathbb{N}\hat{w}^+(Y_{t-}, X_{t-}; z)) - f(X_{t-} - Y_{t-})\mu_+(\mathrm{d}t \times \mathrm{d}z). \tag{2.29}$$

This time, however, the intensity of the counting measure is generally larger and the equivalence is obtained as a result of the thinning procedure.

2.2.3. The Validity of the Master Equation

With this much formalism developed, we may conveniently quote the following result:

Theorem 2.1 ([25] , Chap. 8.3.2, Theorem 3.3). Under Assumption 2.1, and if additionally, for $t \in [0,T]$ it holds that

$$EW(X_t) < \infty, \tag{2.30}$$

then (2.7) is valid for $t \in [0,T]$.

Since the governing Equation (2.7) for the expected value of $f(X_t)$ is a direct consequence of (2.4), we can similarly conclude the following:

Corollary 2.2 Under the assumptions of Theorem 2.1, and if, moreover, in an arbitrary norm $\|\cdot\|$,

$$E\|f(X_t)\| < \infty, \quad t \in [0,T], \tag{2.31}$$

then (2.7) is valid for $t \in [0,T]$.

In stating these results we have suppressed the conditional dependency on the initial state which we for simplicity consider to be some non-random state x_0.

2.3. Concrete Examples

Consider the bi-molecular birth-death system,

$$\left.\begin{array}{l} \varnothing \xrightarrow{k_1} A \\ \varnothing \xrightarrow{k_1} B \\ A+B \xrightarrow{k_2 ab} \varnothing \end{array}\right\}, \tag{2.32}$$

that is, the system is in contact with a large reservoir such that A- and B-molecules are emitted at a constant rate k_1. Additionally, a decay reaction happens with probability k_2 per unit of time whenever two molecules meet. For this example we have the stoichiometric matrix

$$\mathrm{N} = \begin{bmatrix} -1 & 0 & 1 \\ 0 & -1 & 1 \end{bmatrix}$$

and the vector propensity function

$$w(x) = [k_1, k_1, k_2 ab]^{\mathrm{T}},$$

where $x = [x_1, x_2]^{\mathrm{T}} = [a,b]^{\mathrm{T}}$.

For a state $X_t = [A_t, B_t]^{\mathrm{T}}$, define $U_t = A_t - B_t \in \mathbf{Z}$. Itô's formula (2.19) with $f(x) = x_1 - x_2$ yields

$$\mathrm{d}U_t = \mathrm{d}f(X_t) = \int_I -[-1,1,0]\,\hat{w}(X_{t-};z)\,\mu(\mathrm{d}t \times \mathrm{d}z), \tag{2.33}$$

which upon a moments consideration is just the same thing as the model

$$\varnothing \underset{k_1}{\overset{k_1}{\rightleftharpoons}} U, \tag{2.34}$$

that is, a constant intensity discrete random walk process. An explicit solution is the difference between two independent Poisson distributions,

$$U_t = U_0 + \Pi_1(k_1 t) - \Pi_2(k_1 t) \sim \mathrm{N}\left(U_0, 2k_1 t\right), \quad \text{as } t \to \infty, \tag{2.35}$$

where N is a normally distributed random variable of the indicated mean and variance. Hence U_t fluctuates between arbitrarily large and small values as $t \to \infty$.

Reversible Versions

From time to time below we shall be concerned with the following closed version of (2.32), consisting of a single reversible reaction,

$$A + B \underset{k_2 c}{\overset{k_1 ab}{\rightleftharpoons}} C \tag{2.36a}$$

This is clearly a finite system since the number $a+b+2c$ is always preserved. An open version of the same system is

$$A + B \underset{k_2 c}{\overset{k_1 ab}{\rightleftharpoons}} C \underset{k_4}{\overset{k_3 c}{\rightleftharpoons}} \varnothing, \tag{2.36b}$$

and will prove to be a useful example in the stochastic setting since formally, all states in $\mathbf{Z}_+^3$ are reachable. For (2.36a) we have

$$\mathrm{N} \equiv \begin{bmatrix} 1 & -1 \\ 1 & -1 \\ -1 & 1 \end{bmatrix}, \quad w(x) \equiv \left[k_1 ab, k_2 c\right]^{\mathrm{T}}, \tag{2.37}$$

while (2.36b) is represented by

$$\mathrm{N} \equiv \begin{bmatrix} 1 & -1 & 0 & 0 \\ 1 & -1 & 0 & 0 \\ -1 & 1 & 1 & -1 \end{bmatrix}, \quad w(x) \equiv \left[k_1 ab, k_2 c, k_3 c, k_4\right]^{\mathrm{T}}. \tag{2.38}$$

These examples, while very simple to deal with, will provide good counterexamples in both Sections 3 and 4.

3. DETERMINISTIC STABILITY

In this section we shall be concerned with the deterministic drift part of the dynamics (2.17). We are interested in techniques for judging the stability of the time-homogeneous ODE (2.2), the so-called reaction rate equations implied by the rates (2.1). Stability and continuity with respect to initial data are considered in Sections 3.1 and 3.2. The main motivation for this discussion stems from the observation that assumptions that do not hold in this very basic setting are unlikely to hold in the stochastic case. In Section 3.3, techniques for explicitly obtaining all our postulated constants are discussed. A good point in favor of taking the time to describe these techniques is that we have not found such a discussion elsewhere.

Initially we will consider states $x \in \mathbf{R}^D$, but we will soon find it convenient to restrict the treatment to $x \in \mathbf{R}_+^D$. In order to remain valid also in the discrete stochastic setting, however, constructed counterexamples will remain relevant also when restricted to $\mathbf{Z}_+^D$.

3.1. Stability

Many stability proofs can be thought of as comparisons with relevant linear cases. This is the motivation for the well-known Grönwall's inequality which we state in the following two versions.

Lemma 3.1. Suppose that $u'(t) \le A + \alpha u(t)$ for $t \ge 0$. Then

$$u(t) \le u(0)\mathrm{e}^{\alpha t} + \frac{A}{\alpha}\left(\mathrm{e}^{\alpha t} - 1\right). \tag{3.1}$$

The same conclusion holds irrespective of the differentiability of u but with $\alpha \ge 0$ and under the weaker integral condition

$$u(t) \le u(0) + \int_0^t A + \alpha u(s)\,\mathrm{d}s. \tag{3.2}$$

The most immediate way of comparing the growth of solutions to the ODE (2.2) to those of a linear ODE is to require that the norm of the driving function is bounded in terms of its argument;

$$\|F(x)\| \le A + \alpha \|x\|, \tag{3.3}$$

since then by the triangle inequality,

$$\|x(t)\| \le \|x_0\| + \int_0^t \|F(x)\| \mathrm{d}t \le \|x_0\| + \int_0^t A + \alpha \|x(t)\| \mathrm{d}t, \tag{3.4}$$

where Grönwall's inequality applies. Unfortunately, (3.3) is a too strict requirement for our applications.

Proposition 3.2 The bi-molecular birth-death system (2.32) does not satisfy (3.3).

Proof. We compute $\|F(x)\| = \|-\mathrm{N}w(x)\| = \sqrt{2}|k_1 - k_2 ab|$ for a state $x = [a,b]^{\mathrm{T}}$. Hence for a = b = N = 0, 1,··· we have for N large enough that $\|F(x)\| = \sqrt{2}k_2 \cdot N^2 - \sqrt{2}k_1$, which can clearly never be bounded linearly in $\|x\| = \sqrt{2}N$. W

The problem with the simple condition (3.3) is that it does not take the direction of growth into account; the offending quadratic propensity in (2.32) actually decreases the number of molecules. To deal with this, let $x \in \mathbf{R}^D$ be an arbitrary vector defining an "outward" direction. The length of the component of the driving function along this direction is $(x, F(x))$ and in order not to have x driven too strongly out along this ray we may, in view of Grönwall's inequality, naturally require that $(x/\|x\|, F(x)) \le \text{constant} \times \|x\|$ for $\|x\|$ sufficiently large. Equivalently, for any x,

$$(x, F(x)) \le A + \alpha \|x\|^2, \tag{3.5}$$

from which one deduces

$$\frac{\mathrm{d}}{\mathrm{d}t}\frac{\|x\|^2}{2} = (x, F(x)) \le A + \alpha \|x\|^2, \tag{3.6}$$

where Grönwall's inequality applies anew. The assumption (3.5) is weaker than (3.3) since the former implies the latter by the Cauchy-Schwarz inequality. Indeed, as in the proof of Proposition 3.2 it is readily checked that for the bi-molecular birth-death system (2.32), we get $(x, F(x)) = k_1(a+b) - k_2(a+b)ab$ which this time readily can be bounded linearly in terms of $\|x\|^2 = a^2 + b^2$.

Unfortunately, in the case of an infinite state space and strong dependencies between the species the assumption (3.5) is also often unrealistic.

Proposition 3.3 Neither (2.36a) nor (2.36b) admits a bound of the kind (3.5).

Proof. As in the proof of Proposition 3.2 we look at a ray $x^{\mathrm{T}} = (a,b,c) = (N,N,3N)$ parametrized by a non- negative integer N. For (2.36a) we compute $(x, F(x)) = (x, -\mathrm{N}w(x)) = (a+b-c)(k_2c - k_1ab) = k_1N^3 - 3k_2N^2$, which clearly cannot be bounded linearly in $\|x\|^2 = 11N^2$. The same argument applies also to (2.36b). W

This negative result can perhaps best be appreciated as a kind of loss of information about the dependencies between the species in the functional form of the condition (3.5). The number of A- and B-molecules is strongly correlated with the number of C-molecules such that, in fact, in (2.36a) $a+b+2c$ is a preserved quantity. By contrast, in (3.5) the growth of $\|x\|^2$ is estimated from the sum of the growth of the individual elements of x as if they where independent.

A way around this limitation can be found provided that we leave the general case $x \in \mathbf{R}^D$. We therefore specify the discussion to the positive quadrant $x \in \mathbf{R}_+^D$ and assume from now on that it can be shown a priori that the initial data x_0 belongs to this set and that the subsequent trajectory $x(t)$ never leaves it (compare Assumption 2.1).

It then follows that $\|x\|_1 = (\mathbf{1}, x) = \mathbf{1}^T x$, where $\mathbf{1}$ is the vector of length D containing all ones. This vector also defines a suitable "outward" vector for states $x \in \mathbf{R}_+^D$ since solutions to the ODE (2.2) cannot grow without simultaneously growing also in the direction of $\mathbf{1}$.

Again, in view of Grönwall's inequality Lemma 3.1, we tentatively require that $(\mathbf{1}, F(x)) \le \text{constant} \times \|x\|_1$ for $\|x\|_1$ sufficiently large. Equivalently, for any x,

$$(\mathbf{1}, F(x)) \le A + \alpha \|x\|_1, \tag{3.7}$$

implying the bounded dynamics

$$\frac{\mathrm{d}}{\mathrm{d}t}\|x\|_1 = (\mathbf{1}, F(x)) \le A + \alpha \|x\|_1. \tag{3.8}$$

We remark in passing that the criterion (3.7) is sharp in the sense that if the reversed inequality can be shown to be true, then the growth of solutions can be estimated from below.

Example 3.1 As a point in favor of this approach we compute for the bi-molecular birth-death system (2.32), $(\mathbf{1}, F(x)) = (\mathbf{1}, -\mathrm{N}w(x)) = 2k_1 - 2k_2 ab$ which evidently falls under the assumption (3.7) with $(A, \alpha) = (2k_1, 0)$. For the reversible case (2.36a) we similarly get $(\mathbf{1}, F(x)) = -k_1 ab + k_2 c$ such that (3.7) applies with $(A, \alpha) = (0, k_2)$. Finally, and in the same fashion, the open case (2.36b) is seen to be covered by letting $(A, \alpha) = (k_4, (k_2 - k_3) \vee 0)$.

The chosen "outward" vector $\mathbf{1}$ is by no means special. Clearly, any strictly positive vector $\boldsymbol{l}$ may be used in its place since $\|x\|_1$ and $\|x\|_l := \boldsymbol{l}^T x$ are equivalent norms over $\mathbf{R}_+^D$. This is a general and useful observation as it may be used to discard parts of a system that are closed without any restrictions on the associated propensities.

Example 3.2 For the reversible system (2.36a), we have already noted that $a+b+2c$ is a conserved quantity such that the choice $l=[1,1,2]^{\mathrm{T}}$ yields $\mathrm{d}/\mathrm{d}t\|x\|_l=0$. The open case (2.36b) also benefits from this weighted norm in that we get $\mathrm{d}/\mathrm{d}t\|x\|_l\le 2k_4$.

Example 3.3 A slightly more involved model reads as follows:

$$\left.\begin{array}{ll}\varnothing\xrightarrow{k_1}A & A+B\xrightarrow{k_2ab}3C\\ \varnothing\xrightarrow{k_1}B & C\xrightarrow{k_3c}\varnothing\end{array}\right\}$$

This example has been constructed such that the quadratic reaction increases $\|x\|_1$ and hence (3.7) does not apply. However, taking $l=[3,3,2]^{\mathrm{T}}$ we get

$$\frac{\mathrm{d}}{\mathrm{d}t}\|x_t\|_l=6k_1-2k_3c\le 6k_1.$$

This example hints at a general technique for obtaining suitable candidates for the weight vector l. Simply form the matrix N_2 consisting of the columns of N that are affected by superlinear propensities. If a vector $l>0$ annihilating these propensities exists, it can be found in the null-space of $-\mathrm{N}_2^{\mathrm{T}}$, readily available by linear algebra techniques. We omit the details.

3.2. Continuity

For well-posedness of the ODE (2.2) we also need continuity with respect to the initial data. We cannot ask for uniform Lipschitz continuity since $\|F(x)-F(y)\|\le L\|x-y\|$ clearly implies (3.3) which we have already refuted. For the same reason, a uniform one-sided Lipschitz condition $(x-y,F(x)-F(y))\le\lambda\|x-y\|^2$ cannot be assumed to hold since it implies (3.5). The problem here is the global nature of the estimate and it therefore seems to be reasonable to localize this assumption. For instance, one might ask for

$$\left(x-y,F(x)-F(y)\right)\le \lambda_R\left\|x-y\right\|^2 \quad \text{whenever } \left\|x\right\|,\left\|y\right\|\le R, \tag{3.9}$$

presumably with some growth restrictions on λ_R. Although very general, such an analysis is likely to be less informative when it comes to estimating actual constants in later results. We shall therefore consider the following simpler version,

$$\left(x-y,F(x)-F(y)\right)\le\left(M+\mu\left\|x+y\right\|_1\right)\left\|x-y\right\|^2, \tag{3.10}$$

where the form of λ_R has been restricted to better suit the present purposes. Trivially, the norms $\|\cdot\|_1$ and $\|\cdot\|$ are equivalent and hence the specific choice made in (3.10) is just a matter of convenience. Since the idea here is to use a priori bounds on x and y when deriving perturbation bounds, using $\|\cdot\|_1$ (or $\|\cdot\|_l$) is natural.

Theorem 3.4 Suppose that the ODE (2.2) satisfies (3.7) and (3.10) and that initial data $x_0\in\mathbf{R}_+^D$ implies a solution $x(t)\in\mathbf{R}_+^D$. Then for any $t\in[0,T]$ there is a unique such solution $x(t)$. Moreover, define $C(t;x_0,y_0)\equiv M+\mu\left\|x_t+y_t\right\|_1$, where x_t and y_t are two trajectories associated with initial data $x_0,y_0\in\mathbf{R}_+^D$, respectively. Then

$$\begin{aligned}\left\|x(t)-y(t)\right\|&\le\left\|x_0-y_0\right\|\exp\left(\int_0^t C(s;x_0,y_0)\,ds\right)\\&=\left\|x_0-y_0\right\|\left[1+\left(M+\mu\left(\left\|x_0+y_0\right\|_1\right)\right)t+O\left(t^2\right)\right].\end{aligned} \tag{3.11}$$

Proof. Combining (3.7) with Grönwall's inequality we get the a priori estimate

$$\left\|x(t)+y(t)\right\|_1\le\left\|x_0+y_0\right\|_1+\left(\alpha\left\|x_0+y_0\right\|_1+2A\right)\left(e^{\alpha t}-1\right)/\alpha.$$

Hence the (bounded) solution to

$$\frac{\mathrm{d}}{\mathrm{d}t}\|x-y\|^2 = 2\left(x-y, F(x)-F(y)\right) \le 2C(t;x_0,y_0)\|x-y\|^2.$$

is readily found through its integrating factor. The order estimate is a consequence of the fact that

$$\int_0^t \|x_s\|_1 - \|x_0\|_1 \,\mathrm{d}s = \mathrm{O}\left(t^2\right),$$

since the trajectory is continuous. W

3.3. Bounds for Elementary Reactions

As briefly discussed by the end of Section 3.1, finding bounds on A and α in (3.7) as well as a suitable weight-vector l amounts to basic inequalities and some fairly straightforward linear algebra manipulations. In this section we therefore consider precise bounds in (3.10) for the elementary propensities (2.3). Since (3.10) is linear in F, a reasonable approach is to consider linear and quadratic propensities separate (constant propensities trivially satisfy (3.10) with $M = \mu = 0$).

Proposition 3.5 (linear case) Write a set of R linear propensities as $w_r(x) = q_r^{\mathrm{T}} x$, $r = 1, \ldots, R$, each with the corresponding stoichiometric vector N_r. Then $F(x) := -\sum_r \mathrm{N}_r w_r(x)$ satisfies $\left(x-y, F(x)-F(y)\right) \le M\|x-y\|^2$ with $M = \mu_2\left[-\mathrm{N}Q^{\mathrm{T}}\right]$ in terms of the Euclidean logarithmic norm $\mu_2[\cdot]$

and the matrix Q containing the vectors q_r as columns. In particular, in the case of a single linear propensity and, if as is usually the case, $q_j = k\delta_{jn}$ is all-zero except for a single rate constant k in the nth position, then this reduces to $M = k\left(-\mathrm{N}_{r,n} + \|\mathrm{N}_r\|\right)/2$.

Proof. The first assertion is immediate since the smallest such constant M by definition is the logarithmic norm (see e.g. [32]). To compute $\mu_2\left[-\mathrm{N}_r q_r^{\mathrm{T}}\right]$ when q has the form indicated, we determine the extremal eigenvalue of $-\left(\mathrm{N}_r q_r^{\mathrm{T}} + q_r \mathrm{N}_r^{\mathrm{T}}\right)/2$. By the (signed) scaling invariance of the logarithmic norm we may without loss of generality take $k \equiv 1$. The spectral relation for an eigenpair (λ, z) can be written as

$$-\frac{1}{2}\mathrm{N}_{r,j} z_n = \lambda z_j, \quad j = 1, 2, \ldots, D,\ j \neq n,$$

$$-\frac{1}{2}\mathrm{N}_{r,n} z_n - \frac{1}{2}\mathrm{N}_r^{\mathrm{T}} z = \lambda z_n.$$

For non-zero λ the first relation can be solved for z_j. When inserted into the second relation, using that $z_n \neq 0$ (or otherwise $z = 0$), we get a quadratic equation for λ with a single extremal root. $\blacksquare$

Example 3.4 The simple special case in Proposition 3.5 is generally sharp except for when there are linear reactions affecting all species considered in the model. For example, in a one-dimensional state space, the single decay $A \to \varnothing$ with propensity $w_1(a) = ka$ allows the optimal value $M = -k$. In general D-dimensional space, a chain with unit rate constants of the form $A_1 \to A_2 \to \ldots \to A_D \to \varnothing$, or a closed loop in which the last transition is replaced with $A_D \to A_1$, both admit bounds $M \leq 0$ as an inspection of the Gershgorin-discs of $-\left(\mathrm{N} + \mathrm{N}^{\mathrm{T}}\right)/2$ shows.

Other than for those special examples, for the most important linear cases, Table 1 summarizes the bounds as obtained from the special case in Proposition 3.5 (with all reaction constants normalized to unity).

Proposition 3.6 (quadratic case). Write a general quadratic propensity as $w_r(x) = x^{\mathrm{T}} S x$ with S a symmetric matrix. Then $F_r(x) = -\mathrm{N}_r w_r(x)$ satisfies (3.10) with $M = 0$ and

$\mu = \|x+y\|_1^{-1} \mu_2\left[-\mathrm{N}_r (x+y)^{\mathrm{T}} S\right] \le \|\mathrm{N}_r\| \|S\|$. For the special case that $S_{ij} = k\left(\delta_{im}\delta_{jn} + \delta_{jm}\delta_{in}\right)/2$ there holds $\mu \le k \max_{j\in\{m,n\}}\left(-\mathrm{N}_{r,j} + \|\mathrm{N}_r\|\right)/4$.

Proof. Since S is symmetric we have $x^{\mathrm{T}}Sx - y^{\mathrm{T}}Sy = (x+y)^{\mathrm{T}} S(x-y)$. Hence an explicit expression for μ is obtained as follows:

$$\mu \le \sup_{x,y\in\mathbf{Z}_+^D} \|x+y\|_1^{-1} \|x-y\|^{-2}\left(x-y, -\mathrm{N}_r (x+y)^{\mathrm{T}} S(x-y)\right)$$
$$= \sup_{w\in\mathbf{Z}_+^D} \sup_{v\in\mathbf{Z}^D} \|w\|_1^{-1} \|v\|^{-2}\left(v, -\mathrm{N}_r w^{\mathrm{T}} S v\right) \le \sup_{\substack{u\ge 0\\ \|u\|_1 \le 1}} \mu_2\left[-\mathrm{N}_r u^{\mathrm{T}} S\right].$$

The indicated upper bound is derived from the fact that $|\mu_2[\cdot]| \le \|\cdot\|$ [32]. For the useful special case, define first the vector $q = q_1 + q_2$ in terms of $q_{1,j} = k\delta_{jm}\left(x_n + y_n\right)/2$, and $q_{2,j} = k\delta_{jn}\left(x_m + y_m\right)/2$. Using the fact that the logarithmic norm is sub-additive we can reuse the calculation in the proof of Proposition 3.5, $\mathbf{W}$

Example 3.5 The most important quadratic cases are summarized in Table 2. For the dimerizations in the lower half of the table there is also a linear part M in (3.10).

Example 3.6 The bi-molecular birth-death model (2.32) admits the constants $(M,\mu) = \left(0, k_2\left(\sqrt{2}-1\right)/4\right)$ in (3.10). Similarly, the reversible cases (2.36a) and (2.36b) both obeys (3.10) with

$(M,\mu) = \left(\sqrt{3}-1\right)/2 \times (k_2, k_1/2)$. All these results are sharp except for the open case (2.36b) for which one can obtain a slightly smaller constant M by using the general formula stated in Proposition 3.5.

Example 3.7 As a highly prototypical example we consider the following natural extension of the bi- molecular birth-death model (2.32),

$$\left.\begin{array}{cc} \emptyset \underset{k_3 a}{\overset{k_1 V}{\rightleftharpoons}} A & \emptyset \underset{k_3 b}{\overset{k_1 V}{\rightleftharpoons}} B \\ A+B \xrightarrow{k_2 ab/V} \emptyset & \end{array}\right\},$$

where in this example it is informative to consider the dependence on the system's size V. It is straightforward to show the bounds $(A,\alpha) \le (2k_1 V, -k_3)$ in (3.7) and hence that the system is effectively bounded despite being of open character. This is seen from the fact that, for states $\|x\|_1 \ge 2k_1/k_3 \cdot V$, the dynamics is dissipative in the $\|\cdot\|_1$-norm. Furthermore, from Proposition 3.5 and 3.6 we get the sharp bounds $(M,\mu) \le \left(-k_1 V, k_2/V \cdot \left(\sqrt{2}-1\right)/4\right)$ in (3.10). It follows that for states $\{x,y\}$ such that $\|x+y\|_1 \gtrsim 9.7k_1/k_2 \cdot V^2$, the dynamics is contractive in the Euclidean norm. For density dependent propensities we expect that $\|x\| \sim V$ in any norm as V grows, and hence the region of contractivity grows in a relative sense. Intuitively one expects that these results offer an insight into the evolution of the process that is relevant also in the stochastic setting.

4. STOCHASTIC STABILITY

We now consider the properties of the stochastic jump SDE (2.16). For convenience we start by collecting all assumptions in Section 4.1. In the stochastic setting the requirements for existence and uniqueness are slightly stronger than in the deterministic case such that the one-sided bound (3.10) needs to be augmented with an

Table 1. Linear propensities and bounds of M in (3.10).

Reaction	Bound on M
$A \to \emptyset$	0
$A \to B$	$\left(\sqrt{2}-1\right)/2$
$A \to B + C$	$\left(\sqrt{3}-1\right)/2$

Table 2. Quadratic propensities and bounds of M, μ in (3.10).

Reaction	Bound on μ	M
$A+B\to\varnothing$	$\left(\sqrt{2}-1\right)/4$	
$A+B\to C$	$\left(\sqrt{3}-1\right)/4$	
$A+B\to A$	$1/4$	
$A+B\to A+C$	$\sqrt{2}/4$	
$A+A\to\varnothing$	0	2
$A+A\to B$	$\sqrt{5}/2-1$	$\sqrt{5}/2+1$

unsigned version, implying essentially the assumption of at most quadratically growing propensities. We demonstrate that this assumption is reasonable by constructing a model involving cubic propensities and with unbounded second moments. On the positive side we show in Section 4.2 that the assumptions are strong enough to guarantee finite moments of any order during finite time intervals.

We prove existence and uniqueness of solutions to the jump SDE (2.16) in Section 4.3. A sufficient condition for the existence of asymptotic bounds of the pth order moment is given in Section 4.4 where we also derive some stability estimates.

4.1. Working Assumptions

We state formally the set of assumptions on the jump SDE (14) as follows.

Assumption 4.9 For arguments x, $y\in\mathbf{Z}_{+}^{D}$, $F(x):=-\mathrm{N}w(x)$, and weighted norm $\|x\|_{l}:=l^{\mathrm{T}}x$ we assume that

(1) $-l^{\mathrm{T}}\mathrm{N}w(x)\le A+\alpha\|x\|_{l}$ ("bounded growth"),

(2) $\left(-l^{\mathrm{T}}\mathrm{N}\right)^{2}w(x)/2\le B+\beta_{1}\|x\|_{l}+\beta_{2}\|x\|_{l}^{2}$ ("absolutely bounded growth"),

(3) $\left(\mathbf{1}^{\mathrm{T}}\mathrm{N}^{2}\right)\left|w(x)-w(y)\right| \le L\left(1+\|x+y\|_{1}\right)\|x-y\|$,

(4) $\left(x-y,F(x)-F(y)\right) \le \left(M+\mu\|x+y\|_{1}\right)\|x-y\|^{2}$.

The parameters $\{A,B,\beta_1,\beta_2,L\}$ are assumed to be positive (with β_2 possibly zero) but we allow also negative values of $\{\alpha,M,\mu\}$. The vector $\boldsymbol{l}$ is normalized such that $\min_i \boldsymbol{l}_i = 1$; hence the bound $\|x\|_1 \le \|x\|_l$ is sharp.

After the original draft of the current paper was finished, the author became aware of two other papers discussing very similar conditions [33] [34] . In particular, Assumption 4.1 (1), (2) are also found in ([33] , Condition (1). In fact, these very conditions can be shown to be exactly what is needed to apply the earlier and quite general theory found in ([35] , Theorem 7.1).

In Assumption 4.9 (2) the case $\beta_2 = 0$ will merit special attention. For well-posedness it turns out that we will need to require a higher regularity of the initial data when $\beta_2 > 0$ (see Theorem 4.7) and the condition for ergodicity becomes more restrictive (see Theorem 4.9). In practice, $\beta_2 = 0$ implies that opposing quadratic reactions of the type

$$\left.\begin{array}{l} 2X \xrightarrow{x(x-1)} 3X \\ 2X \xrightarrow{x(x-1)} X \end{array}\right\}, \tag{4.1}$$

are impossible. Similarly, when $\beta_1 = 0$ reactions of the type

$$X \xrightarrow{x} 2X \tag{4.2}$$

are excluded.

Note that (2) and (4) are redundant in the sense that they are both implied by (3). However, as we saw in Section 3.3, in (4) it is often possible to find sharper constants M and μ by considering this bound in isolation. Also, although (3) is stronger than (4), it is in particular valid for quadratic propensities as can be seen from the representation used in the proof of Proposition 3.6,

$$\left|w_r(x)-w_r(y)\right| = \left|(x+y)^{\mathrm{T}} S(x-y)\right| \le \|S\|\|x+y\|_1\|x-y\|. \tag{4.3}$$

The Danger with Cubic Propensities

Assumption 4.1 (2) specifies the discussion to propensities with at most quadratic growth, at least when measured in the direction of the weight vector l. To show that this is natural we now demonstrate that additional care should be taken when considering cubic propensities.

Example 4.2 Consider the model

$$\left.\begin{array}{l} 3X \xrightarrow{x(x-1)(x-2)/2} X \\ 3X \xrightarrow{x(x-1)(x-2)} 4X \end{array}\right\},$$

such that the stoichiometric vector is given by $\mathrm{N} = [-2,1]$, and hence that the drift $-\mathrm{N}w(x) = 0$.

Proposition 4.1 For the model in Example 4.2, if $X_0 \geq 3$, then the second moment explodes in finite time.

Proof. Assume that both the second and the third moment are bounded for $t \in [0,T)$ with $T > 0$. From (2.7) we get the governing equation

$$\frac{\mathrm{d}}{\mathrm{d}t} EX_t^2 = E\left[3X_t(X_t-1)(X_t-2)\right],$$

such that the growth of the second moment remains bounded only provided that the third moment remains finite. It is convenient to look at the cumulative third order moment. From (2.7),

$$\frac{\mathrm{d}}{\mathrm{d}t} EC_3(X_t) := \frac{\mathrm{d}}{\mathrm{d}t} EX_t(X_t-1)(X_t-2) = E\left[9C_3(X_t)(X_t-2/3)\right].$$

By the arithmetic-geometric mean inequality, $x-2/3 \geq x-1 \geq \left[x(x-1)(x-2)\right]^{1/3}$, such that by Jensen's inequality,

$$\frac{\mathrm{d}}{\mathrm{d}t} EC_3(X_t) \geq 9E\left[C_3(X_t)^{4/3}\right] \geq 9\left[EC_3(X_t)\right]^{4/3}.$$

We put $u^3 = EC_3(X_t)$ and get the differential inequality

$$\frac{\mathrm{d}(1/u)}{\mathrm{d}t} \le -3,$$

which can be integrated and rearranged to produce the bound

$$E\left[X_t(X_t-1)(X_t-2)\right] \ge \frac{X_0(X_0-1)(X_0-2)}{1-3tX_0(X_0-1)(X_0-2)}. \tag{4.4}$$

Hence the third, and consequently also the second moment explode for some finite t whenever $X_0 \ge 3$. W

Interestingly, we note that if $X_0 = 3$, then the probability that the cubic decay transition occurs first is $1/3$, and if this happens the state of the system will be stuck with a single molecule indefinitely.

4.2. Moment Bounds

In this section we consider general moment bounds derived from (2.20) using localization. To get some guidance, let us first assume that the differential form of Dynkin's formula (2.7) is valid. Since any trajectory $(X_t)_{t\ge 0}$ by the basic Assumption 2.1 will belong to $\mathbf{Z}_+^D$, we may use that $\|X_t\|_l = (l, X_t)$. Hence from (2.7) with $f(x) = (l, x)$ we get that

$$\frac{\mathrm{d}}{\mathrm{d}t} E\|X_t\|_l = (l, F(X_t)) \le A + \alpha\|X_t\|_l, \tag{4.5}$$

by Assumption 4.1 (1). Clearly, the differential form of Grönwall's inequality in Lemma 3.1 applies here. A correct version of this argument unfortunately looses the sign of α.

Proposition 4.2 If Assumption 4.1 (1) is true, then

$$E\|X_t\|_l \le \|X_0\|_l \exp(\alpha_+ t) + A\left(\exp(\alpha_+ t) - 1\right)/\alpha_+,$$

where $\alpha_+ = \alpha \vee 0$.

Here and below we shall make use of the stopping time $\tau_P = \inf_{t\geq 0}\{\|X_t\|_l > P\}$ and define $\hat{t} = t \wedge \tau_P$.

Proof. From (2.20) with $f(x) = (l, x)$ we get that

$$E\|X_{\hat{t}}\|_l = \|X_0\|_l + E\int_0^{\hat{t}}(l, F(X_s))\,\mathrm{d}s \leq \|X_0\|_l + E\int_0^t A + \alpha_+ \|X_{\hat{s}}\|_l\,\mathrm{d}s, \tag{4.6}$$

By the integral form of Grönwall's inequality in Lemma 3.1 we deduce in terms of $Y_t := X_{t\wedge\tau_P}$ that

$$E\|Y_t\|_l \leq \|X_0\|_l \exp(\alpha_+ t) + A(\exp(\alpha_+ t) - 1)/\alpha_+ \tag{4.7}$$

such that the same bound holds for X_t by letting $P \to \infty$. W

We attempt a similar treatment for obtaining bounds in mean square. Assuming tactically that (2.7) is valid, writing $\|x\|^2 = x^{\mathrm{T}}x$ we get after some work that

$$\frac{\mathrm{d}}{\mathrm{d}t}E\|X_t\|^2 = E\left[\mathbf{1}^{\mathrm{T}}\mathrm{N}^2 w(X_t) - 2X_t^{\mathrm{T}}\mathrm{N}w(X_t)\right] \tag{4.8}$$

where $\mathrm{N}_{ij}^2 \equiv (\mathrm{N}_{ij})^2$. We expect from Grönwall's inequality that $E\|X_t\|^2$ grows at most exponentially with αt whenever

$$\mathbf{1}^{\mathrm{T}}\mathrm{N}^2 w(x) - 2x^{\mathrm{T}}\mathrm{N}w(x) \leq A + \alpha\|x\|^2. \tag{4.9}$$

However, this tentative condition is often violated in practice since the second term $-x^{\mathrm{T}}\mathrm{N}w(x) = (x, F(x))$, and since we already know from Proposition 3.3 that this quantity does not admit bounds in terms of $\|x\|^2$ even for very simple problems.

More realistic conditions arise when seeking to bound $\|X_t\|_l^2$ instead.

Proposition 4.3 If for some constants γ and C,

$$\left(\boldsymbol{l}^{\mathrm{T}}\mathrm{N}\right)^2 w(x) - 2\boldsymbol{l}^{\mathrm{T}}\mathrm{N}w(x)\|x\|_{\boldsymbol{l}} \le C + \gamma\|x\|_{\boldsymbol{l}}^2, \tag{4.10}$$

$\left(\boldsymbol{l}^{\mathrm{T}}\mathrm{N}\right)^2$ understood elementwise), then $E\|X_t\|_{\boldsymbol{l}}^2 \le \|X_0\|_{\boldsymbol{l}}^2 \exp(\gamma_+ t) + C(\exp(\gamma_+ t) - 1)/\gamma_+$.

The proof of Proposition 4.3 follows the same pattern as for Proposition 4.2, but using this time

$f(x) = \|x\|_{\boldsymbol{l}}^2 = (\boldsymbol{l}, x)^2$ in (2.20). The condition (4.10) is typically more realistic than (4.9) since we recognize the term $-\boldsymbol{l}^{\mathrm{T}}\mathrm{N}w(x)\|x\|_{\boldsymbol{l}} = (\boldsymbol{l}, F(x))\|x\|_{\boldsymbol{l}}$, which under the evidently reasonable Assumption 4.1 (1) is $\le \left(A + \alpha\|x\|_{\boldsymbol{l}}\right)\|x\|_{\boldsymbol{l}}$. It follows that if $\left(\boldsymbol{l}^{\mathrm{T}}\mathrm{N}\right)^2 w(x)$ grows at most quadratically with $\|x\|_{\boldsymbol{l}}$, then this assumption is sufficient to yield bounds in mean square. Stated formally,

Proposition 4.4 Under Assumption 4.1 (1) and (2) the condition (4.10) of Proposition 4.3 is true with $\gamma = 2\beta_2 + 2\alpha + 2$ and $C = 2B + \beta_1^2 + A^2$.

Proof. This is straightforward: we get by the assumptions and Hölder's inequality,

$$\left(\boldsymbol{l}^{\mathrm{T}}\mathrm{N}\right)^2 w(x) - 2\boldsymbol{l}^{\mathrm{T}}\mathrm{N}w(x)\|x\|_{\boldsymbol{l}} \le 2B + 2\beta_1\|x\|_{\boldsymbol{l}} + 2\beta_2\|x\|_{\boldsymbol{l}}^2 + 2\left(A + \alpha\|x\|_{\boldsymbol{l}}\right)\|x\|_{\boldsymbol{l}},$$

where an application of Young's inequality yields the indicated bounds. W

As a strong point in favor of our running assumptions we now demonstrate that the above reasoning can be generalized: these two conditions implies finite time stability in any order moment. We note that in a recent manuscript [34] , related conditions for the same results are proposed.

Theorem 4.5 (Moment estimate). Under Assumption 4.1 (1) and (2), for any integer $p \ge 1$,

$$E\|X_t\|_{\boldsymbol{l}}^p \le \left(\|X_0\|_{\boldsymbol{l}}^p + 1\right)\exp(Ct) - 1, \tag{4.11}$$

where $C > 0$ is a constant depending on the assumptions and on p.

The proof of Theorem 4.5 and some later results will simplify using the following bound.

Lemma 4.6 Let $H(x) \equiv (x+y)^p - x^p$ with $x \in \mathbf{R}_+$ and $y \in \mathbf{R}$. Then for integer $p \geq 1$ we have the bounds

$$H(x) \leq pyx^{p-1} + 2^{p-4}p(p-1)y^2\left[x^{p-2} + |y|^{p-2}\right], \tag{4.12}$$

$$|H(x)| \leq p|y|2^{p-2}\left[x^{p-1} + |y|^{p-1}\right]. \tag{4.13}$$

Proof. Both results follow from Taylor expansions;

$$H(x) = pyx^{p-1} + \frac{p(p-1)}{2}y^2[x + \theta_1 y]^{p-2},$$

$$|H(x)| = p|y||x + \theta_2 y|^{p-1},$$

respectively, where $\theta_{1,2} \in [0,1]$. Using the triangle inequality and the elementary inequality $(a+b)^p \leq 2^{p-1}(a^p + b^p)$ the lemma is proved. W

Proof of Theorem 4.5. Using (2.20) with $f(X_t) \equiv [l^{\mathrm{T}}X_t]^p$ we get

$$E\|X_{\hat{t}}\|_l^p = \|X_0\|_l^p + E\int_0^{\hat{t}} \underbrace{\sum^{R} w_r(X_s)\left[\left[l^{\mathrm{T}}(X_s - \mathrm{N}_r)\right]^p - \left[l^{\mathrm{T}}X_s\right]^p\right]}_{=:G(X_s)} ds. \tag{4.14}$$

Using Lemma 4.6 (4.12) and Assumption 4.1 (1) and (2) we obtain

$$G(x) \leq p(A + \alpha\|x\|_l)\|x\|_l^{p-1} + 2^{p-3}p(p-1)\left(B + \beta_1\|x\|_l + \beta_2\|x\|_l^2\right)\left(\|x\|_l^{p-2} + \Delta^{p-2}\right),$$

where $\Delta := \left\| l^{\mathrm{T}} \mathrm{N} \right\|_{\infty}$. Expanding and using Young's inequality with exponents $\{p/(p-1), p/(p-2)\}$ and conjugate exponents $\{p, p/2\}$,

$$E\left\|X_{\hat{t}}\right\|_{l}^{p} \leq \left\|X_{0}\right\|_{l}^{p} + \int_{0}^{\hat{t}} C\left(1 + E\left\|X_{\hat{s}}\right\|_{l}^{p}\right) \mathrm{d}s,$$

for some constant C which thus depends on the assumptions. Applying Grönwall's inequality and letting $P \to \infty$ we obtain the stated result.

4.3. Existence and Uniqueness

We shall now prove that the jump SDE (2.16) under Assumption 4.1 has a uniquely defined and locally bounded solution. To this end and following ([8] , Section 3.1.2), we introduce the following spaces of path-wise locally bounded processes:

$$S_{\mathrm{F}}^{p,\mathrm{loc}}\left(\mathbf{Z}_{+}^{D}\right) = \Big\{ X(t,\omega) : X \text{ is } \mathrm{F}_{t}\text{-adapted and } \mathbf{Z}_{+}^{D}\text{-valued such that}$$

$$E \sup_{t \in [0,T]} \left\| X(t,\omega) \right\|_{1}^{p} < \infty \text{ for } \forall T < \infty \Big\}. \tag{4.15}$$

Theorem 4.7 (Existence). Let X_t be a solution to (2.16) under Assumption 4.1 (1) and (2) with $\beta_2 = 0$.

Then if $\left\|X_0\right\|_l^p < \infty$, $\{X_t\}_{t \geq 0} \in S_{\mathrm{F}}^{p,\mathrm{loc}}\left(\mathbf{Z}_{+}^{D}\right)$. If $\beta_2 > 0$ then the conclusion remains under the additional

requirement that $\left\|X_0\right\|_l^{p+1} < \infty$.

Proof. Below we let C denote a positive constant which may be different on each occasion used. As before we use the stopping time $\tau_P = \inf_{t \geq 0} \left\{ \left\|X_t\right\|_l > P \right\}$ and put $\hat{t} = t \wedge \tau_P$. We get from Itô's formula (with G defined in (4.14))

$$\left\|X_{\hat{t}}\right\|_{l}^{p} = \left\|X_{0}\right\|_{l}^{p} + \int_{0}^{\hat{t}} G(X_{s}) \mathrm{d}s + \int_{0}^{\hat{t}} \int_{I} \sum_{r=1}^{R} \hat{w}_{r}(X_{s-}; z) \left[\left[l^{\mathrm{T}} (X_{s-} - \mathrm{N}_{r}) \right]^{p} - \left[l^{\mathrm{T}} X_{s-} \right]^{p} \right] (\mu - m)(\mathrm{d}s \times \mathrm{d}z).$$

Since the propensities are bounded for bounded arguments (Assumption 2.1), using the stopping time we find that the jump part is absolutely integrable and hence a local martingale $M_{\hat{t}}$. We estimate its quadratic variation under Assumption 4.1 (2),

$$E[M]_{\hat{t}}^{1/2} \le E\left[\int_0^{\hat{t}}\int_I \sum_{r=1}^{R} \hat{w}_r(X_s;z)\left|\left[l^{\mathrm{T}}(X_s-\mathrm{N}_r)\right]^p-\left[l^{\mathrm{T}}X_s\right]^p\right|\mu(\mathrm{d}s\times\mathrm{d}z)\right]$$

$$= E\left[\int_0^{\hat{t}}\int_I \sum_{r=1}^{R} \hat{w}_r(X_s;z)W(X_s)\left|\left[l^{\mathrm{T}}(X_s-\mathrm{N}_r)\right]^p-\left[l^{\mathrm{T}}X_s\right]^p\right|\mathrm{d}s\times\mathrm{d}z\right]$$

$$= E\left[\int_0^{\hat{t}} \sum_{r=1}^{R} w_r(X_s)\left|\left[l^{\mathrm{T}}(X_s-\mathrm{N}_r)\right]^p-\left[l^{\mathrm{T}}X_s\right]^p\right|\mathrm{d}s\right]$$

$$\le E\left[\int_0^{\hat{t}} \sum_{r=1}^{R} p\left|l^T\mathrm{N}_r\right| w_r(X_s)2^{p-2}\left[\left\|X_s\right\|_l^{p-1}+\left|l^{\mathrm{T}}\mathrm{N}_r\right|^{p-1}\right]\mathrm{d}s\right] \tag{4.16}$$

$$\le E\left[\int_0^{\hat{t}} C\left(B+\beta_1\left\|X_s\right\|_l+\beta_2\left\|X_s\right\|_l^2\right)\left[\left\|X_s\right\|_l^{p-1}+\Delta^{p-1}\right]\mathrm{d}s\right] \tag{4.17}$$

$$\le E\int_0^{\hat{t}} C\left(1+\left\|X_s\right\|_l^p+\beta_2\left\|X_s\right\|_l^{p+1}\right)\mathrm{d}s \tag{4.18}$$

where $\Delta=\left\|l^{\mathrm{T}}\mathrm{N}\right\|_\infty$. In (4.16) Lemma 4.6 (4.13) was applied and Assumption 4.1 (2) entered in (4.17). Assume first that $\beta_2=0$. Then for the drift part we have already constructed a suitable bound in Theorem 4.5 such that

$$\left\|X_{\hat{t}}\right\|_l^p \le \left\|X_0\right\|_l^p+\int_0^{\hat{t}} C\left(1+\left\|X_s\right\|_l^p\right)\mathrm{d}s+\left|M_{\hat{t}}\right|.$$

Taking supremum and expectation values we get from Burkholder's inequality ([6] , Chap. IV.4) that

$$E\sup_{s\in[0,\hat{t}]}\left\|X_s\right\|_l^p \le \left\|X_0\right\|_l^p+\int_0^{\hat{t}} C\left(1+E\sup_{s'\in[0,s]}\left\|X_{s'}\right\|_l^p\right)\mathrm{d}s.$$

Writing $\left\|X\right\|_l^p(t)\equiv\sup_{s\in[0,t]}\left\|X_s\right\|_l^p$ we conclude that

$$E\|X\|_l^p(t\wedge\tau_p)\le\|X_0\|_l^p+\int_0^t C\left(1+E\|X\|_l^p(s\wedge\tau_p)\right)\mathrm{d}s.$$

By Grönwall's inequality we have thus shown that $E\|X\|_l^p(t\wedge\tau_p)$ can be bounded in terms of the initial data and time t. The result now follows by letting $P\to\infty$ and using Fatou's lemma.

Next assume that $\beta_2>0$. Then we have to rely more directly on Theorem 4.5 in (4.18),

$$E[M]_{\hat t}^{1/2}\le\int_0^{\hat t}C\left(1+E\|X_s\|_l^{p+1}\right)\mathrm{d}s\le\left(\mathrm{e}^{C\hat t}-1\right)\left(\|X_0\|_l^{p+1}+1\right),$$

where, although there is now a dependency on $\|X_0\|^{p+1}$, the rest of the argument carries through.

For the case that the initial data X_0 is non-deterministic we see that the general quadratic case Assumption 4.1 (2) with $\beta_2>0$ requires a one order higher moment of the initial data in order for a solution in $S_{\mathrm{F}}^{p,\mathrm{loc}}\left(\mathbf{Z}_+^D\right)$ to exist.

Theorem 4.8 (Uniqueness) Let Assumption 4.1 (1)-(4) hold true. Then any two paths X_t and Y_t coupled according to the description in Section 2.2.2 with $X_0=Y_0$ are equal.

We shall be using the observation that, for $x\in\mathbf{Z}_+^D$, we have that $\|x\|_1\le\|x\|^2$ (referred below to as the "integer inequality").

Proof. Under the same stopping time as before we get from Itô's formula using the coupling described in Section 2.2.2 that

$$\begin{aligned}E\|X_{\hat t}-Y_{\hat t}\|^2&=E\int_0^{\hat t}2\left(X_s-Y_s,F(X_s)-F(Y_s)\right)+\left(\mathbf{1}^{\mathrm{T}}\mathrm{N}^2\right)|w(X_s)-w(Y_s)|\,\mathrm{d}s\\&\le E\int_0^{\hat t}2\left(M+\mu\|X_s+Y_s\|_1\right)\|X_s-Y_s\|^2+\left\|\mathbf{1}^{\mathrm{T}}\mathrm{N}^2\right\|_\infty L\left(1+\|X_s+Y_s\|_1\right)\|X_s-Y_s\|\,\mathrm{d}s.\end{aligned}$$

From the integer inequality we find that there is a constant $C \geq 0$ depending on P such that

$$E\left\|(X-Y)(t \wedge \tau_P)\right\|^2 \leq \int_0^{t \wedge \tau_P} CE\left\|X_s - Y_s\right\|^2 \mathrm{d}s \leq \int_0^t CE\left\|(X-Y)(s \wedge \tau_P)\right\|^2 \mathrm{d}s.$$

Using that $E\left\|X_0 - Y_0\right\|^2 = 0$ and Grönwall's inequality we conclude that the only solution is the zero solution. Letting $P \to \infty$ and using Fatou's lemma the statement is therefore proved. W

In a certain sense the previous result is trivial; from the Poisson representation (2.8) we see that up to the first explosion, a path is uniquely determined from an initial state and a series of Poisson distributed events. However, and as we shall see below, the above proof is prototypical for more involved situations. An example would be when devising hybrid approximations in continuous state space. Indeed, in the above proof, note that if the integer inequality did not hold we would naturally have to rely on the Cauchy-Schwartz inequality instead. With $u(t)^2 = E\left\|X_t - Y_t\right\|^2$ this leads to bounds of the typical kind

$$u(t)^2 \leq \int_0^t C\left(u(s) + u(s)^2\right) \mathrm{d}s, \tag{4.20}$$

for which $u(t) = \exp(Ct/2) - 1$ is an admissible solution. This observation shows that the integer inequality as used in the proof is crucial; without it the integral inequality (4.19) admits growing solutions.

4.4. Stability

Although Theorem 4.5 shows that any moments are bounded in finite time, a relevant question from the modeling point of view is whether the first few moments remain bounded indefinitely. We give a result to this effect which relies on the existence of solutions in $S_{\mathrm{F}}^{p,\mathrm{loc}}\left(\mathbf{Z}_+^D\right)$ which implies that the differential form (2.7) of Dynkin's formula may be used (cf. Corollary 2.2) such that in turn the differential Grönwall inequality applies. We mention anew that a very similar result has recently appeared in ([33] , Theorem 2).

Theorem 4.9 (Ergodicity) Under Assumption 4.1 (1), (2), suppose that

$$\alpha + \beta_2 (p-1) \le -\kappa_p < 0. \tag{4.21}$$

Then for integer $p \ge 1$, $E\|X_t\|_t^p$ remains bounded as $t \to \infty$.

Proof. The case $\beta_2 > 0$ is slightly more complicated to obtain so we shall concentrate on this. We omit the case $p = 1$ since it follows from (4.5) under the present assumptions. The idea of the proof is to asymptotically bound $E\left(C + \|X_t\|_t\right)^p$ with a certain positive constant $C = C(p)$ to be decided upon below. By (2.7) we get with $Z_t = C + \|X_t\|_t$,

$$\begin{aligned}\frac{\mathrm{d}EZ_t^p}{\mathrm{d}t} &= E\sum_{r=1}^{R} w_r(X_t)\left[\left[Z_t - l^{\mathrm{T}}\mathrm{N}_r\right]^p - Z_t^p\right] \\ &= E\sum_{r=1}^{R} w_r(X_t)\left[-l^{\mathrm{T}}\mathrm{N}_r p Z_t^{p-1} + \frac{\left(-l^{\mathrm{T}}\mathrm{N}_r\right)^2}{2} p(p-1)\left(Z_t - \theta_r l^{\mathrm{T}}\mathrm{N}_r\right)^{p-2}\right],\end{aligned} \tag{4.22}$$

by Taylor's formula for some $\theta_r \in [0,1]$. Using the assumptions we get the bound

$$\frac{\mathrm{d}EZ_t^p}{\mathrm{d}t} \le pE\left[\left(A + \alpha\|X_t\|_t\right)Z_t^{p-1} + (p-1)\left(B + \beta_1\|X_t\|_t + \beta_2\|X_t\|_t^2\right)\left(Z_t + \|l^{\mathrm{T}}\mathrm{N}\|_\infty\right)^{p-2}\right]. \tag{4.23}$$

For the first term in (4.23) we get from the scaled Young's inequality with exponent $p/(p-1)$ and conjugate exponent p that

$$\left(A + \alpha\|X_t\|_t\right)Z_t^{p-1} = \alpha Z_t^p + (A - \alpha C)Z_t^{p-1} \le \alpha Z_t^p + \frac{\varepsilon(p-1)}{p} Z_t^p + \frac{\varepsilon^{1-p}}{p}|A - \alpha C|^p,$$

for some $\varepsilon > 0$. As for the second term in (4.23) we first estimate for $\beta_2 > 0$

$$B + \beta_1\|X_t\|_t + \beta_2\|X_t\|_t^2 \le \beta_2\left(\|X_t\|_t + B/\beta_1\right)\left(\|X_t\|_t + \beta_1/\beta_2\right).$$

Next by the arithmetic-geometric mean inequality we get

$$\left(\|X_t\|_l + B/\beta_1\right)\left(\|X_t\|_l + \beta_1/\beta_2\right)\left(Z_t + \|l^{\mathrm{T}}\mathrm{N}\|_\infty\right)^{p-2}$$
$$\leq \left(\frac{B}{p\beta_1} + \frac{\beta_1}{p\beta_2} + \frac{p-2}{p}\left(\|l^{\mathrm{T}}\mathrm{N}\|_\infty + C\right) + \|X_t\|_l\right)^p = Z_t^p,$$

provided that we choose C as the solution to the equation

$$\frac{B}{p\beta_1} + \frac{\beta_1}{p\beta_2} + \frac{p-2}{p}\left(\|l^{\mathrm{T}}\mathrm{N}\|_\infty + C\right) = C. \tag{4.24}$$

Taken together we thus have

$$\frac{\mathrm{d}EZ_t^p}{\mathrm{d}t} \leq p\left(-\kappa_p + \frac{\epsilon(p-1)}{p}\right)EZ_t^p + \epsilon^{1-p}|A - \alpha C|^p.$$

Since $\kappa_p > 0$ we may pick a small enough ϵ such that the bracketed expression remains negative. By Grönwall's inequality this then proves the result with $\beta_2 > 0$. To prove the case $\beta_2 = 0$ the same idea of proof applies and results in

$$\frac{\mathrm{d}EZ_t^p}{\mathrm{d}t} \leq p\left(-\kappa_p + \frac{\epsilon(p-1)}{p} + \frac{(p-1)^2}{p}\beta_1\right)EZ_t^p + \epsilon^{1-p}\left(|A - \alpha C|^p + \beta_1(p-1)\right),$$

for a certain new constant C satisfying an equation similar to (4.24). W

We next aim at deriving some stability estimates with respect to perturbations in the reaction coefficients. An early account of this was given by Kurtz in [31] , see also [18] for a recent discussion in a bounded setting.

Given the linear dependence on the coefficients k_r, $r = 1, \ldots, 4$ in the elementary reactions (2.3) a suitable model seems to be that a perturbation $k_r \to k_r + \delta_r$ in a propensity $w_r(x)$ spreads linearly in a relative sense,

$$\left|w_r(x, k_r) - w_r(x, k_r + \delta_r)\right| \leq \text{constant} \times \delta_r w_r(x, k_r). \tag{4.25}$$

We make this formal by requiring that

$$\left(\mathbf{1}^{\mathrm{T}}\mathrm{N}^2\right)\left|w(x)-w_\delta(x)\right| \le \delta\left\|w(x)\right\|_1 \le \delta C\left(1+\left\|x\right\|_1^2\right), \tag{4.26}$$

where δ is a suitable measure of the total perturbation vector and where the perturbed propensity vector function is given by

$$w_\delta(x) \equiv \left[w_1(x, k_1+\delta_1), \mathrm{L}\ , w_R(x, k_R+\delta_R)\right]^{\mathrm{T}}. \tag{4.27}$$

The existence of an absolute constant C in (4.26) follows from Assumption 4.1 (3). We further conveniently assume that the entire statement of Assumption 4.1 carries over to the perturbed system, and for convenience we shall also assume that all constants are the same. By the triangle inequality and Assumption 4.1 (3) we obtain from (4.26) the bound

$$\left(\mathbf{1}^{\mathrm{T}}\mathrm{N}^2\right)\left|w(x)-w_\delta(y)\right| \le L\left(1+\left\|x+y\right\|_1\right)\left\|x-y\right\|+\delta C\left(1+\left\|y\right\|_1^2\right) \le C\left(\delta+\left\|x-y\right\|^2\right), \tag{4.28}$$

with C some constant and where the simplification in (4.28) assumes an a priori bound (e.g. stopping time) $\left\|x+y\right\|_1 \le P$ and additionally requires the integer inequality.

The starting point for the analysis will be Itô's formula under the coupling described in Section 2.2.2. The techniques used below generalize well to pth order moment estimates, but for ease of exposition we let $p=2$.

Hence under the model for coefficient perturbations (4.26)-(4.27) we have that

$$\begin{aligned}\left\|X_t-Y_t\right\|^2 &= \int_0^t 2\left(X_s-Y_s, F(X_s)-F_\delta(Y_s)\right)+\left(\mathbf{1}^{\mathrm{T}}\mathrm{N}^2\right)\left|w(X_s)-w_\delta(Y_s)\right|\mathrm{d}s \\ &+\int_0^t\int_l 2\left(X_{s-}-Y_{s-}, -\mathrm{N}\left(\hat{w}^\delta(X_{s-},Y_{s-};z)-\hat{w}^\delta(Y_{s-},X_{s-};z)\right)\right) \\ &+\left\|-\mathrm{N}\left(\hat{w}^\delta(X_{s-},Y_{s-};z)-\hat{w}^\delta(Y_{s-},X_{s-};z)\right)\right\|^2(\mu_\delta-m_\delta)(\mathrm{d}s\times\mathrm{d}z).\end{aligned} \tag{4.29}$$

Theorem 4.10 (Continuity). Let two trajectories X_t and Y_t be given, with the same initial data and coupled according to the discussion in Section 2.2.2. Let the propensities for Y_t be perturbed by δ as indi- cated in (4.26), (4.27). Then

$$\lim_{\delta\to 0+} E\left\|X_t-Y_t\right\|^2 = 0. \tag{4.30}$$

Proof. We use the stopping time $\tau_P = \inf_{t\geq 0}\left\{\|X_t + Y_t\|_1 > P\right\}$ and put $\hat{t} = t \wedge \tau_P$. From (4.29) we get

$$E\|X_{\hat{t}} - Y_{\hat{t}}\|^2 = \int_0^{\hat{t}} E\left[2\left(X_s - Y_s, F(X_s) - F_\delta(Y_s)\right) + \left(\mathbf{1}^{\mathrm{T}}\mathrm{N}^2\right)\left|w(X_s) - w_\delta(Y_s)\right|\right]\mathrm{d}s$$
$$\leq \int_0^{\hat{t}} E\left[2\left(M + \mu\|X_s + Y_s\|_1\right)\|X_s - Y_s\|^2 + 2\delta\left(\mathbf{1}^{\mathrm{T}}\mathrm{N}^2\right)^{1/2}\|w(Y_s) - w_\delta(Y_s)\|\|X_s - Y_s\| + C\left(\delta + \|X_s - Y_s\|^2\right)\right]\mathrm{d}s, \tag{4.31}$$

where (4.28) was used. Simplifying further for a bounded δ we get

$$E\left\|(X - Y)(t \wedge \tau_P)\right\|^2 \leq \int_0^{t\wedge\tau_P} C\left(\delta + E\|X_s - Y_s\|^2\right)\mathrm{d}s$$
$$\leq \int_0^t C\left(\delta + E\left\|(X - Y)(s \wedge \tau_P)\right\|^2\right)\mathrm{d}s.$$

Using that $X_0 = Y_0$ and Grönwall's inequality we conclude that

$$E\|X_{\hat{t}} - Y_{\hat{t}}\|^2 \leq \delta\left(\exp(Ct) - 1\right).$$

To get rid of the stopping time we write in terms of indicator functions,

$$E\|X_t - Y_t\|^2 = E\left[\|X_t - Y_t\|^2\left[t < \tau_P\right] + \|X_t - Y_t\|^2\left[t \geq \tau_P\right]\right]$$
$$\leq E\|X_{\hat{t}} - Y_{\hat{t}}\|^2 + \left(E\|X_t - Y_t\|^4\right)^{1/2}\left(\mathbf{P}\left[t \geq \tau_P\right]\right)^{1/2}. \tag{4.32}$$

Using $\mathbf{P}\left[t \geq \tau_P\right] = \mathbf{P}\left[\sup_{s\in[0,t]}\|X_s + Y_s\|_1 > P\right]$ we get from Markov's inequality and the previous estimate

$$E\|X_t - Y_t\|^2 \leq \delta\left(\exp(Ct) - 1\right) + \left(E\|X_t - Y_t\|^4\right)^{1/2} P^{-1/2}\left(E\sup_{s\in[0,t]}\|X_s + Y_s\|_1\right)^{1/2}. \tag{4.33}$$

Relying on the existence result in Theorem 4.7 we find that for any given $\varepsilon > 0$ we can select P (and hence also C) such that the right term is $< \varepsilon/2$. We can next find $\delta_0 > 0$ such that for all $\delta \leq \delta_0$, also the left term is $< \varepsilon/2$. Hence for all $\delta \leq \delta_0$, $E\|X_t - Y_t\|^2 < \varepsilon$ as claimed. W

As a by-product of the proof we see that if the process is bounded, then for P large enough the probability in (4.32) is zero.

Corollary 4.11 (Perturbation estimate, bounded version) If in Theorem 4.10, the processes X_t and Y_t are bounded, then for a constant $C>0$,

$$E\|X_t - Y_t\|^2 \le \delta(\exp(Ct)-1). \tag{4.34}$$

The constant C in (4.34) can be bounded explicitly by inspection of (4.28) and (4.31).

For an unbounded system it is apparently much more difficult to obtain explicit estimates. However, by controlling also the martingale part we can strengthen Theorem 4.10 in another direction.

Theorem 4.12 (Continuity/sup) Under the same assumptions as Theorem 4.10 we have that

$$\lim_{\delta\to 0+} E \sup_{s\in[0,t]} \|X_s - Y_s\|^2 = 0. \tag{4.35}$$

Proof. The quadratic variation of the martingale part in (4.29) can be bounded as

$$\begin{aligned}
E[M]_{\hat{t}}^{1/2} &\le E\Big[\int_0^{\hat{t}}\int_I \big|2\big(X_{s-}-Y_{s-}, -\mathrm{N}\big(\hat{w}^\delta(X_{s-},Y_{s-};z)-\hat{w}^\delta(Y_{s-},X_{s-};z)\big)\big) \\
&\quad + \big\|\mathrm{N}\big(\hat{w}^\delta(X_{s-},Y_{s-};z)-\hat{w}^\delta(Y_{s-},X_{s-};z)\big)\big\|^2\big|\,\mu_\delta(\mathrm{d}s\times\mathrm{d}z)\Big] \\
&\le E\Big[\int_0^{\hat{t}}\int_I \Big(2\|X_s-Y_s\|\big\|\mathrm{N}\big(\hat{w}^\delta(X_s,Y_s;z)-\hat{w}^\delta(Y_s,X_s;z)\big)\big\| \\
&\quad + \big\|\mathrm{N}\big(\hat{w}^\delta(X_s,Y_s;z)-\hat{w}^\delta(Y_s,X_s;z)\big)\big\|^2\Big) W^\delta(X_s,Y_s)\,\mathrm{d}s\times\mathrm{d}z\Big] \\
&= E\Big[\int_0^{\hat{t}}\Big(2\|X_s-Y_s\|\big(\mathbf{1}^T\mathrm{N}^2\big)^{1/2}+\big(\mathbf{1}^T\mathrm{N}^2\big)\Big)\big|w(X_s)-w_\delta(Y_s)\big|\,\mathrm{d}s\Big] \\
&\le \int_0^{\hat{t}} C\big(\delta + E\|X_s-Y_s\|^2\big)\,\mathrm{d}s,
\end{aligned}$$

after using (4.28) and the integer inequality anew. For the drift part we may use the corresponding bound developed in the proof of Theorem 4.10. After taking supremum and expectation values of (4.29) and using Burkholder's inequality we therefore arrive at

$$E\|X-Y\|^2(t\wedge\tau_P)\le\int_0^t C\left(\delta+E\|X-Y\|^2(s)\right)ds\le\int_0^t C\left(\delta+E\|X-Y\|^2(s\wedge\tau_P)\right)ds\le\delta\left(\exp(Ct)-1\right)$$

by Grönwall's inequality and using the notation $\|X\|(t)\equiv\sup_{s\in[0,t]}\|X_t\|$. We now rely on the same strategy as in the proof of Theorem 4.10 to similarly arrive at

$$E\|X-Y\|^2(t)\le\delta\left(\exp(Ct)-1\right)+\left(E\|X-Y\|^4(t)\right)^{1/2}P^{-1/2}\left(E\|X+Y\|_1(t)\right)^{1/2},$$

and the conclusion follows as before.

5. CONCLUSIONS

We have proposed a theoretical framework consisting of a priori assumptions and estimates for problems in stochastic chemical kinetics. The assumptions are strong enough to guarantee well-posedness for a large and physically relevant class of problems. Long time estimates and limit results for perturbations in rate constants have been studied to exemplify the theory. The assumptions are constructive in the sense that explicit techniques for obtaining all postulated constants have either been worked out in detail or at least indicated. We have seen that the case $\beta_2=0$ in Assumption 4.1 (2) is particularly promising from the analysis point of view in that the conditions for existence in Theorem 4.7 and the ergodicity in Theorem 4.9 both can be formulated naturally.

In the course of motivating our setup we have seen that most problems do not admit global Lipschitz constants and that one-sided versions do not provide a better alternative. Another conclusion worth highlighting is that it pays off to consider jump SDEs in a fully discrete setting in that there are potential complications in proving uniqueness in continuous state space. A practical implication is that care should be exercised when forming continuous approximations to these types of jump SDEs.

For future work we intend to re-visit certain classical results from the perspective of the framework developed herein; for example, thermodynamic limit results, time discretization strategies, and quasi-steady state approximations—all of which have a practical impact in a range of applications.

ACKNOWLEDGEMENTS

The author likes to express his sincere gratitude to Takis Konstantopoulos for several fruitful and clarifying discussions. Early inputs on this work were also gratefully obtained from Henrik Hult, Ingemar Kaj, and Per Lötstedt.

Stefan Engblom This work was supported by the Swedish Research Council within the UPMARC Linnaeus center of Excellence.

REFERENCES

1. Lestas, I., Vinnicombe, G. and Paulsson, J. (2010) Fundamental Limits on the Suppression of Molecular Fluctuations. Nature, 467, 174-178.
2. Raj, A. and van Oudenaarden, A. (2008) Nature, Nurture, or Chance: Stochastic Gene Expression and Its Consequences. Cell, 135, 216-226.
3. Samoilov, M.S. and Arkin, A.P. (2006) Deviant Effects in Molecular Reaction Pathways. Nature Biotechnology, 24, 1235-1240.
4. Gillespie, D.T. (1976) A General Method for Numerically Simulating the Stochastic Time Evolution of Coupled Chemical Reactions. Journal of Computational Physics, 22, 403-434.
5. Gillespie, D.T. (1992) A Rigorous Derivation of the Chemical Master Equation. Physica A, 188, 404-425.
6. Protter, P.E. (2005) Stochastic Integration and Differential Equations. Number 21 in Stochastic Modelling and Applied Probability. 2nd Edition, Version 2.1, Springer, Berlin.
7. Applebaum, D. (2004) Lévy Processes and Stochastic Calculus, Volume 93 of Cambridge Studies in Advanced Mathematics. Cambridge University Press, Cambridge.
8. Situ, R. (2005) Theory of Stochastic Differential Equations with Jumps and Applications. Mathematical and Analytical Techniques with Applications to Engineering. Springer, New York.
9. Øksendal, B. and Zhang, T. (2007) The Itô-Ventzell Formula and Forward Stochastic Differential Equations Driven by Poisson Random Measures. Osaka Journal of Mathematics, 44, 207-230.
10. Hausenblas, E. (2007) SPDEs Driven by Poisson Random Measure with Non Lipschitz Coefficients: Existence Results. Probability Theory and Related Fields, 137, 161-200.
11. Marinelli, C., Prévôt, C. and Röckner, M. (2010) Regular Dependence on Initial Data for Stochastic Evolution Equations with Multiplicative Poisson Noise. Journal of Functional Analysis, 258, 616-649.
12. Marinelli, C. and Röckner, M. (2010) Well-Posedness and Asymptotic Behavior for Stochastic Reaction-Diffusion Equations with Multiplicative Poisson Noise. Electronic Journal of Probability, 15, 1529-1555.
13. Filipović, D., Tappe, S. and Teichmann, J. (2010) Jump-Diffusions in Hilbert Spaces: Existence, Stability and Numerics. Stochastics, 82, 475-520.
14. van Kampen, N.G. (2004) Stochastic Processes in Physics and Chemistry. 2nd Edition, Elsevier, Amsterdam.
15. Higham, D.J. and Kloeden, P.E. (2005) Numerical Methods for Nonlinear Stochastic Differential Equations with Jumps. Numerische Mathematik, 101, 101-119.

16. Li, T.J. (2007) Analysis of Explicit Tau-Leaping Schemes for Simulating Chemically Reacting Systems. Multiscale Modeling & Simulation, 6, 417-436.
17. Engblom, S. (2009) Parallel in Time Simulation of Multiscale Stochastic Chemical Kinetics. Multiscale Modeling & Simulation, 8, 46-68.
18. Anderson, D.F. (2012) An Efficient Finite Difference Method for Parameter Sensitivities of Continuous Time Markov Chains. SIAM Journal on Numerical Analysis, 50, 2237-2258.
19. Kawamura, A. (2009) Lipschitz Continuous Ordinary Differential Equations Are Polynomial-Space Complete. 24th Annual IEEE Conference on Computational Complexity, Paris, 15-18 July 2009, 149-160.
20. Hutzenthaler, M., Jentzen, A. and Kloeden, P.E. (2011) Strong and Weak Divergence in Finite Time of Euler's Method for Stochastic Differential Equations with Non-Globally Lipschitz Continuous Coefficients. Proceedings of the Royal Society A, 467, 1563-1576.
21. Chen, W.Y. and Bokka, S. (2005) Stochastic Modeling of Nonlinear Epidemiology. Journal of Theoretical Biology, 234, 455-470.
22. Ewens, W.J. (2004) Mathematical Population Genetics I. Theoretical Introduction, Volume 27 of Interdisciplinary Applied Mathematics. 2nd Edition, Springer, New York.
23. Escudero, C., Buceta, J., de la Rubia, F.J. and Lindenberg, K. (2004) Extinction in Population Dynamics. Physical Review E, 69, 021908.
24. Ethier, S.N. and Kurtz, T.G. (1986) Markov Processes: Characterization and Convergence. Wiley Series in Probability and Mathematical Statistics. John Wiley & Sons, New York.
25. Brémaud, P. (1999) Markov Chains: Gibbs Fields, Monte Carlo Simulation, and Queues. Number 31 in Texts in Applied Mathematics. Springer, New York.
26. Plyasunov, S. (2005) On Hybrid Simulation Schemes for Stochastic Reaction Dynamics.
27. Brémaud, P. (1981) Point Processes and Queues: Martingale Dynamics. Springer Series in Statistics. Springer, New York.
28. Daley, D.J. and Vere-Jones, D. (2003) An Introduction to the Theory of Point Processes, Volume I: Elementary Theory and Methods. 2nd Edition, Springer, New York.
29. Kurtz, T.G. (1978) Strong Approximation Theorems for Density Dependent Markov Chains. Stochastic Processes and Their Applications, 6, 223-240.
30. Rathinam, M., Sheppard, P.W. and Khammash, M. (2010) Efficient Computation of Parameter Sensitivities of Discrete Stochastic Chemical Reaction Networks. Journal of Chemical Physics, 132, 034103.
31. Kurtz, T.G. (1982) Representation and Approximation of Counting Processes. In: Fleming, W.H. and Gorostiza, L.G., Eds., Advances in Filtering and Optimal Stochastic Control, Vol. 42, Lecture Notes in Control and Information Sciences, Springer, Berlin, 177-191.
32. Ström, T. (1975) On Logarithmic Norms. SIAM Journal on Numerical Analysis, 12, 741-753.
33. Briat, C., Gupta, A. and Khammash, M. (2014) A Scalable Computational Framework for Establishing Long-Term Behavior of Stochastic Reaction Networks. PLoS Computational Biology, 10, e1003669.
34. Rathinam, M. (2014) Moment Growth Bounds on Continuous Time Markov Processes on Non-Negative Integer Lattices. To appear in Quart. Appl. Math.
35. Meyn, S.P. and Tweedie, R.L. (1993) Stability of Markovian Processes III: Foster-Lyapunov Criteria for Continuous- Time Processes. Advances in Applied Probability, 25, 518-548.

CHAPTER 7

Convergence of Invariant Measures of Truncation Approximations to Markov Processes

Andrew G. Hart[1], Richard L. Tweedie[2]

[1]*Centro de Modelamiento Matemático, Facultad de Ciencias Físicas y Matemáticas, Universidad de Chile, Santiago, Chile*

[2]*Division of Biostatistics, School of Public Health, University of Minnesota, Minneapolis, USA*

ABSTRACT

Let Q be the Q-matrixof an irreducible, positive recurrent Markov process on a countable state space. We show that, under a number of conditions, the stationary distributions of the n × n north-west corner augmentations of Q converge in total variation to the stationary distribution of the process. Twoconditions guaranteeing such convergence include exponential ergodicity and stochastic monotonicity of the process. The same also holds for processes dominated by a stochastically monotone Markov process. In addition, we shall show that finite perturbations of stochastically monotone processes may be viewed as being dominated by a stochastically monotone process, thus extending the scope of these results to a larger class of processes. Consequently, the augmentation method provides an attractive, intuitive method for approximating the stationary distributions of a large class of Markov processes on countably infinite state spaces from a finite amount of known information.

Keywords: Invariant Measure; Truncation Approximation; Augmentation; Exponential Ergodicity; Stochastic Monotonicity; Markov Process

1. INTRODUCTION

Let $Q=(q(i.j),i,j\in S)$ be the stable, conservative *Q-matrix* of a continuous-time Markov process on a countable state space $S=\{0,1,2,\cdots\}$ The *Q-matrix* satisfies

$$\begin{aligned}&Q(i,j)\geq 0,\text{ if } j\neq i,\\&0\leq q(i):=-Q(i,i)<\infty,\text{if } j=i,\text{and}\\&\sum_{j\in S}Q(i,j)=0 \text{ for all } i\in S\end{aligned}$$

In addition, we assume that Q is regular, which means there exists no non-trivial, non-negative solution

$x=(x(j),j\in S)$ to $\sum_{j\in S}Q(i,j)x(j)=kx(i);0\leq x(i)\leq 1,i\in S$

for some (and then all) $k>0$.

Under these assumptions, the state transition probabilities of the process are given by the unique Q-function $F=(F^t(i,j),i,j\in S,t\geq 0)$ which satisfies the Kolmogorov backward equations, $\frac{\mathrm{d}}{\mathrm{d}t}F^t=QF^t,t\geq 0,F^0=I.$

The object F, which is also called a transition function, is a family of $S\times S$ matrices indexed over the reals which constitutes an analytic semi-group. an analytic semi-group is characterised by three properties: F^0 is the identity matrix, the row sums of F^t are less than or equal to unity and F^{s+t} is equal to the matrix product F^sF^t for all $s,t \geq 0$. This last property, known as the Chapman-Kolmogorov equation, implies $F^t = (F^1)^t$. Thus, even though $F^t$is generally thought of as the matrix of state transition probabilities at time t, it serves as an analogue to the t-th power of the transition matrix of a discrete-time Markov chain on the state space S. Consequently, using the superscript to denote F as a function of t should not cause any confusion. While on the subject of notation, we should mention that we are using a standard notation common in the literature of continuous-time Markov processes on general state spaces. In the discrete state space setting, this notation causes matrices to look like functions of two variables (or kernels) while measures and vectores appear to be functions over the state space. We have elected to follow this notation in an endeavour to reduce the number of subscripts and superscripts in the sequel.

Note that in the conservative setting posed here, regularity of Q is equivalent to honesty and uniqueness of the transition function, that is, $\sum_{j \in S} F^t(i,j) = 1$ for all $i \in S$ and $t \geq 0$.

The state space S is irreducible if $F^T(i,j) > 0$ for all $i,j \in S$ and $t \geq 0$. On such a state space, a Markov process is said to be positive recurrent or ergodic if $F^t(i,j) \to \pi(j) > 0$ for all $i,j \in S$ as $t \to \infty$. For a positive recurrent process, it can be shown (for example, see Theorem 5.1.6 in [1]) that the $\lim \pi = (\pi_i, i \in S)$ satisfies

$$\sum_{j \in s} \pi(i) = 1 \text{ and } \sum_{j \in S} \pi(i) F^t(i,j) = \pi(j),$$

$$j \in S, t \geq 0 \quad (1)$$

More generally, any measure π satisfying (1) is called an invariant or stationary measure for the process. If, in addition, the measure has mass 1, it is referred to as a stationary or invariant distribution. Any measure satisfying (1) with "$=$" replaced by "$\geq$" is called a subinvariant measure for F. Conversely, if F has a stationary distribution π, then the process is positive recurrent and $F^t(i,j) \to \pi(j)$ as $t \to \infty$.

In this paper, we are interested in approximating π using the $n \times n$ north-west corner truncations of Q. The analogous problem for discrete-time Markov chains has been studied in [2-7]. The final reference contains a review of the literature on the discrete-time version of the truncation problem. Some properties of truncation in continuous-time Markov processes were studied in [8,9].

Truncations of Q are submatrices of Q defined by

$_{(n)}Q = \left({}_{(n)}Q(i,j), i,j \in S_n \right)$, where $_{(n)}Q(i,j) := Q(i,j)$

and $(S_n, n \in \mathbb{N})$ is an increasing sequence of subsets of S such that $\bigcup_{n \in \mathbb{N}} S_n = S$.

The truncation $_{(n)}Q$ is not conservative. By adding the discarded transition rates to $_{(n)}Q$, we may produce a conservative Q-matrix $\overline{_{(n)}Q}$ which generates a uniquehonest, finite, continuous-time Markov process. For example, we may choose to perform linear augmentation, where the aggregate of the transition rates outside of S_n is dispersed amongst the states in S_n according to some probability measure $_{(n)}\alpha = \left({}_{(n)}\alpha(i), i \in S_n \right)$. Then, the

n-th order augmentation $\overline{_{(n)}Q}$ is given by

$$\overline{_{(n)}Q}(i.j) = \begin{cases} Q(i,j) + {}_{(n)}\alpha(j) \sum_{k \notin S_n} Q(i,k), \text{ if } i,j \in S_n, \\ 0, \text{ otherwise.} \end{cases}$$

An important example of this is where we only augment a single column, say h, in which case $_{(n)}\alpha$ is the Dirac measure at h and we obtain The n-th order augmentation $_{(n)}Q_h$ as

$$_{(n)}Q(i.j)=\begin{cases}Q(i,j)+\delta_{jh}\sum_{k\notin S_n}Q(i,k),\text{ if } i,j\in S_n,\\ 0,\text{ otherwise.}\end{cases}$$

Here, δ_{jh} denotes the kronecker delta.

Linear augmentation obtains exactly one irreducible, closed class $A_n \subseteq S_n$ together with zero or more open classes from which A_n is accessible. Since A_n is closed,

$\overline{_{(n)}Q}$ is conservative on A_n and so the minimal

$\overline{_{(n)}Q}$-function $\overline{_{(n)}F}$ is honest and positive recurrent on

A_n. Finiteness of S_n ensures that the remaining open classes are transient. Hence, there exists a unique invariant measure for $\overline{_{(n)}F}$. We shall be mainly concerned with $_{(n)}Q_h$ where either $h=0$ or $h=n$. The minimal $\overline{_{(n)}Q}$-function will be denoted $_{(n)}F_h$ while

$_{(n)}\pi_h=\left(_{(n)}\pi_h(i),i\in S_n\right)$ will be its invariant probability measure.

Two obvious questions now arise. Firstly, when does

$$_{(n)}\pi_h \xrightarrow{w} \pi \text{ as } n\to\infty? \tag{2}$$

Here, we use $\xrightarrow{w}$ to denote convergence in total variation norm. Secondly, how quickly does this convergence occur? This paper considers the first question. We shall present augmentation strategies for approximating invariant distributions for two classes of Markov processes via $_{(n)}\pi_h$ for n large. The classes are:

- Markov processes which satisfy

$$\left|F^t(i,j)-\pi(j)\right|=O\left(e^{-at}\right),i,j\in S,t\geq 0,$$

for some $\alpha > 0$. Such processes are called exponentially ergodic.

- Stochastically monotone Markov processes, which have the property that

$$\sum_{j \geq n} F^t(i,j) \leq \sum_{j \geq n} F^t(k,j)$$

for all $i \leq k, i,k,n \in S$ and $t \geq 0$, and processes dominated by stochastically monotone processes.

Parallelling results for discrete-time chains in [7], we shall also show that Markov processes constructed from finite perturbations of stochastically monotone processes are always dominated by some other stochastically monotone process. This extends the class of processes for which our results are applicable.

In the next section, we begin by showing that the limit of the ${}_{(n)}\pi'$ is unique when it exists. Then, Section 3 considers exponentially ergodic Markov processes while Section 4 studies stochastically monotone Markov processes and their above-mentioned variations.

Finally, some concluding remarks are made in Section 5.

2. PRELIMINARIES

The problem of proving that ${}_{(n)}\pi_h \xrightarrow{w} \pi$ may be broken into two parts. Firstly we must show that ${}_{(n)}\pi_h$ converges weakly to some limit, say $\overline{\pi}$, and secondly, that $\overline{\pi} \equiv \pi$. We consider the latter in this section.

Theorem 2.1 Consider a sequence of linearly augmented Q-matrices$\left({}_{(n)}Q_h\right)$ derived from Q and let ${}_{(n)}F_h$ be the minimal ${}_{(n)}Q_h$-function. Then

$$_{(n)}F_h^t(i,j) \mapsto F^t(i,j) \text{ as } n \to \infty, i,j \in S. \tag{3}$$

Proof: Let ${}_{(n)}F$ denote the minimal ${}_{(n)}Q$-function.

Firstly, observe that ${}_{(n)}F^t(i,j) \le {}_{(n)}F_h^t(i,j)$ for all $i,j \in S$ and $t \ge 0$. This can be seen inductively using the backward integral recurrences for ${}_{(n)}F$ and ${}_{(n)}F_h$. The argument parallels the proof of Theorem 2.2.14 in [1] which states that

$$ {}_{(n)}F^t(i,j) \nearrow F^t(i,j) \text{ as } n \to \infty \tag{4} $$

for all $i,j \in S, t \ge 0$

Next, since ${}_{(n)}F_h$ is honest and ${}_{(n)}F$ is dishonest, we see that

$$ {}_{(n)}F^t(i,j) \le {}_{(n)}F_h^t(i,j) \le {}_{(n)}F^t(i,j) + \delta_n^t(i) \tag{5} $$

for $i,j \in S_n, t \ge 0, n \ge 1$, where

$$ \delta_n^t(i) := 1 - \sum_{k \in S_n} {}_{(n)}F^t(i,k), i \in S_n, t \ge 0 \tag{6} $$

Applying (4) to (6) together with monotone convergence shows that $\delta_n^t(i)$ monotonically decreases to 0 as $n \to \infty$. Taking limits in n on both sides of (5) then completes the proof.

Remark 2.2 Although we have only considered linear augmentations, the statement and proof of Theorem 2.1 is in fact valid for any sequence of augmentations

$\left(\overline{{}_{(n)}F^t}\right)$.

Since the transition function ${}_{(n)}F_h$ is finite, it is positive recurrent on some subset of S_n. Hence it possesses a unique stationary distribution ${}_{(n)}\pi_h$ and

$$ \sum_{j \in S_n} {}_{(n)}\pi_h(i)\, {}_{(n)}F_h^t(i,j) = {}_{(n)}\pi_h(j) \tag{7} $$

for $j \in S_n, t \geq 0$. Positive recurrence establishes anequivalence between the stationary distributions for ${}_{(n)}F_h^t$ and invariant distributions for ${}_{(n)}Q_h$. An invariant distribution for an arbitrary Q-matrix$Q$ is any probability measure π such that $\sum_{i \in S} \pi(i) Q(i,j) = 0$ for all

$j \in S$. So, ${}_{(n)}\pi_h$ uniquely satisfies

$\sum_{i \in S_n} {}_{(n)}\pi_h(i)\, {}_{(n)}Q_h(i,j) = 0$ for all $j \in S_n$

Let us assume for the moment that ${}_{(n)}\pi_h$ converges weakly to some limit measure $\bar{\pi} = (\bar{\pi}(i), j \in S)$. We require that $\bar{\pi} \equiv \pi$. Weak convergence to $\bar{\pi}$ implies that $\bar{\pi}$ is a probability distribution. By taking the limit infimum on both sides of (7) and applying Fatou's Lemma, we have

$$
\begin{aligned}
&\sum_{j \in S} \bar{\pi}(i) F^t(i,j) \\
&\leq \liminf_{n \to \infty} \sum_{j \in S} {}_{(n)}\pi_h(i)\, {}_{(n)}F_h^t(i,j) \\
&= \liminf_{n \to \infty} {}_{(n)}\pi_h(j) = \bar{\pi}(j)
\end{aligned}
$$

for $j \in S, t \geq 0$. The measure $\bar{\pi}$ is therefore a subinvariant probability measure for F. However, F is positive recurrent and hence, by Theorem 4 in [10], $\bar{\pi}$ is both invariant and the unique probability measure satisfying (1). Hence, $\bar{\pi} \equiv \pi$.

3. EXPONENTIAL ERGODICITY

Let Q be the Q-matrix of a positive recurrent Markov process χ on S. Consider an increasing sequence of sets $(S_n, n \in \mathbb{N})$ such that $\bigcup_{n \in \mathbb{N}} S_n = S$ and $0 \in S_n$

for all n. Let ${}_{(n)}Q$ be the truncation of Q corresponding to S_n. In this section, we shall consider augmentations ${}_{(n)}Q_0$ obtained by linearly augmenting ${}_{(n)}Q$ in column 0. We shall prove that exponential ergodicity of the Markov process is sufficient for ${}_{(n)}\pi \xrightarrow{w} \pi$ as $n \to \infty$ where ${}_{(n)}\pi$ is taken to be ${}_{(n)}\pi_0$, the invariant distribution for ${}_{(n)}Q_0$. In order to do this, we shall require the notion of a V-norm. Let $V = (V(i), i \in S)$ be an arbitrary vector (function) such that $1 \leq V(i) \leq \infty$ for all $i \in S$. In future, we abreviate this to $V \geq 1$. The V - norm of a signed measure n is then

$$\|V\|_V := \sup_{f:|f|\leq V} |v(f)| = \sum_{j\in S} |v(i)| V(i)$$

If $A = [A(i,J) i \in S, J \in B(S))$ is a matrix, then the V-norm of A is

$$|||A|||_V := \sup_{i\in S} \frac{\|A(i,\cdot)\|_V}{V(i)}$$

Rather than working with the Q-matrix augmentationsdirectly, we will use the β-resolvents associated with these. The β-resolvent of a continuoust-time Markov process is the stochastic matrix

$R_\beta = (R_\beta(i,j), i, j \in S)$ given by,

$R_\beta(i,j) := \int_0^\infty \beta e^{-\beta t} F^t(i.j)\, dt,\ i,\ j \in S,\ \beta > 0.$ We note that R_β satisfies the resolvent forms of both the backward and forward equations which are $R_\beta = I + \beta^{-1} Q R_\beta$ and $R_\beta = I + \beta^{-1} R_\beta Q$ respectively.

Since Q is regular, R_β is the unique solution to the resolvent form of the backward equations. Let ${}_{(n)}R_\beta$ and ${}_{(n)}R_{\beta,h}$ denote the unique β -

resolvents of ${}_{(n)}F$ and ${}_{(n)}F_h$ respectively. Here, ${}_{(n)}F$ is the minimal ${}_{(n)}Q$ - funcction while ${}_{(n)}F_h$ denotes the minimal ${}_{(n)}Q_h$ - function. Since S_n is a finite set, ${}_{(n)}Q_h$, ${}_{(n)}F_h$ and ${}_{(n)}R_{\beta,h}$ have the same invariant distribution ${}_{(n)}\pi_h$. The same is true for F and R_β, which share the distribution π.

In the sequel, we shall have need of the following corollary to Theorem 2.1.

Corollary 3.1

i. For all $i, j \in S_n, \beta > 0$ and $n \in \mathbb{N}$,

$$
\begin{aligned}
{}_{(n)}R_\beta(i,j) &\le {}_{(n)}R_{\beta,h}(i,j) \\
&\le {}_{(n)}R_\beta(i,j) + \rho_{\beta,n}(i)
\end{aligned}
\tag{8}
$$

where $\rho_{\beta,n}(i) := 1 - \sum_{j \in S_n(n)} R_\beta(i,j)$; and

ii. ${}_{(n)}R_{\beta,h}(i,j) \to R_\beta(i,j)$ as $n \to \infty$.

Proof: Part i is obtained by integrating both sides of (5) with respect to be $\beta e^{-\beta t} dt$. Part ii then follows by taking limits in (8) and observing that

${}_{(n)}R_\beta(i,j) \nearrow R_\beta(i,j)$ and $\rho_{\beta,n}(i) \searrow 0$ as $n \to \infty$

Next, the various "drift to C" conditions introduced in [11] will play an important role in allowing us to pass between the continuous-time process and the discretetime β -resolvent chain. The drift conditions require the notion of a petite set in both continuoustime processes and discrete-time chains. Let $\mathcal{B}(S)$ denote the Borel σ -algebra on S. Then, A set $C \in \mathcal{B}(S)$ is a petite set in the continuous-time setting if there exists a probability distribution a on $(R^+, \mathcal{B}(R^+))$ and a non-trivial positive measure V_a such that

$$K_a(i,J) := \int_0^\infty F^t(i,J)\, a(dt) \ge V_a(J)$$

for $i \in C, J \in S(\mathcal{B})$, where $F'(i,J)\sum_{j\in J} F'(i,j)$.

Petite sets for discrete-time chains are defined analogously. According to Theorem 5.1 in [11], the following three drift conditions are equivalent, although the petite set C and function V may differ in each instance.

$(\mathcal{D}_T)$: Drift for T-skeletons. For some $T > 0$, there exist constants $\lambda(S)$ bounded for all $s \in (0,T]$ with $\lambda(T) < 1, b < \infty$, together with a petite set $C \in \mathcal{B}(S)$ and a function $V_T \geq 1$ such that

$$\sum_{j\in S} F^S(i,j)V_T(j) \leq \lambda(s)V_T(i) + b\mathbb{I}_C(i),$$

for $i \in S, s \leq T$. We use $\mathbb{I}_C(i)$ to denote the indicator function of the set C which is 1 if $i \in C$ and 0 otherwise.

$(\breve{\mathcal{D}}_\beta)$: Drift for β-resolvents. For some

$\lambda \in (0,1), b < \infty, \beta > 0$, a petite set $C \in \mathcal{B}(S)$ and a function

$$\sum_{j\in S} R_\beta(i,j)V_\beta(j) \leq \lambda V_\beta(i) + b\mathbb{I}_C(i), i \in S.$$

$(\tilde{\mathcal{D}})$: Drift for the Q-matrix. For constants $b, c > 0,$ a petite set $C \in \mathcal{B}(S)$ and a function $\tilde{V} \geq 1,$

$$\sum_{j\in S} q(i,j)\tilde{V}(j) \leq -c\tilde{V}(i) + b\mathbb{I}_C(i), i \in S.$$

An irreducible continuous-time Markov process X is V-uniformly ergodic if, for some invariant probability kernel $\pi, |||F^t - \pi|||_V \to 0 \text{ as } t \to \infty$. In the special case where $V \equiv 1$, the chain is said to be uniformly ergodic or strongly ergodic: For all $i \in S, \|F^t - \pi\|_1 \to 0$ as $t \to \infty$

where, by an abuse of notation, we use π to denote the invariant transition kervnel $\pi(i,j)=\pi(j)$ for all $i,j\in S.$

The following theorem collects together a number of results on exponential and V -uniform ergodicity of Markov processes from the literature.

Theorem 3.2 Let X be an irreducible, aperiodic continuous-time Markov process on S. The following conditions are equivalent.

i. One of the drift conditions $(\mathcal{D}_T),(\breve{\mathcal{D}}_\beta)$ or $(\tilde{\mathcal{D}})$ holds, in which case they all hold, but not necessarily with the same petite set C ;

ii. For all $T>0$, the T-skeleton chain is geometrically ergodic;

iii. For all $\beta>0$, the β -resolvent chain is geometrically ergodic;

iv. X is exponentially ergodic.

v. X is V -uniformly ergodic for some $V\geq 1$.

In particular, it is V_T -uniformly ergodic, V_β -uniformly ergodic and $\tilde{V}$ -uniformly ergodic where V_T,V_β and $\tilde{V}$ satisfy $(\mathcal{D}_T),(\breve{\mathcal{D}}_\beta)$ and $(\tilde{\mathcal{D}})$ respectively.

Proof:

ii $\Leftrightarrow$ iii $\Leftrightarrow$ iv. This was proved in Theorem 5.3 of [11].

i $\Leftrightarrow$ iv. Theorem 5.1 of [11] shows that

$(\mathcal{D}_T),(\breve{\mathcal{D}}_\beta)$ and $(\tilde{\mathcal{D}})$ satisfy a solidarity property in that either all of them hold or none hold. Next, fix $N\geq 1$ and set $C=\{1,2,\cdots,N\}$ which is trivialy petite Since it is finite. In $(\tilde{\mathcal{D}})$, set $c=\lambda$ and $V(i)=cy_i+1$, where

N,λ and the y_i's are those appearing in Theorem 3 of [12]. Finally, an appropriate relabelling of the states in

S reveals $(\tilde{\mathcal{D}})$ to be equivalent to the necessary and sufficient condition for exponential ergodicity givenin Part (ii) of Theorem 3 in [12].

Consequently X is exponentially ergodic if and only if $(\tilde{\mathcal{D}})$ orany of the other drift criteria holds.

i $\Rightarrow$ v. Theorem 5.2 in [11] says that any of $(\mathcal{D}_T),(\breve{\mathcal{D}}_\beta)$ or $(\tilde{\mathcal{D}})$ is sufficient for X to be V-uniformly ergodic where V is either V_T, V_β or $\tilde{V}$ respectively.

v $\Rightarrow$ ii. If X is V-uniformly ergodic for some $V \geq 1$, then so to is the T-skeleton for any $T > 0$ and an application of Theorem 16.0.1 in [13] shows that

$\left\|\left|F^{Tn} - \pi\right|\right\|_V \leq M\rho^n$ for some $M < \infty$ and $\rho \in (0,1)$.

Geometric ergodicity of the T-skeleton F^{Tn} then follows from the definition of the V-norm.

Next, suppose that the Markov process X is exponenttially ergodic. From Theorem 3.2, there exist constants $0 < c, d < \infty$ and a function $V \geq 1$ such that

$$\sum_{j \in S} Q(i,j)V(j) \leq -cV(i) + d\mathbb{I}_C(i),$$
$$i \in S. \tag{9}$$

Without loss of generality, we may take $C = \{0\}$ and assume that $V(0) = 1$. The state space can always be relabelled to accommodate this convention. Then, since $V(k) \geq V(0) = 1$ for all $k \neq 0$, the augmented Q-matrices $\left({}_{(n)}Q_0\right)$ each satisfy

$$\sum_{j \in S_n} {}_{(n)}Q_0(i,j)V(j) \leq -cV(i) + d\mathbb{I}_0(i),$$
$$i \in S_n.$$

Multiplying both sides by ${}_{(n)}R_{\beta,0}$ and re-arranging, we obtain

$$\sum_{j\in S_n} {}_{(n)}R_{\beta,0}(i,j)V(j)$$

$$\le \frac{\beta}{\beta+c}V(i)+\frac{d}{\beta+c}\,{}_{(n)}R_{\beta,0}(i,0),$$

for $i\in S_n$

.

Now, choose $\varepsilon < c/(\beta+c)$ such that

${}_{(N)}R_{\beta}(0,0)\ge\delta := \varepsilon(\beta+c)/d>0$ for some

$N=N(\beta,\varepsilon)\in\mathbb{N}$. This is always possible since R_β is a strictly positive matrix (in particular, $R_\beta(0,0)>0$ and, as noted in the proof of Corollary 3.1,

${}_{(n)}R_{\beta}(0,0)\nearrow R_\beta(0,0)$ as $n\to\infty$. Therefore,

${}_{(n)}R_{\beta,0}(0,0)\ge {}_{(n)}R_{\beta}(0,0)\ge\delta$. So, in addition to R_β

being strongly aperiodic, we see that ${}_{(n)}R_{\beta,0}$ is strongly aperiodic for all $n\ge N$. A transition matrix is strongly aperiodic if it is primative and possesses a non-zero diagonal entry.

Define $C_n := \{k\in S_n : {}_{(n)}R_{\beta,0}(k,0)\ge\delta\}$. By Proposition 5.5.4 in [13], the set C_n is petite since the singleton set $\{0\}$ is trivially petite and $\{0\}$ is uniformly accessible from C_n under the resolvent chain ${}_{(n)}R_{\beta,0}$, that is,

${}_{(n)}R_{\beta,0}(k,0)$ is bounded away from 0 for all $k\in C_n$. By the definition of C_n, we have ${}_{(n)}R_{\beta,0}(i,0)<\delta$ for

$i \in S_n \setminus C_n$. On the other hand, ${}_{(n)}R_{\beta,0}(i,0) \le 1$ for $i \in C_n$, since ${}_{(n)}R_{\beta,0}$ is a stochastic matrix. Hence we have

$$\sum_{j \in S_n} {}_{(n)}R_{\beta,0}(i,j)V(j) \tag{10}$$

$$\le \frac{\beta}{\beta+c}V(i) + \frac{d}{\beta+c}\,{}_{(n)}R_{\beta,0}(i,0)$$
$$\le \frac{\beta}{\beta+c}V(i) + \frac{d}{\beta+c}\mathbb{I}_{C_n}(i) + \varepsilon \mathbb{I}_{S_n \setminus C_n}(i)$$
$$\le \frac{\beta}{\beta+c}V(i) + \frac{d}{\beta+c}\mathbb{I}_{C_n}(i) + \varepsilon V(i)$$
$$= \lambda' V(i) + b' \mathbb{I}_{C_n}(i), i \in S_n, n \ge N, \tag{11}$$

where $\lambda' := \beta/(\beta+c)+\varepsilon < 1$ and $b' := d/(\beta+c) > 0$.

Note that $0 \in C_n$ for all n large enough.

Next, set $P \equiv {}_{(n)}R_{\beta,0}, S = S_n, C = C_n, \alpha = \{0\}, \lambda_C = \lambda'$ and $b_C = b_\alpha^* = b_C^* = b'$ in Theorem 6.1 of [14]. It can be seen that the conditions of the theorem are satisfied and so there exists some $V' \ge 1$ such that

$$\sum_{j \in S_n} {}_{(n)}r_{\beta,0}(i,j)V'(j) \le \lambda V'(i) + b\mathbb{I}_0(i), i \in S_n,$$

where $\lambda = \dfrac{\lambda' + b'/\delta}{1+b'/\delta} = \dfrac{\beta + \varepsilon(\beta+c) + d/\delta}{\beta+c+d/\delta}$ and

$b = (1+1/\delta)b' = \mathrm{d}(1+1/\delta)/(\beta+c)$. Furthermore, we have

$$\left\| {}_{(n)}R_{\beta,0}(i,\cdot) - {}_{(n)}\pi_0(\cdot) \right\|_V \le MV(i)\rho^m, m \in \mathbb{N},$$

where ${}_{(n)}\pi_0$ is the unique invariant distribution for ${}_{(n)}Q_0$, and $M < \infty$ and $\rho < 1$ are completely determined by β, c, d and δ. Note that this is true for every $n \geq N$ so that the rate of convergence ρ is independent of the truncation size. In addition, by applying the preceding argument directly to Q instead of ${}_{(n)}Q_0$, we also have $\left\| R_\beta^m(i,\cdot) - \pi(\cdot) \right\|_V \leq MV(i)\rho^m$ for all m, sinceby assumption, (9) holds and $R_\beta(0,0) \geq \delta > 0$. Thus, not only are R_β and ${}_{(n)}R_{\beta,0}$ V-uniformly ergodic, they are geometrically ergodic with the same convergence rate ρ.

We can now prove the main result of this section.

Theorem 3.3 Let X be an exponentially ergodic, continuous-time Markov chain on a countable, irreducible state space S. Let π and ${}_{(n)}\pi_0$ be the invariant distributions for Q and ${}_{(n)}Q_0$ respectively. Then,

$\left\| {}_{(n)}\pi_0 - \pi \right\| \to 0$ as $n \to \infty$.

Proof: Choose an arbitrary number $m \geq 2$. From the triangle inequality, we have

$$\begin{aligned}
&\left\| {}_{(n)}\pi_0 - \pi \right\| \\
&\leq \left\| R_\beta^m(i,\cdot) - \pi \right\| + \left\| {}_{(n)}R_{\beta,0}^m(i,\cdot) - {}_{(n)}\pi_0 \right\| \\
&+ \left\| R_\beta^m(i,\cdot) - {}_{(n)}R_{\beta,0}^m(i,\cdot) \right\| \\
&\leq 2MV(i)\rho^m + \left\| R_\beta^m(i,\cdot) - {}_{(n)}R_{\beta,0}^m(i,\cdot) \right\|.
\end{aligned} \tag{13}$$

As was pointed out in [7], if A and B are two stochastic matrices, then

$$\begin{aligned}&\left\|A^m(i,\cdot)-B^m(i,\cdot)\right\|\\&\leq \sum_{0\leq s\leq m-1, w\in S} A^S(i,w)\left\|A(w,\cdot)-B(w,\cdot)\right\|,\end{aligned} \tag{14}$$

for $m \geq 2$. Applying this to the last term in (13), we obtain

$$\begin{aligned}&\left\|R_\beta^m(i,\cdot)-{}_{(n)}R_{\beta,0}^m(i,\cdot)\right\|\\&\leq \sum_{0\leq s\leq m-1, w\in S} R_\beta^s(i,w)\delta_n(w),\end{aligned} \tag{15}$$

where

$$\begin{aligned}\delta_n(i) &:= \left\|R_\beta(i,\cdot)-{}_{(n)}R_{\beta,0}(i,\cdot)\right\|\\&= \sum_{j\in S}\left|R_\beta(i,j)-{}_{(n)}R_{\beta,0}(i,j)\right|.\end{aligned}$$

Now, since $\delta_n(i) \to 0$ as $n \to \infty$ for all $j \in S$, we can use dominated convergence to conclude that the third term in (13) vanishes as n tends to infinity. Thus,

$$\lim_{n\to\infty}\left\|{}_{(n)}\pi_0-\pi\right\| \leq 2MV(i)\rho^m,$$

for $m \in \mathbb{N}$, and since m was chosen arbitrarily, $\lim_{n\to\infty}\left\|{}_{(n)}\pi_0-\pi\right\|=0$.

Example

Let $\lambda, \mu > 0$ and define Q by

$$Q(i,j)=\begin{cases}\lambda, \text{if } i\geq 0, j=i+1,\\ \mu, \text{if } j=0 \text{ and } i>1,\\ \lambda, \text{if } j=i=0,\\ -(\lambda+\mu), \text{if } j=i>0,\\ 0, \text{otherwise,}\end{cases}$$

The process with this Q-matrix is essentially a renewal process with renewal times marked by visits to state 0. Each renewal time consists of a geometric number of exponential times of mean $1/\lambda$ followed by an exponential time of mean $1/\mu$. At each jump, the process passes from state i to state $i+1$ with probability $\gamma = \lambda/(\lambda+\mu)$ and falls back to state 0 with probability $1-\gamma$. the state space is clearly irreducible and the process has a geometric stationary distribution π, where $\pi(i) = (1-\gamma)\gamma^i$. Existence of the stationary distribution ensures positive recurrence.

Next, let the vector $\tilde{V}$ be given by $\tilde{V}(i) = (1+\delta)^i$, where $\delta \in (0, \mu/\lambda)$. Also define $b = \mu$ and Set

$C = \{ i \in S : \mu - (\lambda\delta - \mu)(1+\delta)^i > -c \}$, where c is a small positive number. Then, the drift condition $(\tilde{D})$ holds for the specified $\tilde{V}, C, b$ and c. The process is therefore exponentially ergodic by Theorem 3.2. Further, all the conditions of Theorem 3.3 are satisfied. Thus, we can construct augmentations ${}_{(n)}Q_0$ on corresponding sets $S_n = \{0, 1, \cdots, n\}$ and use their invariant distributions ${}_{(n)}\pi_0$ to approximate π.

We can confirm this by solving

$\sum_{i=0}^{n} {}_{(n)}\pi_0(i)\, {}_{(n)}Q_0(i,j) = 0,\ j = 0, 1, \cdots, n$ with

$\sum_{i=0}^{n} {}_{(n)}\pi_0(i) = 1$. We have ${}_{(n)}\pi_0(i) = (1-\gamma)\gamma^i / (1-\gamma^n)$

for $i = 0, 1, \cdots, n$, from which it is evident that

(${}_{(n)}\pi_0(i) \searrow \pi(i)$ as $n \to \infty$. Convergence in total variation follows by the same argument used later in the proof of Theorem 4.2.

4. STOCHASTIC MONOTONICITY

In this section, we develop results for stochastically monotone Markov processes. Our key result says that stochastic monotonicity of the process is sufficient for (2) to hold under arbitrary linear augmentation. The remaining results extend this to larger classes of Markov processes. While our methods generally parallel those employed in [6] TO study the same problem in discrete-time Markov chains, itt is necessary to take greater care constructing the augmentations in the continuous-time setting.

Let μ and ν be two non-trivial measures. Then, ν stochastically dominates μ if $\sum_{k\geq n}\mu(k)\leq\sum_{k\geq n}\nu(k)$ for all $n\in\mathbb{N}$, in which case we write $\mu\preceq\nu$. If F and $\tilde{F}$ are two transition functions, we say that $\tilde{F}$ stochastically dominates F (written $F\preceq\tilde{F}$) if, for all $t\geq 0,\ F^t(i,\cdot)\tilde{F}^t(i,\cdot)$ for all $i\in S$. A more strict classification is stochastic comparability. The transition functions F and $\tilde{F}$ are stochastically comparable if $F^t(i,\cdot)\preceq\tilde{F}^t(m,\cdot)$ for all $t\geq 0$ and $i,m\in S$ with $i\leq m$. We use the notation $F\sqsubseteq\tilde{F}$ to mean that F and $\tilde{F}$ are stochastically comparable. A stochastically monotone Markov process is one whose transition function is stochastically comparable to itself. Thus, if F is stochastically dominated by a transition function $\tilde{F}$ which itself is stochastically monotone, then F and $\tilde{F}$ are stochasticallly comparable. Clearly, $F\sqsubseteq\tilde{F}$ implies $F\preceq\tilde{F}$.

The following theorem is the key to obtaining sufficient conditions for (2) to hold in continuous time. It characterises stochastic comparability and monotonicity in terms of Q-matrix structure and is a special case of a more general result which was proved in [15] (also see Theorem 7.3.4 in [1] for an account). The reader is directed to the last two citations for the proof.

Theorem 4.1 ([15] and [1, Chapter 7.3])

i. Let Q and $\tilde{Q}$ be two conservative Q-matrices. Their corresponding minimal transition functions F and $\tilde{F}$ are stochastically comparable iff, whenever $i \leq m,$ and k is such that either $k \leq i$ or $k > m,$ then

$$\sum_{j \geq k} Q(i,j) \leq \sum_{j \geq k} \tilde{Q}(m,j). \tag{16}$$

ii. Let Q be a conservative Q-matrix. Its minimal Q-function F is stochastically monotone iff, Whenever $i \leq m$ and k is such that either $k \leq i$ or $k > m,$ then

$$\sum_{j \geq k} Q(i,j) \leq \sum_{j \geq k} Q(m,j). \tag{17}$$

As a consequence of this result, we shall speak of stochastically monotone Q-matrices and of two Q-matrices as being stochastically comparable, etc. This abuse of terminology should not cause any confusion.

If $\tilde{F}$ is an irreducible, positive recurrent transition function and it stochastically dominates another irreducible transition function $F,$ then F is also positive recurrent. Furthermore, if π and $\tilde{\pi}$ denote the stationary distributions of F and $\tilde{F}$ respectively, then $\pi \preceq \tilde{\pi},$ which can be seen by letting $t \to \infty$ in

$$\sum_{j \geq k} F^t(i,j) \leq \sum_{j \geq k} \tilde{F}^t(i,j), \forall i,k \in S.$$

In fact, we may say something stronger than this. If F is reducible and contains a collection of closed ireducible classes $(C_i),$ each C_i is positive recurrent with invariant probability measure $\pi_i.$ Since F is dominated by $\tilde{F}$ on $C_i,$ it follows that $\pi_i \preceq \tilde{\pi}.$ Now, any invariant measure on S for Q can be written as a linear combination of the π_i's; that is, $\pi = \sum_i a(i)\pi_i$ for some probability measure $a = (a(i), i = 1,2,\cdots)$.

Therefore, $\pi \preceq \tilde{\pi}$ for all invariant distributions π.

Throughout the rest of this section, we shall use the north-west corner truncations of Q, that is, truncations of the form $S_n = \{0,1,2,\cdots,n\}$ for $n = 0,1,2,\cdots$

4.1. Stochastically Monotone Processes

Let Q be the Q-matrix of a positive recurrent, stochastically monotone Markov process F. By construction, the $n \times n$ north-west corner truncations of Q augmented in the nth column are stochastically monotone. Since ${}_{(n)}Q_n$ is conservative on a finite set S_n, ${}_{(n)}F_n$ has precisely one positive recurrent class, which contains n and is a subset of or equal to S_n. Its limiting distribution ${}_{(n)}\pi_n$ satisfies $\sum_{i \in S_n} {}_{(n)}\pi_n(i) \, {}_{(n)}Q_n\pi(i,j) = 0$

for all $j \in S$. From Theorem 4.1, we also see that ${}_{(n)}F_n$ is stochastically monotone.

Let $\overline{{}_{(n)}Q}$ be an arbitrary augmentation of ${}_{(n)}Q$ and note that $\overline{{}_{(n)}Q}$ is stochastically comparable with ${}_{(n)}Q_n$. As per our comments above, the minimal $\overline{{}_{(n)}Q}$-function $\overline{{}_{(n)}F}$ is positive recurrent on one or more irreducible subsets of S_n and hence any invariant distribution for

$\overline{{}_{(n)}Q}$, say ${}_{(n)}\pi$, is stochastically dominated by ${}_{(n)}\pi_n$.

Now, let us extend ${}_{(n)}Q_n$ and $\overline{{}_{(n)}Q}$ to S as follows:

$$
{}_{(n)}Q_n(i,j)=\begin{cases} Q(i,j), \text{if } j<n, \\ \sum_{k\geq n} Q(i,k), \text{if } j=n \text{ and } i\leq n, \\ \sum_{k=n}^{i-1} Q(i,k), \text{if } i>n \text{ and } j=n, \\ \sum_{k\geq i} Q(i,k), \text{if } i>n \text{ and } j=i, \\ 0, \text{otherwise}, \end{cases} \tag{18}
$$

and

$$
\overline{{}_{(n)}Q}(i,j)=\begin{cases} \overline{{}_{(n)}Q}(i,j), \text{if } 0\leq i,j\leq n, \\ \sum_{k=n}^{i-1} Q(i,k), \text{if } i>n \text{ and } j=0, \\ \sum_{k\geq i} Q(i,k), \text{if } i>n \text{ and } j=i, \\ 0, \text{otherwise}. \end{cases}
$$

We also extend ${}^{(n)}\pi_n$ and ${}^{(n)}\pi$ to S by appending a countably infinite number of 0's to each, so that ${}^{(n)}\pi_n(i)={}_{(n)}\pi(i)=0$ for all $i>n.$ Note that ${}^{(n)}\pi_n$

(resp. ${}^{(n)}\pi$) remains invariant for ${}_{(n)}Q_n$ (resp. $\overline{{}_{(n)}Q}$).

Moreover, since the minimal ${}_{(n)}Q_n$ -function ${}_{(n)}F_n$ is positive recurrent on some subset of $\{0,1,\cdots,n\}$ containing n and transient elsewhere in $S,$

the measure ${}^{(n)}\pi_n$ is the limiting distribution of

${}_{(n)}F_n : {}_{(n)}F^t_n(i,j)\rightarrow {}_{(n)}\pi_n(j)$ as $t\rightarrow\infty$ for all

$i,j \in S.$ Similarly, ${}_{(n)}\pi$ is the limiting distribution for the minimal $\overline{{}_{(n)}Q}$-function $\overline{{}_{(n)}F}$ when given an appropriate initial distribution.

The stochastically monotone matrix ${}_{(n)}Q_n$ dominates

$\overline{{}_{(n)}Q}$ while ${}_{(n)}Q_n$ and ${}_{(n+1)}Q_{n+1}$ are stochastically comparable for all $n=0,1,2,\cdots$ So too are ${}_{(n)}Q_n$ and

Q for all $n.$ Thus, $\overline{{}_{(n)}Q} \sqsubseteq {}_{(n)}Q_n \sqsubseteq {}_{(n+1)}Q_{n+1} \sqsubseteq Q.$ An application of Part i of Theorem 4.1 then shows that $\overline{{}_{(n)}F} \sqsubseteq {}_{(n)}F_n \sqsubseteq {}_{(n+1)}F_{n+1} \sqsubseteq F$ for all $n \in \mathbb{N}.$ Consquently,

$$\overline{{}_{(n)}\pi} \preceq {}_{(n)}\pi_n \preceq {}_{(n+1)}\pi_{n+1} \preceq \pi, \tag{19}$$

where π is the unique stationary distribution for $F.$ The sequence $\left({}_{(n)}\pi\right)$ is therefore tight and so

${}_{(n)}\pi(j) \to \pi_j$ for all $j \in S$ as $n \to \infty.$ The same is true for $\left({}_{(n)}\pi_n\right).$

From (19), we observe that

$0 \le \sum_{i \ge k}\left(\pi(i) - {}_{(n)}\pi_n(i)\right) \le \sum_{i \ge k}\left(\pi(i) - {}_{(n)}\pi(i)\right),$ for all

$k \in S,$ and so ${}_{(n)}\pi_n$ is at least as good an approximation to p as ${}_{(n)}\pi$. Thus, any invariant measure derived from a north-west corner truncation of Q augmented in its last column is optimal for approximating π.

As was pointed out in [7], the pointwise convergence of measures on a countable set can easily be extended to convergence in total variation. We therefore have the following result.

Theorem 4.2 Let Q be the Q-matrix of a positive recurrent, stochastically monotone Markov process on $S.$ Let p be the stationary distribution of the minimal Q-function F and denote the invariant distribution of an arbitrary $n \times n$ north-west corner augmentation

$\overline{{}_{(n)}Q}$ by ${}_{(n)}\pi$. Furthermore, let ${}_{(n)}Q_n$ be the $n \times n$

north-west corner truncation augmented in column n and take ${}_{(n)}\pi_n$ to be its invariant distribution. Then,

$\left\| {}_{(n)}\pi - \pi \right\| \to 0$ as $n \to \infty$. The same is true of the sequence $\left({}_{(n)}\pi_n \right)$ which is the optimal approximation in the sense that its tail mass more closely approximates that of π.

Proof: The fact that, for all $j \in S$, ${}_{(n)}\pi(j) \to \pi(j)$ and ${}_{(n)}\pi_n(j) \to \pi(j)$ as $n \to \infty$ was established in the preceding discussion. So too was the optimality of ${}_{(n)}\pi_n$ as an approximation to π. To prove convergence in total variation, fix an arbitrary finite $k \in S$. Then, we obtain

$$\begin{aligned}
\left\| {}_{(n)}\pi - \pi \right\| &= \sum_{j \in S} \left| {}_{(n)}\pi(j) - \pi(j) \right| \\
&\leq \sum_{j \leq k} \left| {}_{(n)}\pi(j) - \pi(j) \right| + \sum_{j > k} \left| {}_{(n)}\pi(j) + \pi(j) \right| \\
&\leq \sum_{j \leq k} \left| {}_{(n)}\pi(j) - \pi(j) \right| + 2 \sum_{j > k} \pi(j).
\end{aligned}$$

The analogous statement holds for ${}_{(n)}Q_n$ and the proof is completed by letting first n and then k tend to infinity.

As remarked in [6], ${}_{(n)}\pi(a)$ will be strictly positive for sufficiently large n where a is an arbitrary state in S. Thus, S_n contains a positive recurrent class to which n belongs. Computationally speaking, this means that any invariant distribution will suffice as an approximation to π, provided n is sufficiently large.

Finally, if n is large enough so that ${}_{(n)}Q$ possesses a quasistationary distribution (n)r supported on a nondegenerate irreducible subset of S_n, then the sequence of distributions $\left({}_{(n)}\gamma\right)$ converges weakly to π. We can always find a sequence $(C_n, n = 1, 2, \cdots)$ of irreducible sets such that $C_1 \subseteq C_2 \subseteq \cdots \subseteq S$ and $\bigcup_{n \in \mathbb{N}} C_n = S.$

See Lemma 5.1 in [16] for a proof of this; the analogue for discrete-time Markov chains may be found in [3], Theorem 3.1.

For a finite state Markov process, every quasistationary distribution is equivalent to a probabilitynormalised left eigenvector of its Q-matrix restricted to an ireducible class. In other words, If Q is a nonconservative Q-matrix on a finite state space S containing an ireducible class $C \in S, r = \left(r(i), i \in C\right)$ is a quasistationary distribution on C for the process if and only if $\sum_{i \in C} r(i) = 1$ and, for some $\lambda > 0,$

$$\sum_{j \in S} r(i) Q(i,j) = -\lambda r(j), j \in C.$$

Note that by virtue of λ-theory, $r(i)$ is strictly positive for all $i \in C$. By convention, we extend r to S by setting $r(i) = 0$ for $i \in S \setminus C$. If we then construct the linear augmentation $\bar{Q}$ as

$$\bar{Q}(i,j) = Q(i,j) - \left(\sum_{k \in C} Q(i,k)\right) r(j), \tag{20}$$

$i, j \in S,$ it is not difficult to see that the invariant distribution π for $\bar{Q}$ is unique and equivalent to r.

Now, let us return to the case of a countably infinite state space. Given n large, we may construct $\overline{{}_{(n)}Q}$ from

${}_{(n)}Q$ in the same manner as (20) using a left eigenvector ${}_{(n)}r$ of ${}_{(n)}Q$ supported on an irreducible class C_n. The conditions of Theorem 4.2 are satisfied and so the sequence $\left({}_{(n)}r\right)$ converges in total variation to the invariant distribution of Q. This observation subsumes results concerning the truncation approximation of invariant distributions of birth-death processes and subcritical Markov branching processes, for example, see [16-18]. The convergence of quasistationary distributions of truncations to the invariant distribution of the original process also holds under the weaker conditions we discuss in the next two subsections.

4.2. Processes Dominated by Stochastically Monotone Processes

Now we shall consider a much larger class of Markov processes, namely those whose transition functions are stochastically dominated by a positive recurrent, stochastically monotone process. To begin, let $\tilde{F}$ be the stochastically monotone transition function of an irreducible, positive recurrent Markov process. Suppose that $\tilde{F}$ dominates a transition function F. We shall use Q and $\tilde{Q}$ to denote the corresponding Q-matrices. As noted earlier, F must be positive recurrent and the invariant distributions π and $\tilde{\pi}$, corresponding to F and $\tilde{F}$ respectively, satisfy $\pi \preceq \tilde{\pi}$.

Let ${}_{(n)}Q_n$ and ${}_{(n)}\tilde{Q}_n$ respectively denote the $n \times n$ north-west corner truncations of Q and $\tilde{Q}$ augmented in the nth column. By extending these in the analogous way to (18) and applying Part i of Theorem 4.1, we see that ${}_{(n)}F_n$ and ${}_{(n)}\tilde{F}_n$ are stochastically comparable.

Also, let $\overline{{}_{(n)}Q}$ be an arbitrary augmentation of an $n \times n$ north-west corner truncation of Q and note that

$\overline{{}_{(n)}Q} \sqsubseteq {}_{(n)}\tilde{Q}_n,$ whence $\overline{{}_{(n)}F} \sqsubseteq {}_{(n)}\tilde{F}_n.$ From the previous subsection, $\overline{{}_{(n)}F} \sqsubseteq {}_{(n+1)}\tilde{F}_{n+1} \sqsubseteq \tilde{F}$ for $n = 0,1,2\cdots$

Combining these, we obtain

$\overline{{}_{(n)}F}, {}_{(n)}F_n \sqsubseteq {}_{(n)}\tilde{F}_n \sqsubseteq {}_{(n+1)}\tilde{F}_{n+1} \sqsubseteq \tilde{F}$ which implies that

${}_{(n)}\pi, {}_{(n)}\pi_n \preceq {}_{(n)}\tilde{\pi}_n \preceq {}_{(n+1)}\tilde{\pi}_{n+1} \preceq \tilde{\pi}.$ Thus, the sequence

$\left({}_{(n)}\pi\right)$ is tight and ${}_{(n)}\pi \rightarrow \pi$ componentwise as $n \rightarrow \infty.$

Convergence in total variation follows in the same way as in the proof of Theorem 4.2 and the same is true of

$\left({}_{(n)}\pi_n\right).$ Thus we have proved the following result.

Theorem 4.3 Let Q be the Q-matrix of an irreducible Markov process which is dominated by a positive recurrent, stochastically monotone Markov process. Then,

$\left\|{}_{(n)}\pi - \pi\right\| \rightarrow 0$ as $n \rightarrow \infty$ where p is the unique invariant distribution for Q and ${}_{(n)}\pi,$ for $n \in \mathbb{N},$ constitutes an invariant distribution of an arbitrary $n \times n$ north-west corner augmentation of $Q.$

As the augmentation ${}_{(n)}Q_n$ is a special case of $\overline{{}_{(n)}Q},$

it follows from the theorem that $\left\|{}_{(n)}\pi_n - \pi\right\| \rightarrow 0$ as

$n \rightarrow \infty.$ However, unlike the situation in which F is stochastically monotone, it is not clear which of ${}_{(n)}\pi$ and ${}_{(n)}\pi_n$ provides the better approximation to $\pi.$

4.3. Finitely Perturbed Stochastically Monotone Markov Processes

Finally, we consider an even more general class of Markov processes which was introduced in [7]. We say that Q is a finite perturbation of $\hat{Q}$ if the two Q-matrices differ in at most a finite number of columns. Let $\hat{Q}$ be stochastically monotone and suppose without loss of generality that Q and $\hat{Q}$ differ in the first k columns. Let

$\tau : \min\left\{\hat{Q}(m) : m = 0,1,\cdots,k-1\right\}$ and construct a

Q-matrix $\tilde{Q}$ as follows:

$$\tilde{Q}(i,j) = \begin{cases} -\tau, \text{if } j = i, 0 \le i \le k, \\ \tau + \sum_{m=0}^{k} \hat{Q}(i,m), \text{if } i < k, j = k \\ \sum_{m=0}^{k} \hat{Q}(i,m), \text{if } i \ge k, j = k \\ \hat{Q}(i,j), \text{if } j > k \\ 0, \text{otherwise} \end{cases}$$

Observe that $\tilde{Q}$ is stochastically monotone. This is due firstly to the way in which the first k columns have been constructed from Q, and secondly to the agreement between the remaining columns of Q with the corresponding columns of the stochastically monotone $\hat{Q}$. Now, $\tilde{Q}$ satisfies (17) and so, by Theorem 4.1, the minimal Q-function F and $\tilde{Q}$-function $\tilde{F}$ are stochastically comparable. Direct application of Theorem 4.3 to F and $\tilde{F}$ then yields the following result.

Theorem 4.4 Let Q be a finite perturbation of a Qmatrix $\hat{Q}$ whose minimal $\hat{Q}$-function is irreducible, positive recurrent and stochastically monotone. Also, let p be the unique invariant distribution for Q and denote the invariant distributions of arbitrary $n \times n$ north-west corner augmentations $_{(n)}Q$ by $_{(n)}\pi$. Then,

$$\left\| _{(n)}\pi - \pi \right\| \to 0 \text{ as } n \to \infty$$

4.4. Example

Conrth-death process, whose tridiagonal Q-matrix

$$Q(i,j) = \begin{cases} \lambda_i, \text{if } i \geq 0, j = i+1, \\ \mu_i, \text{if } j \geq 0, i = j+1, \\ \lambda_0, \text{if } j = i = 0, \\ -(\lambda_i + \mu_i), \text{if } j = i > 0, \\ 0, \text{otherwise,} \end{cases}$$

where $(\lambda_i, i \geq 0)$ are strictly positive birth rates and $(\mu_i, i \geq 1)$ are strictly positive death rates. Here, we take the state space S to be the set of non-negative integers. Such processes can be used to model queues having memoriless arrival and service times, simple circuitswitched teletraffic networks and buffers in computer networks, etc.

Let Q be the Q-matrix of an irreducible birth-death process. Then, it can be shown (see [1], Chapter 3) that

Q is regular if and only if $\sum_{i=0}^{\infty} \frac{1}{\lambda_i \tau_i} \sum_{j=0}^{i} \tau_j = \infty$

where $\tau_0 = 1$ and $\tau_i = \frac{\lambda_0 \lambda_1 \cdots \lambda_{i-1}}{\mu_1 \mu_2 \cdots \mu_i}$ for $i \geq 1$. Now for

Q regular, the unique minimal transition function F is positive recurrent if and only if $A = \sum_{i=0}^{\infty} \frac{1}{\lambda_i \tau_i} = \infty$ and $B = \sum_{i=0}^{\infty} \tau_i < \infty$. The stationary distribution for F is

$\pi = (\pi(i), i \in S)$ where $\pi_{(i)} = \tau_i / B$

Now, it is straight forward to verify that Q satisfies (17) and hence, by Theorem 4.1, F is stochastically monotone. furthermore, by Theorem 4.2, we may use $_{(n)}\pi_n$ to approximate π. Letting $_{(n)}Q$ denote the north-west corner truncation on $\{0,1,\cdots,n\}$, augmentation in column n yields the matrix $_{(n)}Q_n$ which differs from $_{(n)}Q$ only in its (n,n) element. More precisely,

$_{(n)}Q_n(n,n) = -\mu_n$ and $_{(n)}Q_n(i,j) = {}_{(n)}Q(i,j) = Q(i,j)$ if either $i < n$ or $j < n$. The stationary distribution $_{(n)}\pi_n$ corresponding to the augmentation $_{(n)}q_n$ is given by $_{(n)}\pi_n(i) = \tau_i \Big/ \sum_{j=0}^{n} \tau_j, 0 \le i \le n$
From this closed form expression, it can immediately be seen that

$_{(n)}\pi_n(i) \searrow \pi(i)$ as $n \to \infty$ since $\sum_{j=0}^{n} \tau_j \nearrow B < \infty$ as $n \to \infty$. Convergence in total variation then follows as in the proof of Theorem 4.2.

5. CONCLUSION

Here we have investigated procedures based on the augmentation of state-space truncations for approximating the stationary distributions of positive recurrent, continuous-time Markov processes on countably infinite state spaces. We have shown that approximation techniques first proposed for application to discrete-time markov chains are also efficacious in the continuous-time setting. Two classes of Markov process were considered: Exponentially ergodic processes and stochastically monotone processes. It was shown that the invariant distributions ${}_{(n)}\pi$ corresponding to the augmented Q-matrix $\overline{{}_{(n)}Q}$ of finite statespace truncations of a Q-matrix converge in total variation to the invariant distribution of the Markov process generated by that Q-matrix. It remains to study the speed of such convergence. An understanding of the convergence rate would enable the truncation size to be selected in order to guarantee that the measure ${}_{(n)}\pi$ approximates π to a desired degree of accuracy.

6. ACKNOWLEDGEMENTS

Andrew G. Hart, Richard L. Tweedie (CMM) Basal CONICYT Program PFB 03 and FONDECYT grant 1070344. AGH would like to thank Servet Martinez for interesting discussion on the truncation of stochastically monotone processes. AGH dedicates this article to co-author Richard Tweedie, who passed away after this work was started.

REFERENCES

1. W. Anderson, "Continuous-Time Markov Chains: An Applications-Oriented Approach," Springer Series in Statistics, Springer-Verlag, New York, 1991.
2. E. Seneta, "Finite Approximations to Infinite Nonnegative Matrices I," Proceedings of the Cambridge Philosophical Society, Vol. 63, No. 4, 1967, pp. 983-992.
3. E. Seneta, "Finite Approximations to Infinite Nonnega tive Matrices II: Refinements and Applications," Proceedings of the Cambridge Philosophical Society, Vol. 64, No. 2, 1968, pp. 465-470.
4. E. Seneta, "Computing the Stationary Distribution for Infinite Markov Chains," Linear Algebra and Its Applications, Vol. 34, 1980, pp. 259-267.
5. D. Gibson and E. Seneta, "Monotone Infinite Stochastic Matrices and Their Augmented Truncations," Stochastic Processes and Their Applications, Vol. 24, No. 2, 1987, pp. 287-292.
6. D. Gibson and E. Seneta, "Augmented Truncations of Infinite Stochastic Matrices," Journal of Applied Probability, Vol. 24, No. 3, 1987, pp. 600-608.
7. R. Tweedie, "Truncation Approximations of Invariant Measures for Markov Chains," Journal of Applied Probability, Vol. 35, No. 3, 1998, pp. 517-536.

8. R. Tweedie, "Truncation Procedures for Nonnegative Matrices," Journal of Applied Probability, Vol. 8, No. 2, 1971, pp. 311-320.
9. R. Tweedie, "The Calculation of Limit Probabilities for Denumerable Markov Processes from Infinitesimal Properties," Journal of Applied Probability, Vol. 10, No. 1, 1973, pp. 84-99.
10. J. Kingman, "The Exponential Decay of Markov Transition Probabilities," Proceedings of the London Mathematical Society, Vol. 13, No. 1, 1963, pp. 337-358.
11. D. Down, S. Meyn and R. Tweedie, "Exponential and Uniform Ergodicity of Markov Processes," Annals of Probability, Vol. 23, No. 4, 1995, pp. 1671-1691.
12. R. Tweedie, "Criteria for Ergodicity, Exponential Ergodicity and Strong Ergodicity of Markov Processes," Journal of Applied Probability, Vol. 18, No. 1, 1981, pp. 122-130.
13. S. Meyn and R. Tweedie, "Markov Chains and Stochastic Stability," Springer-Verlag, London, 1993.
14. S. Meyn and R. Tweedie, "Computable Bounds for Convergence Rates of Markov Chains," Annals of Applied Probability, Vol. 4, No. 4, 1994, pp. 981-1011.
15. B. Kirstein, "Monotonicity and Comparability of TimeHomogeneous Markov Processes with Discrete State Space," Statistics, Vol. 7, No. 1, 1976, pp. 151-168.
16. A. Hart, "Quasistationary Distributions for ContinuousTime Markov Chains," Ph.D. thesis, University of Queensland, Brisbane, 1997.
17. L. Breyer and A. Hart, "Approximations of Quasistationary Distributions for Markov Chains," Mathematical and Computer Modelling, Vol. 31, No. 10-12, 2000, pp. 69-79.
18. M. Kijima and E. Seneta, "Some Results for Quasistationary Distributions of Birth-Death Processes," Journal of Applied Probability, Vol. 28, No. 3, 1991, pp. 503-511.

CHAPTER 8

Construction of Zero Autocorrelation Stochastic Waveforms

Somantika Datta

Department of Mathematics, University of Idaho, Moscow, USA

ABSTRACT

Stochastic waveforms are constructed whose expected autocorrelation can be made arbitrarily small outside the origin. These waveforms are unimodular and complex-valued. Waveforms with such spike like autocorrelation are desirable in waveform design and are particularly useful in areas of radar and communications. Both discrete and continuous waveforms with low expected autocorrelation are constructed. Further, in the discrete case, frames for $\mathbb{C}^d$ are constructed from these waveforms and the frame properties of such frames are studied.

Keywords: Autocorrelation; Frames; Stochastic Waveforms

1. INTRODUCTION

1.1. Motivation

Designing unimodular waveforms with an impulse-like autocorrelation is central in the general area of waveform design, and it is particularly relevant in several applications in the areas of radar and communications. In the former, the waveforms can play a role in effective target recognition, e.g., [1-8]; and in the latter they are used to address synchronization issues in cellular (phone) access technologies, especially code division multiple access (CDMA), e.g., [9,10]. The radar and communications methods combine in recent advanced multifunction RF systems (AMRFS). In radar there are two main reasons that the waveforms should be unimodular, that is, have constant amplitude. First, a transmitter can operate at peak power if the signal has constant peak amplitude—the system does not have to deal with the surprise of greater than expected amplitudes. Second, amplitude variations during transmission due to additive noise can be theoretically eliminated. The zero autocorrelation property ensures minimum interference between signals sharing the same channel.

Constructing unimodular waveforms with zero autocorrelation can be related to fundamental questions in harmonic analysis as follows. Let $\mathbb{R}$ be the real numbers, $\mathbb{Z}$ the integers, $\mathbb{C}$ the complex numbers, and set $\mathbb{T} = \mathbb{R}/\mathbb{Z}$. The aperiodic autocorrelation $A_X : \mathbb{Z} \to \mathbb{C}$ of a waveform $X : \mathbb{Z} \to \mathbb{C}$, is defined as

$$\forall k \in \mathbb{Z},\ A_X[k] = \lim_{N \to \infty} \frac{1}{2N+1} \sum_{m=-N}^{N} X[k+m]\overline{X[m]}. \qquad (1)$$

A general problem is to characterize the family of positive bounded Radon measures F, whose inverse Fourier transforms are the autocorrelations of bounded waveforms X. A special case is when $F \equiv 1$ on $\mathbb{T}$ and X is unimodular on $\mathbb{Z}$. This is the same as when the autocorrelation of X vanishes except at 0, where it takes the value 1. In this case, X is said to have perfect autocorrelation. An extensive discussion on the construction of different classes of deterministic waveforms with perfect autocorrelation can be found in [11]. Instead of aperiodic waveforms that are defined on $\mathbb{Z}$, in some applications, it might be useful to construct periodic waveforms with similar vanishing properties of the autocorrelation function. Let $n \geq 1$ be an integer and $\mathbb{Z}_n$ be the finite group $\{0,1,\cdots,n-1\}$ with addition modulo n. The periodic autocorrelation $A_X : \mathbb{Z}_n \to \mathbb{C}$ of a waveform $X : \mathbb{Z}_n \to \mathbb{C}$ is defined as

$$\forall k = 0,1,2,\cdots,n-1,\ A_X[k] = \frac{1}{n}\sum_{m=0}^{n-1} X[m+k]\overline{X[m]}. \tag{2}$$

It is said that $X : \mathbb{Z}_n \to \mathbb{C}$ is a constant amplitude zero autocorrelation (CAZAC) waveform if each $|X[k]| = 1$ and

$$\forall k = 1,2,\cdots,n-1,\ A_X[k] = \frac{1}{n}\sum_{m=0}^{n-1} X[m+k]\overline{X[m]} = 0.$$

The literature on CAZACs is overwhelming. A good reference on this topic is [3], among many others. Literature on the general area of waveform design include [12-14]. Comparison between periodic and aperiodic autocorrelation can be found in [15].

Here the focus is on the construction of stochastic aperiodic waveforms. Henceforth, the reference to waveforms shall imply aperiodic waveforms unless stated otherwise. These waveforms are stochastic in nature and are constructed from certain random variables. Due to the stochastic nature of the construction, the expected value of the corresponding autocorrelation function is analyzed. It is desired that everywhere away from zero, the expectation of the autocorrelation can be made arbitrarily small. Such waveforms will be said to have almost perfect autocorrelation and will be called zero autocorrelation stochastic waveforms. First discrete waveforms, $X : \mathbb{Z} \to \mathbb{C}$, are constructed such that X has almost perfect autocorrelation and for all $n \in \mathbb{Z},\ |X[n]| = 1.$ This approach is extended to the construction of continuous waveforms, $x : \mathbb{R} \to \mathbb{C}$, with similar spike like behavior of the expected autocorrelation and $|x(t)| = 1$ for all $t \in \mathbb{R}.$ Thus, these waveforms are unimodular. The stochastic and non-repetitive nature of these waveforms means that they cannot be easily intercepted or detected by an adversary. Previous work on the use of stochastic waveforms in radar can be found in [16-18], where the waveforms are only real-valued and not unimodular. In comparison, the waveforms constructed here are complex valued and unimodular. In addition, frame properties of frames constructed from these stochastic waveforms are discussed. This is motivated by the fact that frames have become a standard tool in signal processing. Previously, a mathematical characterization of CAZACs in terms of finite unit-normed tight frames (FUNTFs) has been done in [2].

1.2. Notation and Mathematical Background

Let X be a random variable with probability density function f. Assuming X to be absolutely continuous, the expectation of X denoted by $E(X)$, is $E(X) = \int_{\mathbb{R}} xf(x)\mathrm{d}x.$

The Gaussian random variable has probability density function given by $f(x) = \frac{1}{\sigma\sqrt{2\pi}} e^{-\frac{1}{2}\left(\frac{x-\mu}{\sigma}\right)^2}.$ The mean or expectation of this random variable is μ and the variance, $V(X)$, is σ^2. In this case it is also said that X follows a normal distribution and is written as $X \sim N(\mu, \sigma^2)$.

The characteristic function of X at t, $E\left(e^{itX}\right)$, is denoted by $\phi_X(t)$. For further properties of expectation and characteristic function of a random variable the reader is referred to [19].

Let $\mathcal{H}$ be a Hilbert space and let $V = \{v_k, k \in K\}$, where K is some index set, be a collection of vectors in $\mathcal{H}$. Then V is said to be a frame for $\mathcal{H}$ if there exist constants A and B, $0 < A \leq B < \infty$, such that for any $v \in \mathcal{H}$ $A\|v\|^2 \leq \sum_{k \in K} |\langle v, v_k \rangle|^2 \leq B\|v\|^2.$

The constants A and B are called the frame bounds. Thus a frame can be thought of as a redundant basis. In fact, for a finite dimensional vector space, a frame is the same as a spanning set. If $A = B$, the frame is said to be tight. Orthonormal bases are special cases of tight frames and for these, $A = B = 1$.

If V is a frame for $\mathcal{H}$ then the map $F:\mathcal{H}\rightarrow\ell_2(\mathcal{K})$ given by $F(v)=\{\langle v,v_k\rangle:k\in\mathcal{K}\}$ is called the analysis operator. The synthesis operator is the adjoint map $F^*:\ell_2(\mathcal{K})\rightarrow\mathcal{H},$ given by
$F^*(\{a_k\})=\sum_{k\in\mathcal{K}}a_k v_k.$

The frame operator $\mathcal{F}:\mathcal{H}\rightarrow\mathcal{H}$ is given by $\mathcal{F}=F^*F.$ For a tight frame, the frame operator is just a constant multiple of the identity, i.e., $\mathcal{F}=A\mathcal{I},$ where $\mathcal{I}$ is the identity map. Every $v\in\mathcal{H}$ can be represented as

$$v=\sum_{k\in\mathcal{K}}\left\langle v,\mathcal{F}^{-1}v_k\right\rangle v_k=\sum_{k\in\mathcal{K}}\left\langle v,v_k\right\rangle\mathcal{F}^{-1}v_k.$$

Here $\{\mathcal{F}^{-1}v_k\}$ is also a frame and is called the dual frame. For a tight frame, $\mathcal{F}^{-1}$ is just $\frac{1}{A}\mathcal{F}.$ Tight frames are thus highly desirable since they offer a computationally simple reconstruction formula that does not involve inverting the frame operator. The minimum and maximum eigenvalues of $\mathcal{F}$ are the optimal lower and upper frame bounds respectively [20]. Thus, for a tight frame all the eigenvalues of the frame operator are equal to each other. For the general theory on frames one can refer to [20,21].

1.3. Outline

The construction of discrete unimodular stochastic waveforms, $X:\mathbb{Z}\rightarrow\mathbb{C}$, with almost perfect autocorrelation is done in Section 2. This is first done with the Gaussian random variable and then generalized to other random variables. The variance of the autocorrelation is also estimated. The section also addresses the construction of stochastic waveforms in higher dimensions, i.e., construction of $v:\mathbb{Z}\rightarrow\mathbb{C}^d$, that have almost perfect autocorrelation and are unit-normed, considering the usual norm in $\mathbb{C}^d$. In Section 3 the construction of unimodular continuous waveforms with almost perfect autocorrelation is done using Brownian motion.

As mentioned in Section 1.2, frames are now a standard tool in signal processing due to their effectiveness in robust signal transmission and reconstruction. In Section 4, frames in $\mathbb{C}^d\,(d \ge 2)$ are constructed from the discrete waveforms of Section 2 and the nature of these frames is analyzed. In particular, the maximum and minimum eigenvalues of the frame operator are estimated. This helps one to understand how close these frames are to being tight. Besides, it follows, from the eigenvalue estimates, that the matrix of the analysis operator, F, for such frames, can be used as a sensing matrix in compressed sensing.

2. CONSTRUCTION OF DISCRETE STOCHASTIC WAVEFORMS

In this section discrete unimodular waveforms, $X : \mathbb{Z} \to \mathbb{C}$, are constructed from random variables such that the expectation of the autocorrelation can be made arbitrarily small everywhere except at the origin. First, such a construction is done using the Gaussian random variable. Next, a general characterization of all random variables that can be used for the purpose is given.

2.1. Construction from Gaussian Random Variables

Let $\{Y_\ell\}_{\ell \in \mathbb{Z}}$ be independent identically distributed (i.i.d.)

random variables following a Gaussian or normal distribution with mean 0 and variance σ^2, i.e., $Y_\ell \sim N(0,\sigma^2)$. Define $X : \mathbb{Z} \to \mathbb{C}$ by

$$\forall n \in \mathbb{Z},\ X[n] = e^{\frac{2\pi i}{\varepsilon}\sum_{\ell=-n}^{n} Y_\ell} \tag{3}$$

where i is $\sqrt{-1}$. Thus, for each n, $|X[n]| = 1$ and X is unimodular. The autocorrelation of X at $k \in \mathbb{Z}$ is

$$A_X[k] = \lim_{N\to\infty} \frac{1}{2N+1} \sum_{n=-N}^{N} X[n+k]\overline{X[n]}$$

where the limit is in the sense of probability. Theorem 2.1 shows that the waveform given by (3) has autocorrelation whose expectation can be made arbitrarily small for all integers $k \neq 0.$

Theorem 2.1. Given $\varepsilon > 0,$ the waveform $X:\mathbb{Z} \to \mathbb{C}$ defined in (3) has autocorrelation A_X such that

$$E\left(A_X[k]\right)=\begin{cases}1 & \text{if } k=0\\ e^{-|k|\sigma^2\left(\frac{2\pi}{\varepsilon}\right)^2} & \text{if } k\neq 0.\end{cases}$$

Proof. 1) When $k=0,$

$$A_X[0]=\lim_{N\to\infty}\frac{1}{2N+1}\sum_{n=-N}^{N}X[n]\overline{X[n]}=1.$$

and so $E\left(A_X[0]\right)=1.$

2) Let $k>0.$ One would like to calculate

$$E\left(A_X[k]\right)=E\left(\lim_{N\to\infty}\frac{1}{2N+1}\sum_{n=-N}^{N}X[n+k]\overline{X[n]}\right).$$

Let $g_N(X)=\frac{1}{2N+1}\sum_{n=-N}^{N}X[n+k]\overline{X[n]}.$ Then $|g_N(X)|\leq 1.$

Let $h(X)=1.$ Then for each $N,$

$|g_N(X)|\leq h(X)$ and $E\left(h(X)\right)=1$. Thus, by the Dominated Convergence Theorem [19], which justifies the interchange of limit and integration below, one obtains

$$
\begin{aligned}
E\left(A_X[k]\right) &= E\left(\lim_{N\to\infty}\frac{1}{2N+1}\sum_{n=-N}^{N}X[n+k]\overline{X[n]}\right). \\
&= \lim_{N\to\infty}\frac{1}{2N+1}\sum_{n=-N}^{N}E\left(X[n+k]\overline{X[n]}\right) \\
&= \lim_{N\to\infty}\frac{1}{2N+1}\sum_{n=-N}^{N}E\left(\mathrm{e}^{\frac{2\pi i}{\varepsilon}\sum_{\ell=-n-k}^{n+k}Y_\ell}\,\mathrm{e}^{-\frac{2\pi i}{\varepsilon}\sum_{m=-n}^{n}Y_m}\right) \\
&= \lim_{N\to\infty}\frac{1}{2N+1}\sum_{n=-N}^{N}E\left(\mathrm{e}^{\frac{2\pi i}{\varepsilon}\left(\sum_{\ell=n+1}^{n+k}Y_\ell+\sum_{m=-n-k}^{-n-1}Y_m\right)}\right) \\
&= \lim_{N\to\infty}\frac{1}{2N+1}\sum_{n=-N}^{N}\left[E\left(\mathrm{e}^{\frac{2\pi i}{\varepsilon}Y_1}\right)\right]^{2k} \\
&= \left[E\left(\mathrm{e}^{\frac{2\pi i}{\varepsilon}Y_1}\right)\right]^{2k} = \left[\phi_{Y_1}\left(\frac{2\pi}{\varepsilon}\right)\right]^{2k}
\end{aligned}
$$

where the last line uses the fact that the Y_ℓ s are i.i.d. random variables. Here $\phi_{Y_1}\left(\frac{2\pi}{\varepsilon}\right)$ is the characteristic function at $\frac{2\pi}{\varepsilon}$ of Y_1 which is the same as that for any other Y_ℓ due to their identical distribution. The characteristic function at $\frac{2\pi}{\varepsilon}$ of a Gaussian random variable with mean 0 and variance σ^2 is $\mathrm{e}^{-\frac{\sigma^2}{2}\left(\frac{2\pi}{\varepsilon}\right)^2}$. Thus

$$
E\left(A_X[k]\right) = \left[\mathrm{e}^{-\frac{\sigma^2}{2}\left(\frac{2\pi}{\varepsilon}\right)^2}\right]^{2k} = \mathrm{e}^{-k\sigma^2\left(\frac{2\pi}{\varepsilon}\right)^2}.
$$

3) When $k>0$, a similar calculation for $A_X[-k]$ gives

$$E\left(A_X[-k]\right)=\left[e^{-\frac{\sigma^2}{2}\left(\frac{2\pi}{\varepsilon}\right)^2}\right]^{2k}=e^{-k\sigma^2\left(\frac{2\pi}{\varepsilon}\right)^2}.$$

Together, this shows that given ε and any $k \neq 0,$

$$E\left(A_X[k]\right)=e^{-|k|\sigma^2\left(\frac{2\pi}{\varepsilon}\right)^2}$$

which indicates that the expectation of the autocorrelation at any integer $k \neq 0$ can be made arbitrarily small depending on the choice of ε . □

As shown in Theorem 2.1 the expectation of the autocorrelation can be made arbitrarily small but this is not useful unless one can estimate the variance of the autocorrelation. Denoting the variance of $A_X[k]$ by $V\left(A_X[k]\right)$ one has

$$\begin{aligned}V\left(A_X[k]\right)&=E\left(\left|A_X[k]\right|^2\right)-\left|E\left(A_X[k]\right)\right|^2\\&=E\left(\left|A_X[k]\right|^2\right)-e^{-2|k|\sigma^2\left(\frac{2\pi}{\varepsilon}\right)^2}.\end{aligned}$$

First consider $k>0;$

$$\begin{aligned}\left|A_X[k]\right|^2&=\left(\lim_{N\to\infty}\frac{1}{2N+1}\sum_{n=-N}^{N}X[n+k]\overline{X[n]}\right)\left(\lim_{M\to\infty}\frac{1}{2M+1}\sum_{m=-M}^{M}\overline{X[m+k]}X[m]\right)\\&=\lim_{N\to\infty}\lim_{M\to\infty}\frac{1}{(2N+1)}\frac{1}{(2M+1)}\sum_{n=-N}^{N}\sum_{m=-M}^{M}X[n+k]\overline{X[n]}\,\overline{X[m+k]}X[m]\\&=\lim_{N\to\infty}\lim_{M\to\infty}\frac{1}{(2N+1)}\frac{1}{(2M+1)}\sum_{n=-N}^{N}\sum_{m=-M}^{M}e^{\frac{2\pi}{\varepsilon}i\left(\sum_{j=1}^{k}Y_{-n-j}+\sum_{j=1}^{k}Y_{n+j}-\sum_{j=1}^{k}Y_{-m-j}-\sum_{j=1}^{k}Y_{m+j}\right)}.\end{aligned}$$

By applying the Lebesgue Dominated Convergence Theorem one can bring the expectation inside the double sum to get

$$E\left(\left|A_X[k]\right|^2\right) = \lim_{N\to\infty}\lim_{M\to\infty}\frac{1}{(2N+1)}\frac{1}{(2M+1)}\sum_{n=-N}^{N}\sum_{m=-M}^{M}E\left(e^{\frac{2\pi}{\varepsilon}i\left(\sum_{j=1}^{k}Y_{-n-j}+\sum_{j=1}^{k}Y_{n+j}-\sum_{j=1}^{k}Y_{-m-j}-\sum_{j=1}^{k}Y_{m+j}\right)}\right).$$

The sum

$$\sum_{j=1}^{k}Y_{-n-j}+\sum_{j=1}^{k}Y_{n+j}-\sum_{j=1}^{k}Y_{-m-j}-\sum_{j=1}^{k}Y_{m+j} \tag{4}$$

may have cancelations among terms involving n with terms involving m. Suppose that for a fixed n and m there are $\tilde{k}_{m,n}$ indices that cancel in each of the four sums in (4). Due to symmetry, the same number i.e., $\tilde{k}_{m,n}$ of terms will cancel in each sum. Depending on n and m, $\tilde{k}_{m,n}$ lies between 0 and k, i.e., $0 \le \tilde{k}_{mn} \le k$. For the sake of making the notation less cumbersome, $\tilde{k}_{m,n}$ will from now on be written as $\tilde{k}$. When $m = n$, $\tilde{k} = k$. If $m > n + k$ or $n > m + k$ then $\tilde{k} = 0$. Each sum in (4) has k terms and $\tilde{k}$ of these get cancelled leaving $(k-\tilde{k})$ terms. One can re-index the variables in (4) and write it as

$$\sum_{j=1}^{k}Y_{-n-j}+\sum_{j=1}^{k}Y_{n+j}-\sum_{j=1}^{k}Y_{-m-j}-\sum_{j=1}^{k}Y_{m+j} = \pm Y_{\ell_1} \pm \cdots \pm Y_{l_{4(k-\tilde{k})}}$$

where the sign depends on whether m is less than or greater than n. Thus

$$E\left(\left|A_X[k]\right|^2\right) = \lim_{N\to\infty}\lim_{M\to\infty}\frac{1}{(2N+1)}\frac{1}{(2M+1)}\sum_{n=-N}^{N}\sum_{m=-M}^{M}E\left(e^{\frac{2\pi}{\varepsilon}i\left(\pm Y_{\ell_1} \pm \cdots \pm Y_{l_{4(k-\tilde{k})}}\right)}\right).$$

Due to the independence of the Y_ℓ s, this means

$$E\left(\left|A_X[k]\right|^2\right)=\lim_{N\to\infty}\lim_{M\to\infty}\frac{1}{(2N+1)}\frac{1}{(2M+1)}\sum_{n=-N}^{N}\sum_{m=-M}^{M}\left[\phi_{Y_1}\left(\pm\frac{2\pi}{\varepsilon}\right)\right]^{4(k-\tilde{k})}$$

$$=\lim_{N\to\infty}\lim_{M\to\infty}\frac{1}{(2N+1)}\frac{1}{(2M+1)}\sum_{n=-N}^{N}\sum_{m=-M}^{M}\mathrm{e}^{-\frac{\sigma^2}{2}\left(\frac{2\pi}{\varepsilon}\right)^2 4(k-\tilde{k})}.$$

The minimum is attained for $\tilde{k}=0$ and the maximum at $\tilde{k}=k.$ Thus

$$E\left(\left|A_X[k]\right|^2\right)\leq\lim_{N\to\infty}\lim_{M\to\infty}\frac{1}{(2N+1)}\frac{1}{(2M+1)}\sum_{n=-N}^{N}\sum_{m=-M}^{M}1=1$$

and

$$E\left(\left|A_X[k]\right|^2\right)\geq\lim_{N\to\infty}\lim_{M\to\infty}\frac{1}{(2N+1)}\frac{1}{(2M+1)}\sum_{n=-N}^{N}\sum_{m=-M}^{M}\mathrm{e}^{-\frac{\sigma^2}{2}\left(\frac{2\pi}{\varepsilon}\right)^2 4k}=\mathrm{e}^{-\frac{\sigma^2}{2}\left(\frac{2\pi}{\varepsilon}\right)^2 4k}.$$

This gives

$$0\leq V\left(A_X[k]\right)\leq 1-\mathrm{e}^{-2k\sigma^2\left(\frac{2\pi}{\varepsilon}\right)^2}.$$

A similar calculation can be done for $k<0.$ Thus for $k\neq 0,$

$$0\leq V\left(A_X[k]\right)\leq 1-\mathrm{e}^{-2|k|\sigma^2\left(\frac{2\pi}{\varepsilon}\right)^2}.$$

2.2. Generalizing the Construction to Other Random Variables

So far the construction of discrete unimodular zero autocorrelation stochastic waveforms has been based on Gaussian random variables. This construction can be generalized to many other random variables. The unimodularity of the waveforms is not affected by using a different random variable. The following theorem characterizes the class of random variables that can be used to get the desired autocorrelation.

Theorem 2.2. Let $\{Y_\ell\}_{\ell\in\mathbb{Z}}$ be a sequence of i.i.d. random variables with characteristic function $\phi_Y.$ Suppose that the probability density function of the Y_ℓ s is even and that $\phi_Y(t)$ goes to 0 as t goes to infinity. Then, given ε, the waveform $X:\mathbb{Z}\to\mathbb{C}$ given by

$$X[n]=e^{\frac{2\pi}{\varepsilon}i\sum_{\ell=-n}^{n}Y_\ell}$$

has almost perfect autocorrelation.

Proof. Since the density function of each Y_ℓ is even this means that the characteristic function is real valued [19]. Following the calculation in the proof of Theorem 2.1, the expected autocorrelation of X for $k\neq 0$ is

$$E(A_X[k])=\left[\phi_Y\left(\frac{2\pi}{\varepsilon}\right)\right]^{2|k|}$$

and this goes to zero with ε by the hypothesis. □

Example 2.3. Suppose the Y_ℓ s follow a bilateral distribution that has density $e^{-|x|}$ with $x\in(-\infty,\infty)$ and characteristic function $\phi_Y(t)=\frac{1}{1+t^2}$. Then for $k\neq 0$,

$$E(A_X[k])=\left[\frac{1}{1+\left(\frac{2\pi}{\varepsilon}\right)^2}\right]^{2|k|}$$

and this can be made arbitrarily small with ε.

In the same way as was done in the Gaussian case, for $k>0$,

$$E\left(|A_X[k]|^2\right)$$

$$=\lim_{N\to\infty}\lim_{M\to\infty}\frac{1}{(2N+1)}\frac{1}{(2M+1)}\sum_{n=-N}^{N}\sum_{m=-M}^{M}\left[\phi_{Y_1}\left(\pm\frac{2\pi}{\varepsilon}\right)\right]^{4(k-\bar{k})}$$

$$\leq 1$$

and

$$E\left(\left|A_X[k]\right|^2\right) \geq \left[\frac{1}{1+\left(\frac{2\pi}{\varepsilon}\right)^2}\right]^{4k}.$$

Thus

$$0 \leq V\left(A_X[k]\right) \leq 1-\left(\frac{1}{1+\left(\frac{2\pi}{\varepsilon}\right)^2}\right)^{4|k|}.$$

Example 2.4. Suppose that the Y_ℓ s follow the Cauchy distribution with density function $\frac{1}{\pi\left(1+x^2\right)}$. Note thatdisregarding the constant π, this is the characteristic function of the random variable considered in Example 2.3. The characteristic function of the Y_ℓ s is now $e^{-|t|}$, the same as the distribution function in Example 2.3. For $k \neq 0$,

$$E\left(A_X[k]\right) = \left[\phi_{Y_1}\left(\frac{2\pi}{\varepsilon}\right)\right]^{2|k|} = e^{-\frac{4\pi|k|}{\varepsilon}}$$

which can be made arbitrarily small with ε. Also,

$$0 \leq V\left(A_X[k]\right) \leq 1 - e^{-\frac{8\pi|k|}{\varepsilon}}.$$

2.3. Higher Dimensional Case

Here one is interested in constructing waveforms $v:\mathbb{Z} \to \mathbb{C}^d$, $d \geq 2$. It is desired that v has unit norm and the expectation of its autocorrelation can be made arbitrarily small. One way to construct v is based on the construction of the one dimensional example given in Section 2.1. This is motivated by the higher dimensional construction in the deterministic case [2]. As before, $\{Y_\ell\}_{\ell\in\mathbb{Z}}$ is a sequence of i.i.d. Gaussian random variables with mean zero and variance σ^2. Next, one defines $X[n]=e^{\frac{2\pi}{\varepsilon}i\sum_{\ell=-n}^{n}Y_\ell}$. The waveform $v:\mathbb{Z}\to\mathbb{C}^d$ is then defined as

$$\forall m\in\mathbb{Z},\ v[m]=\frac{1}{\sqrt{d}}\begin{bmatrix}X[m]\\X[m+1]\\\vdots\\X[m+d-1]\end{bmatrix}. \tag{5}$$

In this case, the autocorrelation is given by

$$A_v[k]=\lim_{N\to\infty}\frac{1}{2N+1}\sum_{n=-N}^{N}\langle v[n+k],v[n]\rangle \tag{6}$$

where $\langle\cdot,\cdot\rangle$ is the usual inner product in $\mathbb{C}^d$. The length or norm of any $v[m]$ is thus given by

$$\|v[m]\|^2=\langle v[m],v[m]\rangle.$$

From (5),

$$\|v[m]\|^2=\frac{1}{d}\sum_{n=0}^{d-1}X[m+n]\overline{X[m+n]}=\frac{d}{d}=1.$$

Thus the $v[m]$s are unit-normed. The following Theorem 2.5 shows that the expected autocorrelation of v can be made arbitrarily small everywhere except at the origin.

Theorem 2.5. Given $\varepsilon>0,$ the waveform $v:\mathbb{Z}\to\mathbb{C}^d$ defined in (5) has autocorrelation A_v such that

$$E\left(A_v[k]\right)=\begin{cases}1 & \text{if } \ k=0\\ e^{-|k|\sigma^2\left(\frac{2\pi}{\varepsilon}\right)^2} & \text{if } \ k\neq 0.\end{cases}$$

Proof. As defined in (6),

$$A_v[k]=\lim_{N\to\infty}\frac{1}{2N+1}\sum_{n=-N}^{N}\left\langle v[n+k],v[n]\right\rangle$$

When $k=0$,

$$A_v[0]=\lim_{N\to\infty}\frac{1}{2N+1}\sum_{n=-N}^{N}\left\|v[n]\right\|^2=1.$$

Thus,

$$E\left(A_v[0]\right)=1.$$

For $k\neq 0$, due to (5),

$$\left\langle v[n+k],v[n]\right\rangle=\frac{1}{d}\left\langle\begin{bmatrix}X[n+k]\\ X[n+k+1]\\ \vdots\\ X[n+k+d-1]\end{bmatrix},\begin{bmatrix}X[n]\\ X[n+1]\\ \vdots\\ X[n+d-1]\end{bmatrix}\right\rangle$$
$$=\frac{1}{d}\left(X[n+k]\overline{X[n]}+X[n+k+1]\overline{X[n+1]}+\cdots+X[n+k+d-1]\overline{X[n+d-1]}\right).$$

Consider $k>0$.

$$E\left(A_v[k]\right)=\lim_{N\to\infty}\frac{1}{2N+1}\sum_{n=-N}^{N}E\left(\left\langle v[n+k],v[n]\right\rangle\right)$$
$$=\lim_{N\to\infty}\frac{1}{2N+1}\sum_{n=-N}^{N}E\left(\frac{1}{d}\sum_{m=0}^{d-1}X[n+k+m]\overline{X[n+m]}\right)$$
$$=\lim_{N\to\infty}\frac{1}{2N+1}\sum_{n=-N}^{N}\frac{1}{d}\sum_{m=0}^{d-1}E\left(X[n+k+m]\overline{X[n+m]}\right)$$
$$=\lim_{N\to\infty}\frac{1}{2N+1}\sum_{n=-N}^{N}\frac{1}{d}\sum_{m=0}^{d-1}E\left(e^{\frac{2\pi}{\varepsilon}i\left(Y_{-(n+m+k)}+\cdots+Y_{-(n+m+1)}+Y_{n+m+1}+\cdots+Y_{n+m+k}\right)}\right)$$
$$=\lim_{N\to\infty}\frac{1}{2N+1}\sum_{n=-N}^{N}\frac{1}{d}\sum_{m=0}^{d-1}E\left(e^{\frac{2\pi}{\varepsilon}iY_1}\right)^{2k}=e^{-\sigma^2k\left(\frac{2\pi}{\varepsilon}\right)^2}.$$

Similarly, for $k < 0$, one gets

$$E(A_v[k]) = e^{\sigma^2 k\left(\frac{2\pi}{\varepsilon}\right)^2}.$$

□

Thus the waveform v as defined in this section is unit-normed and has autocorrelation that can be made arbitrarily small.

Remark 2.6. As in the one dimensional construction, it is easy to see that here too the construction can be done with random variables other than the Gaussian. In fact, all random variables that can be used in the one dimensional case, i.e., ones satisfying the properties of Theorem 2.2, can also be used for the higher dimensional construction.

2.4. Remark on the Periodic Case

It can be shown that the periodic case follows the same nature as the aperiodic case. The sequence $X : \mathbb{Z}_n \to \mathbb{C}$ is defined in the same way as in Section 2.1, i.e.,

$$\forall m \in \{0, 1, \cdots, n-1\},\ X[m] = e^{\frac{2\pi}{\varepsilon} i \sum_{\ell=-m}^{m} Y_\ell}$$

where $Y_\ell \sim N(0, \sigma^2)$. Following the definition given in (2), when $k = 0,$

$$A_X[0] = \frac{1}{n} \sum_{m=0}^{n-1} X[m]\overline{X[m]} = 1.$$

When $k \neq 0,$ the expectation of the autocorrelation is

$$E(A_X[k]) = \frac{1}{n} \sum_{m=0}^{n-1} E\left(X[m+k]\overline{X[m]}\right).$$

For $k > 0,$

$$E\left(A_X[k]\right)=\frac{1}{n}\sum_{m=0}^{n-1}E\left(\mathrm{e}^{\frac{2\pi i}{\varepsilon}\sum_{\ell=-m-k}^{m+k}Y_\ell}\,\mathrm{e}^{-\frac{2\pi i}{\varepsilon}\sum_{j=-m}^{m}Y_j}\right)$$

$$=\frac{1}{n}\sum_{m=0}^{n-1}E\left(\mathrm{e}^{\frac{2\pi i}{\varepsilon}\left(\sum_{\ell=m+1}^{m+k}Y_\ell+\sum_{j=-m-k}^{-m-1}Y_j\right)}\right)=\frac{1}{n}\sum_{m=0}^{n-1}\left[E\left(\mathrm{e}^{\frac{2\pi i}{\varepsilon}Y_1}\right)\right]^{2k}$$

$$=\left[E\left(\mathrm{e}^{\frac{2\pi i}{\varepsilon}Y_1}\right)\right]^{2k}=\left[\phi_{Y_1}\left(\frac{2\pi}{\varepsilon}\right)\right]^{2k}=\mathrm{e}^{-k\sigma^2\left(\frac{2\pi}{\varepsilon}\right)^2}$$

where one uses the fact that the Y_ℓ s are i.i.d.. A similar calculation for negative values of k suggests that the autocorrelation can be made arbitrarily small, depending on ε, for all non-zero values of k. Also, as in the aperiodic case, this result can be obtained for random variables other than the Gaussian.

3. CONSTRUCTION OF CONTINUOUS STOCHASTIC WAVEFORMS

In this section continuous waveforms with almost perfect autocorrelation are constructed from a one dimensional Brownian motion.

For a continuous waveform $x:\mathbb{R}\to\mathbb{C}$, the autocorrelation $A_x:\mathbb{R}\to\mathbb{C}$ can be defined as

$$A_x(s)=\lim_{T\to\infty}\frac{1}{2T}\int_{-T}^{T}x(t+s)\overline{x(t)}\,\mathrm{d}t. \tag{7}$$

Let $\{W(t);t>0\}$ be a one dimensional Brownian motion. Then $W(t)$ satisfies

- $W(0)=0$
- $W(t+s)-W(s)\sim N\left(0,\sigma^2 t\right)$
- $W(t_{i+1})-W(t_i)\,0<t_1<\cdots<t_k$, are independent random variables.

Theorem 3.1. Let $W(t)$ be the one dimensional Brownian motion and $\varepsilon > 0$ be given. Define $x:\mathbb{R} \to \mathbb{C}$ by

$$x(t) = \mathrm{e}^{\frac{2\pi}{\varepsilon} i W(t)} \text{ for } t \geq 0,$$

and $x(-t) = x(t)$. Then the autocorrelation of x, A_x, satisfies

$$E(A_x(s)) = \begin{cases} 1 & \text{if } s = 0 \\ \mathrm{e}^{-\frac{\sigma^2}{2}|s|\left(\frac{2\pi}{\varepsilon}\right)^2} & \text{if } s \neq 0. \end{cases}$$

Proof. We would like to evaluate

$$E(A_x(s)) = E\left(\lim_{T\to\infty} \frac{1}{2T} \int_{-T}^{T} x(t+s)\overline{x(t)}\,\mathrm{d}t \right).$$

Let $s > 0$ and let $g_T(s) = \frac{1}{2T} \int_{-T}^{T} x(t+s)\overline{x(t)}\,\mathrm{d}t.$

$$\begin{aligned} E(g_T) &= \frac{1}{2T} \int_{-T}^{T} E\left(x(t+s)\overline{x(t)} \right) \mathrm{d}t \\ &= \frac{1}{2T} \int_{-T}^{T} E\left(\mathrm{e}^{\frac{2\pi}{\varepsilon} i (W(t+s) - W(t))} \right) \mathrm{d}t \\ &= \frac{1}{2T} \int_{-T}^{T} \phi_{W(t+s)-W(t)}\left(\frac{2\pi}{\varepsilon} \right) = \mathrm{e}^{-\frac{\sigma^2}{2} s \left(\frac{2\pi}{\varepsilon}\right)^2} < \infty. \end{aligned}$$

Thus each g_T is integrable and further $|g_T| \leq 1$. Let $h(t) = 1$; $t \in \mathbb{R}$. Then $E(h) = 1$. Therefore, by the Dominated Convergence Theorem, and properties of Brownian motion and characteristic functions, one gets

$$E\left(A_x(s)\right)=E\left(\lim_{T\to\infty}\frac{1}{2T}\int_{-T}^{T}x(t+s)\overline{x(t)}\,\mathrm{d}t\right)$$
$$=\lim_{T\to\infty}\frac{1}{2T}\int_{-T}^{T}E\left(\mathrm{e}^{\frac{2\pi}{\varepsilon}i(W(t+s)-W(t))}\right)\mathrm{d}t$$
$$=\lim_{T\to\infty}\frac{1}{2T}\int_{-T}^{T}\phi_{W(t+s)-W(t)}\left(\frac{2\pi}{\varepsilon}\right)=\mathrm{e}^{-\frac{\sigma^2}{2}s\left(\frac{2\pi}{\varepsilon}\right)^2}$$

which can be made arbitrarily small based on ε. Similarly,

$$E\left(A(-s)\right)=\lim_{T\to\infty}\frac{1}{2T}\int_{-T}^{T}\phi_{W(t)-W(t-s)}\left(-\frac{2\pi}{\varepsilon}\right)$$
$$=\mathrm{e}^{-\frac{\sigma^2}{2}s\left(-\frac{2\pi}{\varepsilon}\right)^2}=\mathrm{e}^{-\frac{\sigma^2}{2}s\left(\frac{2\pi}{\varepsilon}\right)^2}. \quad \square$$

4. CONNECTION TO FRAMES

Consider the mapping $v:\mathbb{Z}\to\mathbb{C}^d$ given by

$$v(k)=\frac{1}{\sqrt{d}}\begin{pmatrix}X[k]\\X[k+1]\\\vdots\\X[k+d-1]\end{pmatrix} \tag{8}$$

where $X[k]=\mathrm{e}^{\frac{2\pi}{\varepsilon}i\sum_{\ell=-k}^{k}Y_\ell}$, as defined in Section 2.1.

Let $M\ge d$ and consider the set $V=\{v(1),v(2),\cdots,v(M)\}$ of M unit vectors in $\mathbb{C}^d$. The matrix

$$F = \frac{1}{\sqrt{d}} \begin{bmatrix} \overline{X[1]} & \overline{X[2]} & \cdots & \overline{X[d]} \\ \overline{X[2]} & \overline{X[3]} & \cdots & \overline{X[d+1]} \\ \vdots & \vdots & \cdots & \vdots \\ \overline{X[M]} & \overline{X[M+1]} & \cdots & \overline{X[M+d-1]} \end{bmatrix}$$

is the matrix of the analysis operator corresponding to V. The frame operator of V is $\mathcal{F} = F^* F$, i.e.,

$$\mathcal{F} = \frac{1}{d} \begin{bmatrix} X[1] & X[2] & \cdots & X[M] \\ X[2] & X[3] & \cdots & X[M+1] \\ \vdots & \vdots & \cdots & \vdots \\ X[d] & X[d+1] & \cdots & X[M+d-1] \end{bmatrix} \begin{bmatrix} \overline{X[1]} & \overline{X[2]} & \cdots & \overline{X[d]} \\ \overline{X[2]} & \overline{X[3]} & \cdots & \overline{X[d+1]} \\ \vdots & \vdots & \cdots & \vdots \\ \overline{X[M]} & \overline{X[M+1]} & \cdots & \overline{X[M+d-1]} \end{bmatrix}$$

.

The entries of $\mathcal{F}$ are given by $\mathcal{F}_{m,m} = \frac{M}{d}$ and for $m \neq n, m > n,$

$$\mathcal{F}_{m,n} = \frac{1}{d}\left(X[m] \quad X[m+1] \quad \cdots \quad X[M+m-1]\right) \begin{pmatrix} \overline{X[n]} \\ \overline{X[n+1]} \\ \vdots \\ \overline{X[M+n-1]} \end{pmatrix}$$

$$= \frac{M}{d}\frac{1}{M}\sum_{\ell=0}^{M-1} X[m+\ell]\overline{X[n+\ell]} = \frac{M}{d}\left(\frac{1}{M}\sum_{\ell=0}^{M-1} \mathrm{e}^{\frac{2\pi}{\varepsilon} i(Y_{-m-\ell}+\cdots+Y_{-n-\ell-1}+Y_{n+\ell+1}+\cdots+Y_{m+\ell})}\right).$$

Note that since $\mathcal{F}$ is self-adjoint, $\mathcal{F}_{m,n} = \bar{\mathcal{F}}_{n,m}$. It is desired that V emulates a tight frame, i.e, $\mathcal{F}$ is close to a constant times the identity, in this case, $\frac{M}{d}$ times the identity. Alternatively, it is desirable that the eigenvalues of $\mathcal{F}$ are all close to each other and close to $\frac{M}{d}$. In this case, due to the stochastic nature of the frame operator, one studies the expectation of the eigenvalues of $\mathcal{F}$.

4.1. Frames in $\mathbb{C}^2$

This section discusses the construction of sets of vectors in $\mathbb{C}^2$ as given by (8). The frame properties of such sets are analyzed. In fact, it is shown that the expectation of the eigenvalues of the frame operator are close to each other, the closeness increasing with the size of the set. The bounds on the probability of deviation of the eigenvalues from the expected value is also derived. The related inequalities arise from an application of Theorem 4.1 [22] below.

Theorem 4.1. (Azuma's Inequality) Suppose that $\{X_k : k = 0,1,2,\cdots\}$ is a martingale and $|X_k - X_{k-1}| < c_k$, almost surely. Then for all positive integers N and all positive reals t,

$$P\left(|X_N - X_0| \geq t\right) \leq 2\mathrm{e}^{\left(-t^2 \Big/ \left(2\sum_{k=1}^{N} c_k^2\right)\right)}.$$

Consider $M \geq 3$ vectors in $\mathbb{C}^2$ i.e., $d = 2$ in (8). Then $v : \mathbb{Z} \to \mathbb{C}^2$ and

$$v(1) = \frac{1}{\sqrt{2}}\begin{pmatrix} X[1] \\ X[2] \end{pmatrix}, v(2) = \frac{1}{\sqrt{2}}\begin{pmatrix} X[2] \\ X[3] \end{pmatrix}, \cdots, v(M) = \frac{1}{\sqrt{2}}\begin{pmatrix} X[M] \\ X[M+1] \end{pmatrix}. \quad (9)$$

Considering the set $V = \{v(1), v(2), \cdots, v(M)\}$, the frame operator of V is

$$\mathcal{F} = \frac{1}{2}\begin{bmatrix} X[1] & X[2] & X[3] & \cdots & X[M] \\ X[2] & X[3] & X[4] & \cdots & X[M+1] \end{bmatrix}\begin{bmatrix} \overline{X[1]} & \overline{X[2]} \\ \overline{X[2]} & \overline{X[3]} \\ \overline{X[3]} & \overline{X[4]} \\ \vdots & \vdots \\ \overline{X[M]} & \overline{X[M+1]} \end{bmatrix}$$

$$\text{or, } \mathcal{F} = \frac{M}{2}\begin{bmatrix} 1 & \frac{1}{M}\sum_{m=1}^{M} X[m]\overline{X[m+1]} \\ \frac{1}{M}\sum_{m=1}^{M} \overline{X[m]}X[m+1] & 1 \end{bmatrix}. \quad (10)$$

Theorem 4.2. 1) Consider the set $V=\{v(1),v(2),\cdots,v(M)\}\in\mathbb{C}^2$, $M\geq 3$, where the vectors $v(n)$ are given by (9). The minimum eigenvalue, $\lambda_{\min}(\mathcal{F})$, and the maximum eigenvalue, $\lambda_{\max}(\mathcal{F})$, of the frame operator of V satisfy

$$\frac{M}{2}(1-\delta)\leq E(\lambda_{\min}(\mathcal{F}))\leq E(\lambda_{\max}(\mathcal{F}))\leq\frac{M}{2}(1+\delta) \quad (11)$$

where $\delta=\sqrt{\frac{1}{M}+\frac{M-1}{M}\mathrm{e}^{-2\sigma^2\left(\frac{2\pi}{\varepsilon}\right)^2}}$.

2) The deviation of the minimum and maximum eigenvalue of $\mathcal{F}$ from their expected value is given, for all positive reals r, by

$$P\left(\left|\lambda_{\min}(\mathcal{F})-E(\lambda_{\min}(\mathcal{F}))\right|>r\right)\leq 2\mathrm{e}^{-\frac{4r^2}{8M^3}},$$

$$P\left(\left|\lambda_{\max}(\mathcal{F})-E(\lambda_{\max}(\mathcal{F}))\right|>r\right)\leq 2\mathrm{e}^{-\frac{4r^2}{8M^3}}.$$

Proof. 1) The frame operator of

$V=\{v(1),v(2),\cdots,v(M)\}$ is given in (10). The eigenvalues of $\frac{2}{M}\mathcal{F}$ are $\lambda_1=1-|\alpha|$ and $\lambda_2=1+|\alpha|$ where $\alpha=\frac{1}{M}\sum_{m=1}^{M}X[m]\overline{X[m+1]}$.

Let

$$\gamma_1=X[1]\overline{X[2]}=\mathrm{e}^{\frac{2\pi}{\varepsilon}i(Y_{-2}+Y_2)},$$

$$\gamma_2=X[2]\overline{X[3]}=\mathrm{e}^{\frac{2\pi}{\varepsilon}i(Y_{-3}+Y_3)},$$

$$\vdots$$

$$\gamma_M=X[M]\overline{X[M+1]}=\mathrm{e}^{\frac{2\pi}{\varepsilon}i\left(Y_{-(M+1)}+Y_{(M+1)}\right)},$$

so that

$$\alpha = \frac{\gamma_1 + \gamma_2 + \cdots + \gamma_M}{M}.$$

Note that for $m \neq n,\ \gamma_m$ and γ_n are independent and so $E\left(\gamma_m \overline{\gamma_n}\right) = E\left(\gamma_m\right) E\left(\overline{\gamma_n}\right)$. Also, since the Y_ℓ s are i.i.d. and the characteristic function of the Y_ℓ s is symmetric,

$\forall\, 1 \leq m \leq M,$

$$E\left(\gamma_m\right) = E\left(\mathrm{e}^{\frac{2\pi}{\varepsilon} i\left(Y_{-(m+1)} + Y_{(m+1)}\right)}\right) = \left\{E\left(\mathrm{e}^{\frac{2\pi}{\varepsilon} iY_1}\right)\right\}^2$$

$$= \mathrm{e}^{-\sigma^2\left(\frac{2\pi}{\varepsilon}\right)^2} = E\left(\overline{\gamma_m}\right)$$

and therefore

$$E\left(\gamma_m \overline{\gamma_n}\right) = \mathrm{e}^{-2\sigma^2\left(\frac{2\pi}{\varepsilon}\right)^2}.$$

Thus

$$E\left(|\alpha|^2\right) = E\left(\alpha\overline{\alpha}\right)$$

$$= \frac{1}{M^2} E\left(\left(\gamma_1 + \gamma_2 + \cdots + \gamma_M\right)\left(\overline{\gamma_1} + \overline{\gamma_2} + \cdots + \overline{\gamma_M}\right)\right)$$

$$= \frac{1}{M^2} E\left(|\gamma_1|^2 + |\gamma_2|^2 + \cdots + |\gamma_M|^2 + \sum_{m \neq n} \gamma_m \overline{\gamma_n}\right)$$

$$= \frac{1}{M} + \frac{1}{M^2} E\left(\sum_{m \neq n} \gamma_m \overline{\gamma_n}\right) = \frac{1}{M} + \frac{1}{M^2} \sum_{m \neq n} E\left(\gamma_m \overline{\gamma_n}\right)$$

$$= \frac{1}{M} + \frac{M-1}{M} \mathrm{e}^{-2\sigma^2\left(\frac{2\pi}{\varepsilon}\right)^2}.$$

The above estimate on $E\left(|\alpha|^2\right)$ implies that

$$E\left(|\alpha|\right) \le \sqrt{E\left(|\alpha|^2\right)} = \sqrt{\frac{1}{M} + \frac{M-1}{M} e^{-2\sigma^2\left(\frac{2\pi}{\varepsilon}\right)^2}}. \tag{12}$$

Since $E(\lambda_1) = 1 - E\left(|\alpha|\right)$ and $E(\lambda_2) = 1 + E\left(|\alpha|\right)$, (12) implies

$$1 - \sqrt{\frac{1}{M} + \frac{M-1}{M} e^{-2\sigma^2\left(\frac{2\pi}{\varepsilon}\right)^2}} \le E(\lambda_1) \le E(\lambda_2) \le 1 + \sqrt{\frac{1}{M} + \frac{M-1}{M} e^{-2\sigma^2\left(\frac{2\pi}{\varepsilon}\right)^2}}.$$

Noting that $\lambda_{\min}(\mathcal{F}) = \frac{M}{2}\lambda_1$ and $\lambda_{\max}(\mathcal{F}) = \frac{M}{2}\lambda_2$, one finally gets, after setting $\delta = \sqrt{\frac{1}{M} + \frac{M-1}{M} e^{-2\sigma^2\left(\frac{2\pi}{\varepsilon}\right)^2}}$,

$$\frac{M}{2}(1-\delta) \le E\left(\lambda_{\min}(\mathcal{F})\right) \le E\left(\lambda_{\max}(\mathcal{F})\right) \le \frac{M}{2}(1+\delta).$$

2) To prove 2) we use the Doob martingale and Azuma's inequality [22]. For $n = 2,\cdots,M+1$, let $Z_{n-1} = Y_{-n} + Y_n$. Here the Doob martingale is the sequence $\{U_0, U_1, \cdots, U_{M-1}\}$ where

$$\text{for } k = 1,\cdots,M-1,$$

$$U_k = E\left(\frac{1}{M}\left|\sum_{j=1}^{M} e^{\frac{2\pi}{\varepsilon} i Z_j}\right| \middle| Z_1, Z_2, \cdots, Z_k\right)$$

and

$$U_0 = E\left(\frac{1}{M}\left|\sum_{j=1}^{M} e^{\frac{2\pi}{\varepsilon} i Z_j}\right|\right).$$

Note that $U_0 = E\left(|\alpha|\right)$ and $U_{M-1} = |\alpha|$. Also,

$$\left|U_k - U_{k-1}\right| \le \left|U_k\right| + \left|U_{k-1}\right| \le 2.$$

So by Azuma's Inequality (see Theorem 4.1)

$$P\left(\left|U_{M-1}-U_0\right|\geq r\right)=P\left(\left|\left|\alpha\right|-E\left(\left|\alpha\right|\right)\right|\geq r\right)$$

$$\leq 2\mathrm{e}^{-r^2\Big/\left(2\sum\limits_{k=1}^{M}2^2\right)}=2\mathrm{e}^{-r^2/8M}.$$

Since $\left|\lambda_1-E(\lambda_1)\right|=\left|\lambda_2-E(\lambda_2)\right|=\left|\left|\alpha\right|-E\left(\left|\alpha\right|\right)\right|,$ this means

$$P\left(\left|\lambda_1-E(\lambda_1)\right|>r\right)\leq 2\mathrm{e}^{-r^2/8M}$$

and

$$P\left(\left|\lambda_2-E(\lambda_2)\right|>r\right)\leq 2\mathrm{e}^{-r^2/8M}.$$

Going back to the actual frame operator $\mathcal{F}$, whose eigenvalues are $\frac{M}{2}\lambda_1$ and $\frac{M}{2}\lambda_2,$ the following estimates hold.

$$P\left(\left|\lambda_{\max}(\mathcal{F})-E\left(\lambda_{\max}(\mathcal{F})\right)\right|>r\right)$$

$$=P\left(\frac{M}{2}\left|\left|\alpha\right|-E\left(\left|\alpha\right|\right)\right|>r\right)$$

$$=P\left(\left|\left|\alpha\right|-E\left(\left|\alpha\right|\right)\right|>\frac{2}{M}r\right)\leq 2\mathrm{e}^{-\frac{4r^2}{8M^3}}$$

and

$$P\left(\left|\lambda_{\min}(\mathcal{F})-E\left(\lambda_{\min}(\mathcal{F})\right)\right|>r\right)$$

$$=P\left(\frac{M}{2}\left|\left|\alpha\right|-E\left(\left|\alpha\right|\right)\right|>r\right)$$

$$=P\left(\left|\left|\alpha\right|-E\left(\left|\alpha\right|\right)\right|>\frac{2}{M}r\right)\leq 2\mathrm{e}^{-\frac{4r^2}{8M^3}}.\ \square$$

Corollary 4.3. The eigenvalues of the frame operator considered in Theorem 4.2 satisfy, for all positive reals r,

$$P\left(\lambda_{\min}(\mathcal{F}) < \frac{M}{2}(1-\delta) - r\right) \leq e^{-4r^2/8M^3},$$

$$P\left(\lambda_{\max}(\mathcal{F}) > \frac{M}{2}(1+\delta) + r\right) \leq e^{-4r^2/8M^3},$$

where $\delta = \sqrt{\frac{1}{M} + \frac{M-1}{M} e^{-2\sigma^2\left(\frac{2\pi}{\varepsilon}\right)^2}}$.

Proof. Due to part 1) of Theorem 4.2

$$\lambda_{\min}(\mathcal{F}) < \frac{M}{2}(1-\delta) - r \Rightarrow \lambda_{\min}(\mathcal{F}) < E(\lambda_{\min}(\mathcal{F})) - r.$$

This implies, as a consequence of part 2) of Theorem 4.2, that

$$P\left(\lambda_{\min}(\mathcal{F}) < \frac{M}{2}(1-\delta) - r\right) \leq P(\lambda_{\min}(\mathcal{F}) < E(\lambda_{\min}(\mathcal{F})) - r) \leq e^{-4r^2/8M^3}.$$

In a similar way, from part 1) of Theorem 4.2,

$$\lambda_{\max}(\mathcal{F}) > \frac{M}{2}(1+\delta) + r \Rightarrow \lambda_{\max}(\mathcal{F}) > E(\lambda_{\max}(\mathcal{F})) + r$$

which implies, as a consequence of part 2) of Theorem 4.2, that

$$P\left(\lambda_{\max}(\mathcal{F}) > \frac{M}{2}(1+\delta) + r\right) \leq P(\lambda_{\max}(\mathcal{F}) > E(\lambda_{\max}(\mathcal{F})) + r) \leq e^{-4r^2/8M^3}. \quad \square$$

Remark 4.4. In Theorem 4.2, as M tends to infinity, the value of δ in (11) can be made arbitrarily small based on the choice of ε. This in turn implies that the two eigenvalues can be made arbitrarily close to each other, with ε. On the other hand, for a fixed M, as ε tends to zero, (11) becomes

$$\frac{M}{2}\left(1-\sqrt{\frac{1}{M}}\right) \le E\left(\lambda_{\min}(\mathcal{F})\right)$$
$$\le E\left(\lambda_{\max}(\mathcal{F})\right) \le \frac{M}{2}\left(1+\sqrt{\frac{1}{M}}\right).$$

4.2. Frames in $\mathbb{C}^d$; $d > 2$

For general d and M, in order to use existing results on the concentration of eigenvalues of random matrices [23, 24], a slightly different construction of the frame needs to be considered. Let $\{Y_{mn}\}_{m,n\in\mathbb{Z}}$ be i.i.d. random variables following a Gaussian distribution with mean zero and variance σ^2. It can be shown that

$$E\left(e^{\frac{2\pi}{\varepsilon}iY_{mn}}\right) = e^{-\frac{\sigma^2}{2}\left(\frac{2\pi}{\varepsilon}\right)^2}$$

and the variance

$$V\left(e^{\frac{2\pi}{\varepsilon}iY_{mn}}\right) = 1 - e^{-\sigma^2\left(\frac{2\pi}{\varepsilon}\right)^2}.$$

One can define the following two dimensional sequence. For $m, n \in \mathbb{Z}$,

$$X_{mn} = e^{\frac{2\pi}{\varepsilon}iY_{mn}} - e^{-\frac{\sigma^2}{2}\left(\frac{2\pi}{\varepsilon}\right)^2}.$$

Consider the mapping $v : \mathbb{Z} \to \mathbb{C}^d$ given by

$$v(\ell) = \frac{1}{\sqrt{d}}\begin{pmatrix} X_{1\ell} \\ X_{2\ell} \\ \vdots \\ X_{d\ell} \end{pmatrix}. \tag{13}$$

As before, let $M \geq d$ and consider the set of M unit vectors $V = \{v(1), v(2), \cdots, v(M)\}$ in $\mathbb{C}^d$. The frame operator of this set is

$$\mathcal{F} = \frac{1}{d}\begin{bmatrix} X_{11} & X_{12} & \cdots & X_{1M} \\ X_{21} & X_{22} & \cdots & X_{2M} \\ \vdots & \vdots & \cdots & \vdots \\ X_{d1} & X_{d2} & \cdots & X_{dM} \end{bmatrix}\begin{bmatrix} \overline{X}_{11} & \overline{X}_{21} & \cdots & \overline{X}_{d1} \\ \overline{X}_{12} & \overline{X}_{22} & \cdots & \overline{X}_{d2} \\ \vdots & \vdots & \cdots & \vdots \\ \overline{X}_{1M} & \overline{X}_{2M} & \cdots & \overline{X}_{dM} \end{bmatrix}.$$

Let

$$A = \frac{1}{\sqrt{d}}\begin{bmatrix} X_{11} & X_{12} & \cdots & X_{1M} \\ X_{21} & X_{22} & \cdots & X_{2M} \\ \vdots & \vdots & \cdots & \vdots \\ X_{d1} & X_{d2} & \cdots & X_{dM} \end{bmatrix} \tag{14}$$

so that $\mathcal{F} = AA^*$. The matrix A has entries with mean zero and variance $\hat{\sigma}^2 = \frac{1}{d}\left(1 - e^{-\sigma^2\left(\frac{2\pi}{\varepsilon}\right)^2}\right)$. According to results in [23], if $\frac{d}{M} \to c$ as $d,\ M \to \infty$, then the smallest and largest eigenvalues of $\mathcal{F}$ converge almost surely to $\hat{\sigma}^2\left(1 - \sqrt{c}\right)^2$ and $\hat{\sigma}^2\left(1 + \sqrt{c}\right)^2$, respectively.

Theorem 4.5. Let $s_1(A) \leq s_2(A) \leq \cdots \leq s_d(A)$ be the singular values of the matrix A given by (14). Then the following hold.

1) Given ε_0, there is a large enough d such that

$$P\left(s_d(A) \geq \hat{\sigma}\left(1 + \sqrt{\frac{d}{M}}\right) + \varepsilon_0 + r\right) \leq 2e^{-r^2 d/16}. \tag{15}$$

2)

$$P\left(s_1(A) \leq c_1\right) \leq e^{-c_2 M}, \tag{16}$$

where c_1 and c_2 are universal positive constants.

Proof. Let s_d be the mapping that associates to a matrix A it largest singular value. Equip $\mathbb{C}^{dM}$ with the Frobenius norm

$$\|A\|^2 := Tr\left(AA^*\right) = \sum_{m,n} |A_{mn}|^2 .$$

Then the mapping s_d is convex and 1-Lipschitz in the sense that

$$\left|s_d\left(A\right) - s_d\left(A'\right)\right| \le \|A - A'\|$$

for all pairs $\left(A, A'\right)$ of d by M matrices [24].

We think of A as a random vector in $\mathbb{R}^{2dM}$. The real and imaginary parts of the entries of $\frac{1}{\sqrt{d}}A$ are supported in $\left[-\frac{1}{\sqrt{d}}, \frac{1}{\sqrt{d}}\right]$. Let P be a product measure on $\left[-\frac{1}{\sqrt{d}}, \frac{1}{\sqrt{d}}\right]^{2dM}$. Then as a consequence of the concentration inequality (Corollary 4.10, [24]) we have

$$P\left(\left|s_d\left(A\right) - m\left(s_d\right)\right| \ge r\right) \le 4\mathrm{e}^{-r^2 d/16}$$

where $m\left(s_d\right)$ is the median of $s_d\left(A\right)$. It is known that the minimum and maximum singular values of A converge almost surely to $\hat{\sigma}\left(1-\sqrt{c}\right)$ and $\hat{\sigma}\left(1+\sqrt{c}\right)$, respectively, as d, M tend to infinity and $\frac{d}{M} \to c$. As a consequence, for each ε_0 and M sufficiently large, one can show that the medians belong to the fixed interval

$$\left[\hat{\sigma}\left(1-\sqrt{\frac{d}{M}}\right)-\varepsilon_0, \hat{\sigma}\left(1+\sqrt{\frac{d}{M}}\right)+\varepsilon_0\right]$$ which gives

$$P\left(s_d(A) \geq \hat{\sigma}\left(1+\sqrt{\frac{d}{M}}\right)+\varepsilon_0+r\right) \leq 2e^{-r^2 d/16}.$$

For the smallest singular value we cannot use the concentration inequality as used for s_d since the smallest singular value is not convex. However, following results in [25] (Theorem 3.1) that have been used in [26] in a similar situation as here, one can say that whenever $M > (1+\delta)d$, where δ is greater than a small constant,

$P(s_1(A) \leq c_1) \leq e^{-c_2 M}$ where c_1 and c_2 are positive universal constants.

□

Remark 4.6. Note that the square of the singular values of A are the eigenvalues of $\mathcal{F}$ and so the estimates given in (15) and (16) give insight into the corresponding deviation of the eigenvalues of the frame operator $\mathcal{F}$.

Remark 4.7. (Connection to compressed sensing) The theory of compressed sensing [27-29] states that it is possible to recover a sparse signal from a small number of measurements. A signal $x \in \mathbb{C}^M$ is k-sparse in a basis $\Psi = \{\psi_j\}_{j=1}^M$ if x is a weighted superposition of at most k elements of Ψ. Compressed sensing broadly refers to the inverse problem of reconstructing such a signal x from linear measurements $\{y_\ell = \langle x, \phi_\ell \rangle | \ell = 1, \cdots, d\}$ with $d < M$, ideally with $d \ll M$. In the general setting, one has $\Phi^* x = y$, where Φ is a $M \times d$ sensing matrix having the measurement vectors ϕ_ℓ as its columns, x is a length-M signal and y is a length-d measurement.

The standard compressed sensing technique guarantees exact recovery of the original signal with very high probability if the sensing matrix satisfies the Restricted Isometry Property (RIP). This means that for a fixed k, there exists a small number δ_k, such that

$$\left(1-\delta_k\right)\|x\|_{\ell_2}^2 \leq \left\|\Phi^* x\right\|_{\ell_2}^2 \leq \left(1+\delta_k\right)\|x\|_{\ell_2}^2,$$

for any k-sparse signal x. By imitating the work done in [26] (Lemmas 4.1 and 4.2), it can be shown, due to Theorem 4.5, that matrices A of the type given in (14) satisfy

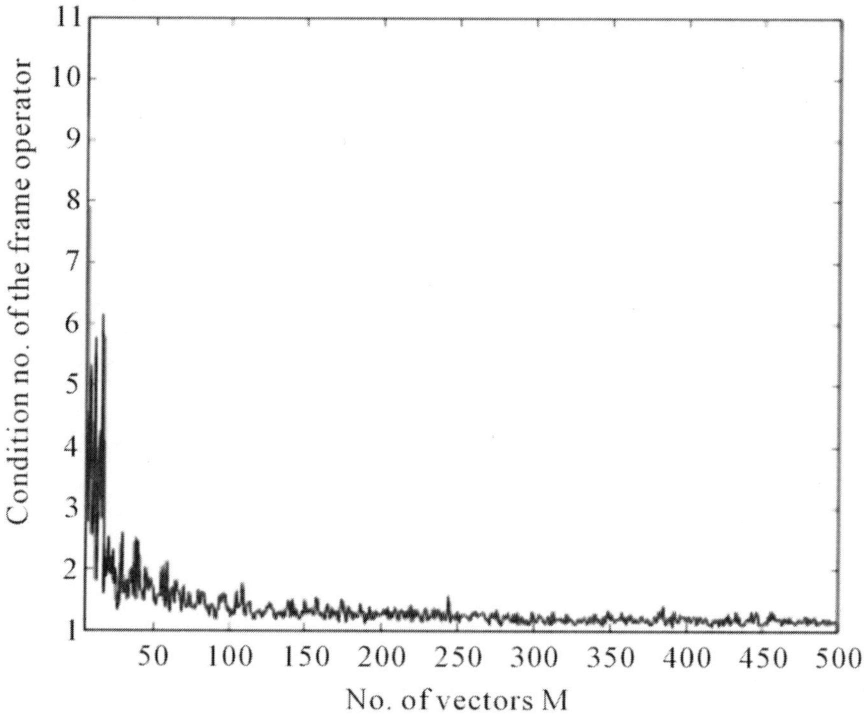

Figure 1. Behavior of the condition number of the frame operator with increasing size of the frame; $\varepsilon = 0.0001$, $d = 3$, $\sigma = 1$.

the RIP condition and can therefore be used as measurement matrices in compressed sensing. These matrices are different from the traditional random matrices used in compressed sensing in that their entries are complexvalued and unimodular instead of being real-valued and not unimodular.

Example 4.8. This example illustrates the ideas in this subsection. First consider M = 5 and d = 3 so that there are 5 vectors in $\mathbb{C}^3$. Taking from a normal distribution with mean 0 and variance $\sigma = 1$, a realization of the matrix $\left[Y_{mn}\right]_{1\le m\le 3, 1\le n\le 5}$ is

$$\begin{bmatrix} -0.0353 & 0.5004 & -0.6299 & -0.1472 & 0.4003 \\ -0.4804 & -0.9344 & 0.4220 & -0.9509 & 0.2783 \\ -0.8609 & -0.4822 & -0.4680 & -0.9509 & 1.2284 \end{bmatrix}.$$

Then taking $\varepsilon = 0.001$, $A = \frac{1}{\sqrt{3}}\left[e^{\frac{2\pi}{\varepsilon} i Y_{mn}}\right]$ is

$$A = \frac{1}{\sqrt{3}}\begin{bmatrix} -0.27-0.96i & -0.89+0.44i & 0.85+0.52i & 0.33-0.94i & -0.24+0.97i \\ -0.92-0.39i & -0.89-0.46i & 0.99+0.05i & 0.93+0.37i & -0.47+0.88i \\ 0.74+0.68i & 0.16-0.99i & 0.99+0.09i & -0.30-0.95i & -0.74+0.67i \end{bmatrix}.$$

The condition number, ratio of the maximum and minimum eigenvalues, of $\mathcal{F} = 4.8667$. As the number of vectors M is increased, the condition number gets closer to 1. **Figure 1** shows the behavior of the condition number with the increase in the number of vectors.

5. CONCLUSION

The construction of discrete unimodular stochastic waveforms with arbitrarily small expected autocorrelation has been proposed. This is motivated by the usefulness of such waveforms in the areas of radar and communications. The family of random variables that can be used for this purpose has been characterized. Such construction been done in one dimension and generalized to higher dimensions. Further, such waveforms have been used to construct frames in $\mathbb{C}^d$ and the frame properties of such frames have been studied. Using Brownian motion, this idea is also extended to the construction of continuous unimodular stochastic waveforms whose autocorrelation can be made arbitrarily small in expectation.

6. ACKNOWLEDGEMENTS

Somantika Datta The author wishes to acknowledge support from AFOSR Grant No. FA9550-10-1-0441 for conducting this research. The author is also grateful to Frank Gao and Ross Richardson for their generous help with probability theory.

REFERENCES

1. L. Auslander and P. E. Barbano, "Communication Codes and Bernoulli Transformations," Applied and Computational Harmonic Analysis, Vol. 5, No. 2, 1998, pp. 109- 128.
2. J. J. Benedetto and J. J. Donatelli, "Ambiguity Function and Frame Theoretic Properties of Periodic Zero Autocorrelation Waveforms," IEEE Journal of Selected Topics in Signal Processing, Vol. 1, 2007, pp. 6-20.
3. T. Helleseth and P. V. Kumar, "Sequences with Low Correlation," Handbook of Coding Theory, North-Holland, Amsterdam, 1998, pp. 1765-1853.
4. N. Levanon and E. Mozeson, "Radar Signals," Wiley Interscience, New York, 2004.
5. M. L. Long, "Radar Reflectivity of Land and Sea," Artech House, 2001.
6. W. H. Mow, "A New Unified Construction of Perfect Root-of-Unity Sequences," Proceedings of IEEE 4th International Symposium on Spread Spectrum Techniques and Applications, Sun City, 22-25 September 1996, pp. 955-959.
7. F. E. Nathanson, "Radar Design Principles: Signal Processing and the Environment," SciTech Publishing Inc., Mendham, 1999.
8. G. W. Stimson, "Introduction to Airborne Radar," SciTech Publishing Inc., Mendham, 1998.
9. S. Ulukus and R. D. Yates, "Iterative Construction of Optimum Signature Sequence Sets in Synchronous CDMA Systems," IEEE Transactions on Information Theory, Vol. 47, No. 5, 2001, pp. 1989-1998.
10. S. Verdú, "Multiuser Detection," Cambridge University Press, Cambridge, 1998.
11. J. J. Benedetto and S. Datta, "Construction of Infinite Unimodular Sequences with Zero Autocorrelation," Advances in Computational Mathematics, Vol. 32, No. 2, 2010, pp. 191-207.
12. D. Cochran, "Waveform-Agile Sensing: Opportunities and Challenges," IEEE International Conference on Acoustics, Speech, and Signal Processing, Pennsylvania, 18-23 March 2005, pp. 877-880.
13. M. R. Bell, "Information Theory and Radar Waveform Design," IEEE Transactions on Information Theory, Vol. 39, No. 5, 1993, pp. 1578-1597.
14. S. P. Sira, Y. Li, A. Papandreou-Suppappola, D. Morrell, D. Cochran and M. Rangaswamy, "Waveform-Agile Sensing for Tracking," IEEE of Signal Processing Magazine, Vol. 26, No. 1, 2009, pp. 53-64.
15. H. Boche and S. Stanczak, "Estimation of Deviations between the Aperiodic and Periodic Correlation Functions of Polyphase Sequences in Vicinity of the Zero Shift," IEEE 6th International Symposium on Spread Spectrum Techniques and Applications, Parsippany, 6-8 September 2000, pp. 283-287.
16. R. Narayanan, "Through Wall Radar Imaging Using UWB Noise Waveforms," IEEE International Conference on Acoustics, Speech and Signal Processing, Beijing, 31 March-4 April 2008, pp. 5185-5188.

17. R. M. Narayanan, Y. Xu, P. D. Hoffmeyer and J. O. Curtis, "Design, Performance, and Applications of a Coherent Ultra-Wideband Random Noise Radar," Optical Engineering, Vol. 37, No. 6, 1998, pp. 1855-1869.
18. D. Bell and R. Narayanan, "Theoretical Aspects of Radar Imaging Using Stochastic Waveforms," IEEE Transactions on Signal Processing, Vol. 49, No. 2, 2001, pp. 394-400.
19. A. F. Karr, "Probability of Springer Texts in Statistics," Springer-Verlag, New York, 1993.
20. O. Christensen, "An Introduction to Frames and Riesz Bases," Birkhäuser, 2003.
21. I. Daubechies, "Ten Lectures on Wavelets," CBMS-NSF Regional Conference Series in Applied Mathematics, 1992.
22. W. Hoeffding, "Probability Inequalities for Sums of Bounded Random Variables," Journal of American Statistical Association, Vol. 58, No. 301, 1963, pp. 13-30.
23. Z. D. Bai, "Methodologies in Spectral Analysis of LargeDimensional Random Matrices: A Review," Statistica Sinica, Vol. 9, No. 3, 1999, pp. 611-677.
24. M. Ledoux, "The Concentration of Measure Phenomenon," Mathematical Surveys and Monographs, American Mathematical Society, 2001.
25. A. E. Litvak, A. Pajor, M. Rudelson and N. TomczakJaegermann, "Smallest Singular Value of Random Matrices and Geometry of Random Polytopes," Advances in Mathematics, Vol. 195, No. 2, 2005, pp. 491-523.
26. E. Candès and T. Tao, "Near Optimal Signal Recovery from Random Projections: Universal Encoding Strategies?" IEEE Transactions on Information Theory, Vol. 52, No. 12, 2006, pp. 5406-5425.
27. E. Candès, "Compressive Sampling," Proceedings of the International Congress of Mathematicians, Madrid, 22-30 August 2006, pp. 1433-1452.
28. E. Candès, J. Romberg and T. Tao, "Robust Uncertainty Principles: Exact Signal Reconstruction from Highly Incomplete Frequency Information," IEEE Transactions on Information Theory, Vol. 52, No. 2, 2006, pp. 489-509.
29. D. L. Donoho, "Compressed Sensing," IEEE Transactions on Information Theory, Vol. 52, No. 4, 2006, pp. 1289- 1306.

NOTES

*This work was supported by AFOSR Grant No. FA9550-10-1-0441.

CHAPTER 9

Multiyear Discrete Stochastic Programming with a Fuzzy Semi-Markov Process

C. S. Kim[1], Richard M. Adams[2], Dannele E. Peck[3]

[1]*Economic Research Service, U.S. Department of Agriculture, Washington, DC, USA*

[2]*Department of Applied Economics, Oregon State University, Corvallis, OR, USA*

[3]*Department of Agricultural and Applied Economics, University of Wyoming, Laramie, WY, USA*

ABSTRACT

Drought conditions at a given location evolve randomly through time and are typically characterized by severity and duration. Researchers interested in modeling the economic effects of drought on agriculture or other water users often capture the stochastic nature of drought and its conditions via multiyear, stochastic economic models. Three major sources of uncertainty in application of a multiyear discrete stochastic model to evaluate user preparedness and response to drought are: (1) the assumption of independence of yearly weather conditions, (2) linguistic vagueness in the definition of drought itself, and (3) the duration of drought. One means of addressing these uncertainties is to re-cast drought as a stochastic, multiyear process using a "fuzzy" semi-Markov process. In this paper, we review "crisp" versus "fuzzy" representations of drought and show how fuzzy semi-Markov processes can aid researchers in developing more robust multiyear, discrete stochastic models.

Keywords: Drought, Discrete Stochastic Economic Modeling, Fuzzy Logic, Fuzzy Markov Process, Fuzzy Semi-Markov Process

1. INTRODUCTION

The impacts of drought on water supplies and those dependent on such supplies have long been an important topic for economists and policy makers. Climate projections for key hydrologic inputs—seasonal precipitation, snowpack storage, evaporative loss, and frequency and severity of drought—are used to anticipate future stresses on water quantity and variability. Water managers use these projections to allocate seasonal water sup plies across various classes of users. For example, public irrigation districts in the western U.S. typically issue water allocation declarations in March, based on hydrologic forecasts, to help irrigators plan for the coming irrigation season. In periods of sharply reduced water supplies, such as recent multiyear droughts in California (**Figure 1**), such allocation decisions must reflect a range of competing demands, including in stream flow requirements to protect associated species and ecosystems.

Discrete stochastic programming (DSP) has been widely used as an optimization method to assess the benefits and costs of alternative water allocation decisions during varying degrees of drought [1]-[3]. DSP is a useful technique based on traditional probability theory. To illustrate water supply uncertainty in a discrete stochastic programming framework, first consider the following single-year linear programming model:

$$\text{Max } Z = c'x \,, \tag{1.1}$$

subject to:

$$Ax \leq b \tag{1.2}$$

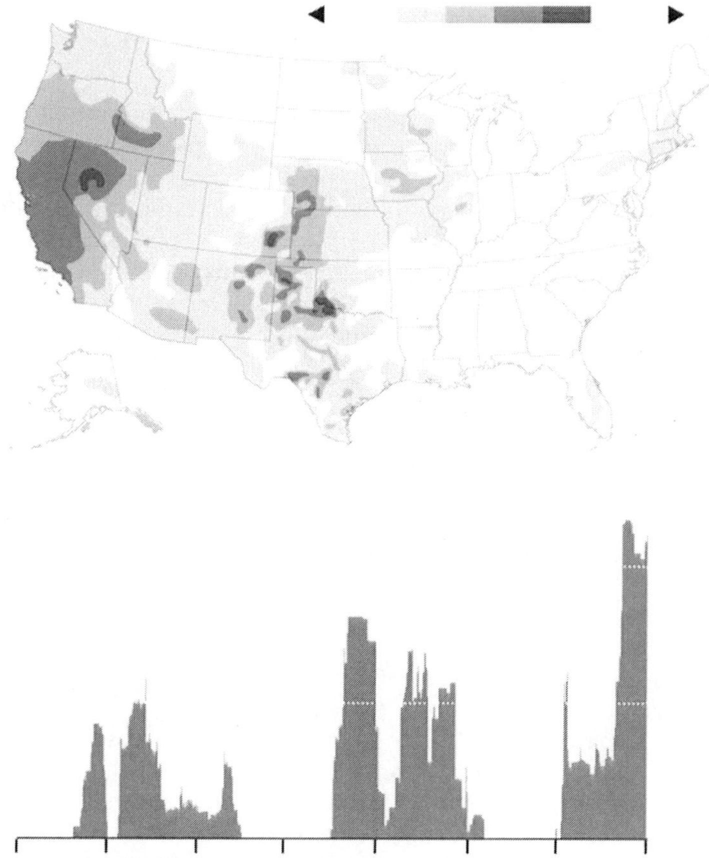

Figure 1. Severity and duration of drought in California (Source: The National Drought Miti- gation Center—USDA).

where $x = (xj)$ is a vector of control variables with $x \geq 0$, $c = (cj)$ is an $(n \times 1)$ column vector of constants, $A = (aij)$ is an $(m \times n)$ matrix of constants, and $b = (bi)$ is an $(m \times 1)$ column vector of constants. The corresponding discrete stochastic linear programming problem, first proposed by Cocks [4], introduces uncertainty into one or more model parameters, such that they satisfy the conditions of a probability distribution:

$$P\left[\left(A,b,c\right)=\left(A_k,b_k,c_k\right)\right]=P_k \quad \left(k=1,2,\cdots,K\right), \tag{1.3}$$

where $\sum_{k=1}^{K} P_k = 1$.

Most previous studies that use DSP to address uncertainty caused by randomness of hydrologic inputs assume a small number of discrete states of nature, sometimes just a binary set (e.g., "drought" = 24 acre-inches of water; "normal" = 40 acre-inches, as in **Figure 2**). This type of linguistic vagueness of the term "drought" can be illustrated by considering the Palmer Drought Severity Index (PDSI), employed by the National Oceanic and Atmospheric Administration (NOAA) to communicate drought conditions to the public. The PDSI comprises five states of drought, including *mild, moderate, severe, extreme,* and *exceptional*. Each state of drought comprises a range of temperature and precipitation conditions. The same quantity of precipitation may therefore result in different states of drought, depending on temperature conditions. A given water allotment does not always result in the same state of drought; instead, it has the potential to lead to various states of drought (*i.e.*, it has "grades of membership" in different states of drought). Which state of drought should actually be assigned to a given water allotment (or *vice versa*) depends on many other complex factors, including but not limited to temperature, available water content of the soil, and other conditions not easily incorporated into an economic model. Correct assignment of a state of drought is important because severe, extreme, or exceptional drought implies a significant reduction in crop production, whereas mild or moderate drought implies a much smaller reduction. An inadequate number of states of nature in a programming model may produce management recommendations that are not robust when they are applied to a real-world setting that has numerous states of nature (e.g., a continuum of water allocations) [5]. There is another *linguistic vagueness* in the definition of "drought", specifically when modelers attempt to assign this term to a specific quantity of irrigation water (e.g., "drought" = 24 acre-inches of water, not 20 or 26, but exactly 24).

In addition to linguistic vagueness, the agricultural economics literature has rarely discussed uncertainty about the *duration* of drought. Peck and Adams [3] developed a multiyear, discrete stochastic model of irrigation water use on a multi-crop farm that faces uncertain surface-water supplies due to the possibility of drought. They then used the model to identify optimal drought preparedness and response plans and measure associated economic benefits that result as the severity and duration of drought is revealed over a six-year period. While they confirmed that the effects of two consecutive years of drought are greater than the sum of two single-year (non-consecutive) droughts, due to inter-annual crop-rotation constraints, they assumed just two discrete (*i.e.*, crisply-defined) states of nature ("normal" = 40 acre-inches per acre; "dry" = 24 acre-inches per acre) and also assumed independence between individual years' states of nature (*i.e.*, the occurrence of drought in one year did not change the probability of drought in subsequent years). They left more sophisticated approaches to these three modeling issues—linguistic vagueness, assumption of independence of yearly water allocation, and more complex assumptions about uncertainty of drought duration—for future research.

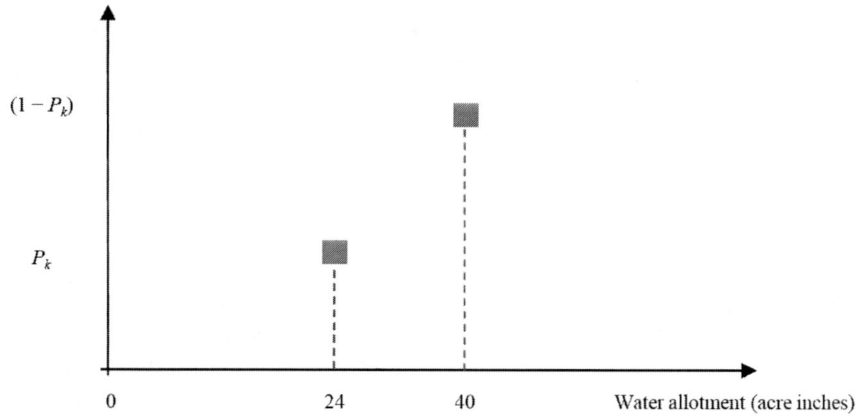

Figure 2. Probabilistic crisp bounds for water allotments.

Uncertainty caused by *linguistic vagueness* has been dealt with more recently using fuzzy logics [6]. A principal difference between discrete-stochastic and fuzzy optimization approaches is in the way uncertainty is modeled: discrete stochastic programming represents random parameters through discrete or continuous probability functions, whereas fuzzy programming treats random parameters as fuzzy numbers and hence constraints as fuzzy sets [7]. Fuzzy programming can be classified into two types.1 Possibilistic fuzzy programming considers uncertainties in the objective function coefficients, *c*, in Equation (1.1), as well as in the constraint coefficients, *A* and *b*, in Equation (1.2). Flexible fuzzy programming considers uncertainties associated only with right-hand-side constraints, *b* [8]-[10].

Furthermore, discrete stochastic programming requires the constraints to be exactly satisfied in each state of nature. For example, the model cannot choose a crop plan that consumes more than the exact quantity of water (24 or 40 acre-inches) that is ultimately revealed to be available. The most commonly used fuzzy mathematical programming method for this type of linguistic vagueness is to transform a crisp programming model into a series of crisp programming problems [11]. Meanwhile, the second type of linguistic vagueness has been dealt with using interval fuzzy programming approaches [12].

However, there are a few shortcomings of fuzzy programming. For example, although it can be used to evaluate the economic effects of drought severity, it is not capable of dealing with randomness in the *duration* of drought (or other random phenomena) [13].

To overcome the shortcomings of the discrete stochastic programming approach [including: (1) crisp definitions for each state of nature, (2) assumed independence of states of nature across multiple periods, e.g., if the probability of drought in a single year is 1/3, then the probability of a two-year drought is simply 1/3 × 1/3 or 1/9), and (3) inability to accommodate uncertainty about future states of nature, *i.e.*, uncertainty about drought duration], we first provide in Section 2 a definition of drought, and then discuss uncertainty caused by linguistic vagueness in the term "drought". We also introduce readers to fuzzy logic for drought severity, where a continuous range of possible water quantities are assigned various grades of membership in a set of drought severity. In section 3, we introduce a semi-Markov process, which allows more complex assumptions about both the probability (or possibility) of transitioning from one state of nature to the next (e.g., the conditional probability (or possibility) of drought occurring next year, given drought occurred this year, is not necessarily equal to the unconditional probability of drought in a single year), and the duration between two transitions. Fuzzy logic and semi-Markov processes are then combined into a fuzzy semi-Markov process, which enables us to capture uncertainty in both drought *severity* and *duration* simultaneously. This fuzzy semi-Markov process can ultimately be incorporated in a discrete stochastic programming framework to overcome the shortcomings mentioned earlier. Finally, in Section 4, we present numerical examples of conditional possibility of drought duration, given a specific state of drought severity. These numerical examples highlight potential implications of using fuzzy semi-Markov processes rather than traditional probability-based representations of multiyear drought.

2. LINGUISTIC VAGUENESS AND FUZZY LOGIC

To evaluate the impacts of drought conditions and associated preparedness and response plans, a clear definition of drought must be provided. One definition of drought is the case in which irrigation water supplied is less than irrigation water demanded, due to inadequate rainfall, snow pack, or other weather conditions. As the difference between irrigation water demanded and supplied increases, severity of drought intensifies along a continuous gradient. While the characterization of drought varies across studies, the following definition of drought pro- vided by Yevjevich [14] has been widely used [6] [15] [16] :

$$y_i = \begin{cases} 0 & \text{if } S_i < S_0 \quad (i = 1,2,3,\cdots) \\ 1 & \text{if } S_i \geq S_0 \end{cases} , \tag{2.1}$$

where S_0 is a constant water allotment threshold (such as 40 acre-inches Peck and Adams [3] used for water allotment under normal weather condition) and S_i is the ith severity state. A Bernoulli variable y_i plays a significant role in estimation of the holding time (i.e., duration) probability mass function of drought in later section of semi-Markov chains.

When economic models represent drought in binary terms (i.e., a water allocation either qualifies or does not qualify as drought), this overly-simplified or deceivingly-crisp (as opposed to fuzzy) measurement of drought severity can cause inefficient resource allocation. Unlike this crisp set (in which an element is either a member of the set or not), fuzzy sets allow elements to be included through a degree of membership, as expressed by a membership function, thus relaxing the binary state assumption [6] [17] . Introduced by Zadeh [18] , fuzzy logic and fuzzy set theory have since been widely adopted to deal with linguistic vagueness in various mathematical optimization models, including fuzzy dynamic programming [19] [20] , and optimal fuzzy control [21] .

$$f_s(w_i) = \begin{cases} 1 & \text{if } w_i \geq w_0 \\ 1 - \dfrac{(w_0 - w_i)}{\alpha} & \text{if } w_0 - \alpha \leq w_i < w_0 . \\ 0 & \text{otherwise} \end{cases} \tag{2.2}$$

3. FUZZY MARKOV AND SEMI-MARKOV PROCESSES

Definition 2. Let $W = \{w_1, w_2, \cdots, w_n\}$ be a finite state-space of possible water allotments from an irrigation water district. Let $(S_n)_{n \in N}$ be random variables taking values in W. Now, let a fuzzy subset of W be defined as a set of ordered pairs $\left(S_i, f_S^m(S_i)\right)$, where $f_S^m(S_i)$ is the grade of membership of S_i in W in the mth transition.

$$f_S^{(m+1)}(S_j) = \max_{i \in S}\left\{\min\left(f_S^m(S_i), P_{ij}\right)\right\} \quad j \in S \qquad ,(3.2)$$

where the fuzzy-state grade of membership, $f_S^m(S_i)$, can be calculated as follows:

$$f_S^m(S_i) = \max_{l \in S}\left[\min\left(f_S^0(S_l), P_{li}^m\right)\right] \quad i \in S \qquad ,(3.3)$$

and the powers of the fuzzy transition matrix, P_{li}^m, are defined by:

$$P_{li}^m = \max_{k \in S}\left[\min\left(P_{lk}, P_{ki}^{m-1}\right)\right], \qquad (3.4)$$

where $P_{li}^1 = P_{li}$ and P_{li}^0 is an identity matrix at the initial time period.

A few empirical studies have been conducted to compare efficiencies between the use of classic Markov chains based on conventional crisp set theory and fuzzy Markov chains in the context of stochastic programming models. Mousavi et al. [20] compared a conventional stochastic dynamic programming model (which employed classic Markov chains) and a fuzzy-state stochastic dynamic programming model (which employed fuzzy Markov chains) in the operation of a multipurpose reservoir in Iran. Results from their study show that fuzzy- state stochastic dynamic programming outperforms conventional stochastic dynamic programming in achieving the flood control objective and in overall performance of the reservoir. Chandramouli and Nanduri [27] also compared a conventional stochastic dynamic model and a fuzzy-state stochastic dynamic model in the operation of a multipurpose reservoir in India. They found that the fuzzy-state stochastic dynamic programming model out performed the conventional stochastic dynamic programming model.

3.1. Fuzzy Semi-Markov Process

3.1.1. Homogeneous Fuzzy Semi-Markov Process

where S^n represents the state at the nth transition, T_n is the time of the nth transition and "t" is arrival time. Equation (3.5) expresses the probability of transitioning to state j at arrival time t, given the system has been in state i after n transitions. Unfortunately no solution has been reported for Equation (3.5) in the literature; therefore, an alternative approach for solving Equation (3.5) has been employed by Cancelliere and Salas [15] and Mirakbari and Ganji [6] .

$$h_{ij}\left(d_{ij}\right)=r_{ij}\left(1-r_{ij}\right)^{d_{ij}-1} \qquad \text{for } d_{ij}=1,2,3,\cdots, \tag{3.9}$$

where r_{ij} represents the probability of the duration d_{ij} in the ith drought severity, before moving to the jth

drought severity. This is estimated by $r_{ij}=\frac{1}{\bar{d}_{ij}},$ where $\bar{d}_{ij}$ is the mean holding-time (i.e., duration) in the ith

drought before moving to the jth drought, which can be easily observed from historical data.

Mirakbari and Ganji [6] compared performances between the classic semi-Markov chains and the fuzzy semi- Markov chains by using profust reliability theory to a rangeland system in India.6 Their results indicate that the reliability of rangeland system decisions increased by 22 percent when the fuzzy semi-Markov process is used over the classic semi-Markov process.

3.1.2. Non-Homogeneous Fuzzy Semi-Markov Process

where q_{ij} is the transition probability of the embedded fuzzy Markov chain in the process.

$$\begin{aligned} P_i\left(s,d_i\right) &= P\left[T_{n+1}-T_n \leq d_i \mid S^n = w_{i,n}, T_n = s\right] \\ &= \max_{j\neq i}\left\{\min\left(P_{ij}\left(s\right),h_{ij}\left(s,d_{ij}\right)\right)\right\}. \end{aligned} \tag{3.17}$$

4. NUMERICAL EXAMPLE

We now present a numerical example of multiyear water supply forecasts under the assumption that a pair of two random variables $\{S_n, T_n\}$ follows a homogeneous fuzzy semi-Markov process (i.e., Equation (3.5) through (3.12)).7 The multiyear forecast associated with the severity and duration of droughts can be incorporated into a multiyear discrete stochastic programming model as shown in Equation (3.12).

Table 1. Transition probabilities and holding-time mass functions.

Equation	*Transition probability*	*Observed mean duration*	*Holding-time mass function*
(A.1)	$p_{11} = 0.4$	$\bar{d}_{11} = 3$	$h_{11}(d_{11}) = \left(\frac{1}{3}\right)\left(\frac{2}{3}\right)^{d_{11}-1}$
(A.2)	$p_{12} = 0.4$	$\bar{d}_{12} = 4$	$h_{12}(d_{12}) = \left(\frac{1}{4}\right)\left(\frac{3}{4}\right)^{d_{12}-1}$
(A.3)	$p_{13} = 0.2$	$\bar{d}_{13} = 5$	$h_{13}(d_{13}) = \left(\frac{1}{5}\right)\left(\frac{4}{5}\right)^{d_{13}-1}$
(B.1)	$p_{21} = 0.3$	$\bar{d}_{21} = 3$	$h_{21}(d_{21}) = \left(\frac{1}{3}\right)\left(\frac{2}{3}\right)^{d_{21}-1}$
(B.2)	$p_{22} = 0.5$	$\bar{d}_{22} = 3$	$h_{22}(d_{22}) = \left(\frac{1}{3}\right)\left(\frac{2}{3}\right)^{d_{22}-1}$
(B.3)	$p_{23} = 0.2$	$\bar{d}_{23} = 4$	$h_{23}(d_{23}) = \left(\frac{1}{4}\right)\left(\frac{3}{4}\right)^{d_{23}-1}$
(C.1)	$p_{31} = 0.4$	$\bar{d}_{31} = 4$	$h_{31}(d_{31}) = \left(\frac{1}{4}\right)\left(\frac{3}{4}\right)^{d_{31}-1}$
(C.2)	$p_{32} = 0.3$	$\bar{d}_{32} = 5$	$h_{32}(d_{32}) = \left(\frac{1}{5}\right)\left(\frac{4}{5}\right)^{d_{32}-1}$
(C.3)	$p_{33} = 0.3$	$\bar{d}_{33} = 6$	$h_{33}(d_{33}) = \left(\frac{1}{6}\right)\left(\frac{5}{6}\right)^{d_{33}-1}$

$See Avrachenkov and Sanchez [25], Gildeh and Dadgar [34], Li and He [35], Mousavi *et al.* [20], and Sanchez [36] for fuzzy transition matrix, and Meenakshi and Kaliraja [37] for interval transition matrix.

$$H\left(d_{ij}\right) = \begin{array}{c} \\ \text{N} \\ \text{MM} \\ \text{SEE} \end{array} \begin{array}{c} \begin{array}{ccc} \text{N} & \text{MM} & \text{SEE} \end{array} \\ \begin{pmatrix} 1-\left(\frac{2}{3}\right)^{d_{11}} & 1-\left(\frac{3}{4}\right)^{d_{12}} & 1-\left(\frac{4}{5}\right)^{d_{13}} \\ 1-\left(\frac{2}{3}\right)^{d_{21}} & 1-\left(\frac{2}{3}\right)^{d_{22}} & 1-\left(\frac{6}{7}\right)^{d_{23}} \\ 1-\left(\frac{3}{4}\right)^{d_{31}} & 1-\left(\frac{4}{5}\right)^{d_{32}} & 1-\left(\frac{5}{6}\right)^{d_{33}} \end{pmatrix} \end{array}. \tag{4.3}$$

Comparison

Due to the lagging and long-term effects of drought on vegetation and soil moisture, or on cropping choices due to agronomic constraints (e.g., rotations), the resilience of drought is equal to or longer than drought duration [38] .9 Therefore, economic impacts of individual years of drought may not necessarily be independent [3] . To compare multiyear drought probabilities between a conventional multiyear discrete stochastic program (which typically assumes the previous year's state of nature does not influence the probability of future states of nature) and a multiyear homogeneous fuzzy semi-Markov process, the steady-state probabilities of the fuzzy transition matrix in Equation (4.2) are estimated as follows:

$$\pi = \overset{\quad N \quad\; MM \quad SEE}{\left[\frac{1}{3}, \frac{1}{3}, \frac{1}{3}\right]}, \tag{4.4}$$

where π is a stationary distribution.1[0] The probabilities in Equation (4.4) represent P_k (k = 1, 2, 3) in Equation (1.3). In a conventional discrete stochastic programming model, the probability of severe drought (SEE) in the

first year, as well as in the second year, or any other year, is $\frac{1}{3}$, such that the probability of two consecutive years of severe drought is $\left(\frac{1}{3}\right)^2 = 0.1111$, while the probability of three consecutive years of severe drought is $\left(\frac{1}{3}\right)^3 = 0.0370$.

Table 2. Results of a numerical example.

Weather	Model	Duration d = 1	d = 2	d = 3
N	FsM	$P_1(d = 1) = 0.3333$	$P_1(d = 2) = 0.2222$	$P_1(d = 3) = 0.1481$
	DSPM	0.3333	0.1111	0.0370
MM	FsM	$P_2(d = 1) = 0.3000$	$P_2(d = 2) = 0.2222$	$P_2(d = 3) = 0.1481$
	DSPM	0.3333	0.1111	0.0370
SEE	FsM	$P_3(d = 1) = 0.2500$	$P_3(d = 2) = 0.1875$	$P_3(d = 3) = 0.1406$
	DSPM	0.3333	0.1111	0.0370

N = Normal; MM = Mild/Moderate drought; SEE = Severe/Extreme/Exceptional drought. FsM = Fuzzy semi-Markov; DSPM = Discrete Stochastic Programming Model.

The possibility mass function of the holding-time spent in each weather condition in a fuzzy semi-Markov process is estimated with Equation (3.11) and results are presented in Table 2. In contrast to a conventional discrete stochastic programming model, the possibility of two consecutive years of SEE drought condition is 0.1875, which is 69 percent higher than the probability assumed in a conventional multiyear discrete stochastic programming approach. As the duration of consecutive years of severe drought increases to three years, the possibility of holding time in a fuzzy semi-Markov process is 0.1406, which is 280 percent higher than the probability assumed in a conventional multiyear discrete stochastic programming approach, 0.0370.

Implementation of adequate measures to control or mitigate drought consequences is a major challenge for irrigators and other water users. Our numerical example, while stylized, demonstrates how economists can use fuzzy semi-Markov processes to incorporate uncertainty about both severity of drought (which necessitates fuzzy sets) and duration of a multiyear drought (which necessitates semi-Markov processes) in stochastic modeling by using a fuzzy semi-Markov process. Such model specification may improve representation of the economic effects of drought severity and duration on water users and the efficacy of alternative mitigation actions.

ACKNOWLEDGEMENTS

The views expressed are those of the authors and should not be attributed to USDA or Economic Research Service.

CITE THIS PAPER

C. S. Kim, Richard M. Adams, Dannele E. Peck (2016) Multiyear Discrete Stochastic Programming with a Fuzzy Semi-Markov Process. *Applied Mathematics*,**07**,482-495.

REFERENCES

1. Engelhardt, T. (1983) Water Supply Optimization with Discrete Stochastic Linear Programming. Proceedings of International Crops Research Institute Workshop on the State of the Art and Management Alternatives for Optimizing the Productivity of SAT Alfisols and Related Soils, ICRISAT Center, 1-3 December 1983, Patancheru, India, 79-87.
2. McCarl, B. and Parandvash, G. (1988) Irrigation Development versus Hydroelectric Generation: Can Interruptible Irrigation Play a Role? Western Journal of Agricultural Economics, 13, 267-276.
3. Peck, D.E. and Adams, R.M. (2010) Farm-Level Impacts of Prolonged Drought: Is a Multiyear Event More than the Sum of Its Parts? The Australian Journal of Agricultural and Resource Economics, 54, 43-60.

4. Cocks, K.D. (1968) Discrete Stochastic Programming. Management Science, 15, 72-79.
5. Watanabe, Y. (1987) Errors in Availability Estimation by 2-State Models of 3-State Systems. Reliability Engineering, 18, 223-235.
6. Mirakbari, M. and Ganji, A. (2010) Reliability Analysis of a Rangeland System: The Application of Profust Theory. Stochastic Environmental Research and Risk Assessment, 24, 399-409.
7. Sahinidis, N. (2004) Optimization under Uncertainty: State-of-the-Art and Opportunities. Computers and Chemical Engineering, 28, 971-983.
8. Fullér, R. (1998) Fuzzy Reasoning and Fuzzy Optimization. Department of Operations, Eötvös Loránd University, Budapest.
9. Inuiguchi, M. and Ramit, J. (2000) Possibilistic Linear Programming: A Brief Review of Fuzzy Mathematical Programming and a Comparison with Stochastic Programming in Portfolio Selection Problem. Fuzzy Sets and System, 111, 3-28.
10. Tanaka, H., Okuda, T. and Asai, K. (1974) Decision Making and Information in Fuzzy Events. Bulletin Series A, 23, 193-202.
11. Li, H. and Gong, Z. (2010) Fuzzy Linear Programming with Possibility and Necessity Relation. In: Cao, et al., Eds., Fuzzy Information and Engineering, Springer-Verlag, Berlin Heidelberg, 305-311.
12. Huang, Y.F., Baetz, B.W., Huang, G.H. and Liu, L. (2002) Violation Analysis for Solid Waste Management Systems: An Interval Fuzzy Programming Approach. Journal of Environmental Management, 65, 431-446.
13. Arkov, V., Kulikov, G.G. and Breikin, T.V. (1999) Fuzzy Markov Modeling in Automatic Control of Complex Dynamic Systems. International Conference on Accelerator and Large Experimental Physics Control Systems, 4-8 October 1999, Trieste, 287-289.
14. Yevjevich, V. (1967) An Objective Approach to Definitions and Investigations to Continental Hydrologic Droughts. Colorado State University, Fort Collins, Colorado.
15. Cancelliere, A. and Salas, J.D. (2004) Drought Length Properties for Periodic-Stochastic Hydrologic Data. Water Resources Research, 40, W02503.
16. Llamas, J. and Siddiqui, M. (1969) Runs of Precipitation Series. Hydrology Paper 33, Colorado State University, Fort Collins.
17. Cai, K.Y., Wen, C.U. and Zhang, M.L. (1993) Fuzzy States as a Basis for a Theory of Fuzzy Reliability. Journal of Microelectronics Reliability, 33, 2253-2263.
18. Zadeh, L.A. (1965) Fuzzy Sets. Information and Control, 8, 338-353.
19. Bellman, R.E. and Zadeh, L.A. (1970) Decision-Making in a Fuzzy Environment. Management Science, 17, 141-164.
20. Mousavi, S., Karamouz, K. and Menhadj, M.B. (2004) Fuzzy-State Stochastic Dynamic Programming for Reservoir Operation. Journal of Water Resources Planning and Management, 130, 460-470.
21. Baten, M.A. and Kamil, A.A. (2012) Optimal Fuzzy Control with Application to Discounted Cost Production Inventory Planning Problem. Proceedings of the World Congress on Engineering, Vol. III, London, 30 June-2 July 2012.
22. Khedun, C.P. and Singh, V. (2014) Drought Risk Management in Agriculture: Current Trends in Research. Texas A&M University, College Station.
23. Alves, A., Mota, G., Costa, G. and Feitosa, R. (2012) Estimation of Transition Possibilities for Fuzzy Markov Chains Applied to the Analysis of Multitemporal Image Sequences. Proceedings of the 4th GEOBIA, Rio de Janeiro, 7-9 May 2012.
24. Praba, B., Sujatha, R. and Srikrishna, S. (2009) A Study on Homogeneous Fuzzy Semi-Markov Model. Applied Mathematical Science, 3, 2453-2467.
25. Avranchenkov, K. and Sanche, E. (2002) Fuzzy Markov Chains and Decision-Making. Fuzzy Optimization and Decision Making, 1, 143-159.
26. Schweizer, P.J. (1968) Pertubation Theory and Finite Markov Chains. Journal of Applied Probability, 5, 401-413.

27. Chandramouli, S. and Nanduri, U. (2011) Comparison of Stochastic and Fuzzy Dynamic Programming Models for the Operation of a Multipurpose Reservoir. Water and Environment Journal, 25, 547-554.
28. Bardossy, A. and Plate, E.J. (1991) Modeling Daily Rainfall Using a Semi-Markov Representation of Circulation Pattern Occurrence. Journal of Hydrology, 122, 33-47.
29. Masruroh, N. (2012) Introduction to Discrete Time Semi Markov Process. http://slideplayer.com/slide/4757341/
30. Praba, B., Sujatha, R. and Srikrishna, S. (2009) Fuzzy Reliability Measures of Fuzzy Probabilistic Semi-Markov Model. International Journal of Recent Trends in Engineering, 2, 25-29.
31. Vassiliou, P.C. (2013) Fuzzy Semi-Markov Migration Process in Credit Risk. Fuzzy Sets and Systems, 223, 39-58.
32. De Medici, J.J. and Manca, R. (2012) The Aggregate Claim Amount Discrete Time Semi-Markov Model. Working Paper, Department of Mathematics of Economic Decisions on Finance and Insurance, University of Rome, La Sapienza.
33. Janssen, J., Manca, R. and di Prignano, E.V. (2012) Non Homogeneous Interest Rate Structure in a Semi-Markov Framework. Working Paper. http://www. actuaries.org
34. Gildeh, B.S. and Dadgar, A. (2007) Reducible Fuzzy Markov Chain and Fuzzy Absorption Probability. 1st Joint Congress on Fuzzy and Intelligent Systems, Ferdowsi University of Mashhad, Iran, 29-31August 2007.
35. Li, J. and He, P. (2010) The Application of Fuzzy Markov Chains in the Analysis of Internet Glance Behavior. 3rd IEEE International Conference on Computer Science and Information Technology (ICCSIT), Vol. 7, Chengdu, 9-11 July 2010, 608-611.
36. Sanchez, E. (1976) Resolution of Composite Fuzzy Relation Equation. Information and Control, 30, 38-48.
37. Meenakshi, A.R. and Kaliraja, M. (2010) Regular Interval Valued Fuzzy Matrices. Advances in Fuzzy Mathematics, 5, 7-15.
38. Rossi, G., Benedini, M., Tsakiris, G. and Giakoumakis, S. (1992) On Regional Drought Estimation and Analysis. Water Resources Management, 6, 249-277.

NOTES

[1]Both possibilistic and flexible fuzzy programming can be presented in interval fuzzy programming [12] .

[2]However, we use the terms possibility and probability interchangeably in this paper.

[3]The max-min operations of fuzzy sets are as follows. For fuzzy subsets A and B of a crisp set W ≠ ϕ, the intersection of A and B is defined as: $(A \cap B)(x) = \min\{A(x), B(x)\} = A(x) \wedge B(x)$, $\forall x \in W$. Similarly, the union of A and B is defined as follows: $(A \cup B)(x) = \max\{A(x), B(x)\} = A(x) \vee B(x)$, $\forall x \in W$.

[4]The set of all subsets of W is called the power set of W.

[5]Markov kernel is also called a transition probability function that maps from a measureable space to another space.

[6]Profust reliability theory consists of two parts, the fuzzy part which considers vagueness in rangeland system failure and the probabilistic part which incorporates randomness of rangeland failure [6] [17] .

[7]Multiyear water supply forecast is a forecast in $d_{ij} = 1$, $d_{ij} = 2$, $d_{ij} = 3$, etc. for all drought severities i = 1, 2, ×××, n.

[8]See Avrachenkov and Sanchez [25] , Gildeh and Dadgar [34] , Li and He [35] , Mousavi et al. [20] , and Sanchez [36] for fuzzy transition matrix, and Meenakshi and Kaliraja [37] for interval transition matrix.

[9]Resilience is a measure of the recovery time of the system [38] .

[10]The stationary distribution must satisfy $\lim_{k \to \infty} P^k = 1\pi$, where 1 is a column vector with all entries equals to one

CHAPTER 10

Measuring Tail Dependence for Aggregate Collateral Losses Using Bivariate Compound Shot-Noise Cox Process

Jiwook Jang[1], Genyuan Fu[2]

[1]*Department of Applied Finance & Actuarial Studies, Faculty of Business and Economics, Macquarie University, Sydney, Australia*

[2]*Benelec Pty Ltd., Sydney, Australia*

ABSTRACT

In this paper, we introduce tail dependene measures for collateral losses from catastrophic events. To calculate these measures, we use bivariate compound process where a Cox process with shot noise intensity is used to count collateral losses. A homogeneous Poisson process is also examined as its counterpart for the case where the catastrophic loss frequency rate is deterministic. Joint Laplace transform of the distribution of the aggregate collateral losses is derived and joint Fast Fourier transform is used to obtain the joint distributions of aggregate collateral losses. For numerical illustrations, a member of Farlie-Gumbel-Morgenstern copula with exponential margins is used. The figures of the joint distributions of collateral losses, their contours and numerical calculations of risk measures are also provided.

Keywords: Aggregate Collateral Losses; Bivariate Compound Cox Process; Shot Noise Process; Farlie-Gumbel-Morgenstern Copula; Tail Dependence; Joint Fast Fourier Transform

1. INTRODUCTION

Over the recent years, numerous papers have looked at the modelling of dependence within an insurance portfolio or between insurance portfolios [1-5]. Also in the field of financial risk management, a range of papers on dependence modelling within credit risk and operational risk can be noticed [6-8]. Besides the construction of specific multivariate models, considerable attraction is given to the use of copulas. In particular, within the theory of Lévy processes, Lévy copulas have proven to be useful [9].

Our paper is very much based on insurance applications where two lines of business are hit by a common external event, hence the word "collateral losses" in the title. These joint losses may, for instance be triggered by events such as flood, windstorm, hail, bushfire, earthquake and terrorism. Particular examples concern collateral losses due to 2011 Great Eastern Japan Earthquake, 2010-2011 Queensland floods, 2009 Victorian Bushfires [10], 2005 Hurricane Katrina [11] and 2001 September 11 attack [12].

For the purpose of this paper, we concentrate on a very specific model and show how, within this model several explicit calculations for relevant risk quantities can be performed. The bivariate model we consider has the following structure:

$$L_t^{(1)} = \sum_{i=1}^{N_t} X_i, L_t^{(2)} = \sum_{i=1}^{N_t} Y_i, \tag{1}$$

where $L_t^{(k)}$ is the total loss arising from risk type $k = 1,2$ and N_t is the number of collateral losses up to time t. The random variables X_i and $Y_i, i = 1,2,\cdots,$ denote the individual loss amounts. In this model, the dependence between two random variables $L_t^{(1)}$ and $L_t^{(2)}$ comes from the common arrival process N_t, together with the dependence between the individual losses X_i and Y_i. The latter is modelled using the notion of copula [13]. To be more specific, we assume the loss random variable X_i and Y_i are independent identically distributed with continuous distribution function F_X and F_Y respectively. The joint distribution of the vector (X,Y) is assumed to be of the form $C(F_X, F_Y)$ with a given copula C. The uniqueness of this two stage construction goes back to Sklar's Theorem.

Theorem 1.1. (Sklar's Theorem) Let F be a joint distribution function with margins F_X and F_Y. Then there exists a copula C such that for all x, y in $\overline{\mathbb{R}} = [-\infty, \infty]$,

$$F(x,y) = C(F_X(x), F_Y(y)). \tag{2}$$

If the margins are continuous, then C is unique; otherwise C is uniquely determined on $\text{Ran}F_X \times \text{Ran}F_Y$, where $\text{Ran}F_i = F_i(\overline{\mathbb{R}})$ denotes the range of F_i. Conversely, if C is a copula and F_X and F_Y are univariate distribution functions, then the function F defined in (2) is a joint distribution function with margins F_X and F_Y.

Proof. See Schweizer and Sklar [14] or Nelsen ([13], p. 18).

To deal with stochastic nature of catastrophic loss arrival in practice, we use a Cox process for N_t. The Cox process provides flexibility by letting the intensity not only depend on time but also allowing it to be a stochastic process. Therefore the Cox process can be viewed as a two step randomisation procedure. A process λ_t is used to generate another process N_t by acting its intensity. That is, N_t is a Poisson process conditional on λ_t which itself is a stochastic process.

Losses arising from a catastrophe depend on its intensity. One of the processes that can be used to measure the impact of catastrophic events is the shot noise process. Previous works of insurance application using shot noise process and a Cox process with shot noise intensity can be found in [15-19]. Reference [20] also used a Cox process with shot noise intensity to model operational risk. The shot noise process is particularly useful to loss arrival process as it measures the frequency, magnitude and time period needed to determine the effect of catastrophic events. As time passes, the shot noise process decreases as more and more losses are settled. This decrease continues until another event occurs which will result in a positive jump in the shot noise process. Therefore the shot noise process can be used as the parameter of a Cox process to measure the number of catastrophic losses, i.e. we will use it as an intensity function to generate a Cox process. We will adopt the shot noise process used by Cox & Isham [21]:

$$\lambda_t = \lambda_0 e^{-\delta t} + \sum_{i=1}^{M_t} Z_i e^{-\delta(t-S_i)}$$

where:

- λ_0 is the initial value of λ_t that is carried on from catastrophic events incurred previously;
- $\{Z_i\}_{i=1,2,\cdots}$ is a sequence of independent and identically distributed random variables with distribution function $G(z)(z>0)$ and $E(Z)<\infty$ (i.e. magnitude of contribution of catastrophic event i to intensity);
- $\{S_i\}_{i=1,2,\cdots}$ is the sequence representing the event times of a Poisson process M_t with constant intensity ρ; and
- δ is the rate of exponential decay.

Catastrophic events may take long to materialise so the decay rate may not be exponential. It is assumed to be of this form for a matter of convenience, i.e. closed-form expressions of final results are easily derived. We also make the additional assumption that a Poisson process

M_t and the sequences $\{Z_i\}_{i=1,2,\cdots}$, $\{X_i\}_{i=1,2,\cdots}$ and

$\{Y_i\}_{i=1,2,\cdots}$ are independent of each other.

A Poisson process with loss frequency rate μ is also studied for N_t, that may be considered when catastrophic loss frequency rate is deterministic.

With the above model specifications, we calculate the following relevant risk measures:

$$\lim_{u \nearrow 1} \mathbb{P}\left\{ L_t^{(1)} > F^{-1}_{L_t^{(1)}}(l) \middle| L_t^{(2)} > F^{-1}_{L_t^{(2)}}(l) \right\}, \tag{3}$$

$$\mathbb{E}\left\{ L_t^{(1)} \middle| L_t^{(1)} > F^{-1}_{L_t^{(1)}}(l), L_t^{(2)} > F^{-1}_{L_t^{(2)}}(l) \right\}, \tag{4}$$

$$\lim_{u \nearrow 1} \mathbb{P}\left\{ L_t^{(1)} > F^{-1}_{L_t^{(1)}}(l) \middle| L_t^{(1)} + L_t^{(2)} > F^{-1}_{L_t^{(1)}+L_t^{(2)}}(l) \right\} \tag{5}$$

and

$$\mathbb{E}\left\{ L_t^{(1)} \middle| L_t^{(1)} + L_t^{(2)} > F^{-1}_{L_t^{(1)}+L_t^{(2)}}(l) \right\}. \tag{6}$$

Here $F_{L_t^{(1)}}$, $F_{L_t^{(2)}}$ and $F_{L_t^{(1)}+L_t^{(2)}}$ are the distribution functions of the random variables $L_t^{(1)}$, $L_t^{(2)}$ and $L_t^{(1)} + L_t^{(2)}$ respectively. The quantities (3) and (5) are known as asymptotic upper tail dependence measures and the quantities (4) and (6) are conditional tail expectations. The motivation for calculating these quantities that measure extremal dependence in the upper tail of a bivariate distribution is that insurance industry is more concerned with dependence between extreme losses. For a discussion on the coefficient of tail dependence parameters, see McNeil et al. [22].

In order to evaluate above risk measures, we need to obtain the joint distribution of the aggregate collateral losses $L_t^{(1)}$ and $L_t^{(2)}$. Unfortunately, it is not easy to derive the joint distribution the aggregate collateral losses explicitly. So in Section 2, we derive joint Laplace transform of the distribution of the aggregate collateral losses expressed with a copula function applying the piecewise deterministic Markov processes (PDMPs) theory. For N_t, a shot-noise Cox process and a homogeneous Poisson process are used respectively. Section 3 provides the expressions of the moments, covariance and linear correlation between $L_t^{(1)}$ and $L_t^{(2)}$ at time t for both cases. In Section 4, we present the expressions for joint probabilities of the aggregate collateral losses and their densities at $L_t^{(1)} = 0$ and $L_t^{(2)} = 0$, which are required to improve the accuracy of the distributions of the aggregate collateral losses inverting joint Fast Fourier transforms. We also provide the figures of the joint distribution of the aggregate collateral losses and their contours. In Section 5, we illustrate the calculations of relevant risk measures (3)-(6) using joint Fast Fourier transforms. For numerical illustrations, an exponential distribution for the jump sizes of catastrophic event, a member of Farlie-Gumbel-Morgenstern copula with exponential margins are used throughout the paper. Section 6 shows the sensitivity analysis on the parameters of a shot-noise Cox process from a Poisson process using the risk measure of (4). Concluding remarks are in Section 7.

2. Joint Laplace Transform of the Distribution of the Aggregate Collateral Losses The piecewise deterministic Markov processes theory developed by Davis [23] is a powerful mathematical tool for examining non-diffusion models. From now on, we present definitions and important properties of $L_t^{(1)}$ and $L_t^{(2)}$ with the aid of piecewise deterministic Markov processes theory [16,24,25]. This theory is used to derive joint Laplace transform of the distribution of the aggregate collateral losses $L_t^{(1)}$ and $L_t^{(2)}$.

2.1. Shot-Noise Cox Process

Assuming that the loss arrival process N_t follows a Cox process with shot noise intensity λ_t, the generator of the process $\left(\lambda_t, \Lambda_t, N_t, L_t^{(1)}, L_t^{(2)}, t\right)$ acting on a function $f\left(\lambda, \Lambda, n, l^{(1)}, l^{(2)}, t\right)$ belonging to its domain is given by

$$\begin{aligned}
&Af\left(\lambda,\Lambda,n,l^{(1)},l^{(2)},t\right)\\
&=\frac{\partial f}{\partial t}+\lambda\frac{\partial f}{\partial \Lambda}\\
&+\lambda\left[\int_0^\infty\int_0^\infty f\left(\lambda,\Lambda,n+1,l^{(1)}+x,l^{(2)}+y,t\right)\right.\\
&\qquad\cdot\frac{\partial^2 C\left(F(x),F(y)\right)}{\partial x\partial y}\mathrm{d}x\mathrm{d}y\\
&\qquad\left.-f\left(\lambda,\Lambda,n,l^{(1)},l^{(2)},t\right)\right]\\
&-\delta\lambda\frac{\partial f}{\partial \lambda}+\rho\\
&\cdot\left[\int_0^\infty f\left(\lambda+z,\Lambda,n,l^{(1)},l^{(2)},t\right)\mathrm{d}G(z)-f\left(\lambda,\Lambda,n,l^{(1)},l^{(2)},t\right)\right]
\end{aligned} \tag{7}$$

where $\Lambda_t=\int_0^t \lambda_s \mathrm{d}s$. Now let us find a suitable martingale in order to derive joint Laplace transform of the distribution of the aggregate collateral losses $L_t^{(1)}$ and $L_t^{(2)}$.

Lemma 2.1. Considering constants

$0\le\zeta\le 1, \nu\ge 0, \xi\ge 0$ and $\kappa\ge 0$ then

$$\begin{aligned}
&\zeta^{N_t}\exp\left(-\nu L_t^{(1)}\right)\exp\left(-\xi L_t^{(2)}\right)\\
&\exp\left[-\left\{\zeta\hat{c}(\nu,\xi)-1\right\}\Lambda_t\right]\\
&\times\exp\left(-\kappa\lambda_t e^{\delta t}\right)\exp\left[\rho\int_0^t\left\{1-\hat{g}\left(\kappa e^{\delta s}\right)\right\}\mathrm{d}s\right]
\end{aligned} \tag{8}$$

is a martingale, where

$$\hat{c}(\nu,\xi)=\int_0^\infty\int_0^\infty e^{-\nu x}e^{-\xi y}\frac{\partial^2 C\left(F(x),F(y)\right)}{\partial x\partial y}\mathrm{d}x\mathrm{d}y<\infty$$

and

$$\hat{g}(u) = \int_0^\infty e^{-uz} \mathrm{d}G(z)$$
.

Proof. From (7), $f\left(\lambda, \Lambda, n, l^{(1)}, l^{(2)}, t\right)$ has to satisfy

$Af = 0$ for $f\left(\lambda, \Lambda, n, l^{(1)}, l^{(2)}, t\right)$ to be a martingale. Setting

$$f\left(\lambda, \Lambda, n, l^{(1)}, l^{(2)}, t\right)$$
$$= \zeta^n e^{\psi\Lambda} \exp\left(-\nu l^{(1)}\right) \exp\left(-\xi l^{(2)}\right) \exp\left(-\kappa\lambda e^{\delta t}\right) e^{B(t)}$$

we get the equation

$$B'(t) + \lambda\psi + \lambda\left\{\zeta\hat{c}(\nu,\xi) - 1\right\} + \rho\left\{\hat{g}\left(\kappa e^{\delta t}\right) - 1\right\} = 0 \tag{9}$$

and the solution is

$$\psi = -\left\{\zeta\hat{c}(\nu,\xi) - 1\right\}, B(t) = \rho\int_0^t \left\{1 - \hat{g}\left(\kappa e^{\mathrm{d}s}\right)\right\} \mathrm{d}s \tag{10}$$

by which the result follows.

Using the martingale obtained in Lemma 2.1 and setting $\zeta = 1$ and $\kappa = 0$, we can easily obtain the general form of joint Laplace transform of the distribution of the aggregate collateral losses $L_t^{(1)}$ and $L_t^{(2)}$, i.e.

$$E\left\{ e^{-\nu L_{t_2}^{(1)}} e^{-\xi L_{t_2}^{(2)}} \middle| \Im_t \right\}$$
$$= \exp\left(-\nu L_{t_1}^{(1)}\right) \exp\left(-\xi L_{t_1}^{(2)}\right) \exp\left[\left\{\hat{c}(\nu,\xi) - 1\right\} \int_{t_1}^{t_2} \lambda_s \mathrm{d}s\right], \tag{11}$$

where the conditional expectation E is based on the probability space $(\Omega, \Im, P)$, and the information set

$\Im = \left\{\Im_t\right\}_{t \geq 0}$ with the filtration

$\Im_t = \sigma\left\{\left(\lambda_s, L_s^{(1)}, L_s^{(2)}\right): s \le t\right\}$. Without loss of generalitychange the time scale and assume that $L_0^{(1)} = 0$ and $L_0^{(2)} = 0,$ then it is given by

$$E\left\{e^{-\nu L_t^{(1)}} e^{-\xi L_t^{(2)}} \middle| \lambda_0\right\} = \exp\left[\left\{\hat{c}(\nu,\xi) - 1\right\} \int_0^t \lambda_s \mathrm{d}s\right]. \tag{12}$$

Throughout the paper, we firstly assume that jump sizes of catastrophic event follow an exponential distribution, i.e. $g(z) = \gamma \exp(-\gamma z), z > 0, \gamma > 0$ and that λ_t is stationary. Then using Theorem 2.6 in [16], (12) is given by

$$E\left\{e^{-\nu L_t^{(1)}} e^{-\xi L_t^{(2)}}\right\} = \left(\frac{\gamma e^{-\delta t}}{\gamma + \frac{1-\hat{c}(\nu,\xi)}{\delta}\left(1 - e^{-\delta t}\right)}\right)^{\frac{\rho}{\delta}} \cdot \left(\frac{\gamma + \frac{1-\hat{c}(\nu,\xi)}{\delta}\left(1 - e^{-\delta t}\right)}{\gamma e^{-\delta t}}\right)^{\frac{\gamma\rho}{\delta\gamma + \{1-\hat{c}(\nu,\xi)\}}}. \tag{13}$$

Secondly, as a specific example for C, we use the Farlie-Gumbel-Morgenstern (FGM) copulas given by

$$C(u,v) = uv + \theta uv(1-u)(1-v), \tag{14}$$

where $u \in [0,1],\ v \in [0,1]$ and $\theta \in [-1,1]$. Lastly, to make this calculation somewhat easier, we assume that $F(x) = 1 - e^{-\alpha x}(\alpha > 0, x > 0)$ and $F(y) = 1 - e^{-\beta y}(\beta > 0, y > 0)$. We omit the corresponding expression for the above joint Laplace transform of the distribution of the aggregate collateral losses as it can be easily obtained using the joint distribution function $F(x,y)$ driven by (14), i.e.

$$\hat{c}(\nu,\xi) = 1 - \frac{(\alpha\xi + \beta\nu + \nu\xi)(2\alpha + \nu)(2\beta + \xi) - \theta\alpha\beta\nu\xi}{(\alpha + \nu)(\beta + \xi)(2\alpha + \nu)(2\beta + \xi)}. \quad (15)$$

It will be of interest to examine the joint Laplace transform of the distribution of the aggregate collateral losses $L_t^{(1)}$ and $L_t^{(2)}$ at time t, using other copulas and other margins $F(x)$ and $F(y)$. If $\{Z_i\}_{i=1,2,\cdots}$, which are the magnitude of contribution of catastrophic event to intensity λ_t, are high, we also need to consider heavy-tailed distributions for jump size of catastrophic event, $G(z)$. We also omit the corresponding expressions for the Laplace transform of the distribution of $L_t^{(i)}, i = 1,2$ which are the Laplace transforms of the distribution of the compound Cox process with shot noise intensity λ_t, where its jump sizes follow an exponential distribution [16].

If we set $\theta = 0$ in (15), we have joint Laplace transform of the distribution of the aggregate collateral losses, which is the case that two losses X and Y occur same time from a sharing loss frequency rate λ_t, but their sizes are independent each other. Due to the dependence of collateral losses of X and Y with sharing loss frequency rate λ_t, we can see that

$$E\left\{e^{-\nu L_t^{(1)}} e^{-\xi L_t^{(2)}}\right\} \neq E\left(e^{-\nu L_t^{(1)}}\right) E\left(e^{-\xi L_t^{(2)}}\right) \quad (16)$$

even if $\theta = 0$. If loss X occurs by arrival process $N_t^{(1)}$ with shot noise intensity $\lambda_t^{(1)}$ that has three parameters of $\delta^{(1)}$, $\rho^{(1)}$ and $G\left(z^{(1)}\right)$ and loss Y occurs by arrival process $N_t^{(2)}$ with shot noise intensity $\lambda_t^{(2)}$ that has three parameters of $\delta^{(2)}$, $\rho^{(2)}$ and $G\left(z^{(2)}\right)$ and everything is independent each other, we can have the joint Laplace transform of the distribution of aggregate losses $L_t^{(1)}$ and $L_t^{(2)}$ at time t, that is the product of the Laplace transforms of the distribution of $L_t^{(i)}, i=1,2$.

2.2. Homogeneous Poisson Process

Let us now assume that the loss arrival process N_t follows a homogeneous Poisson process with loss frequency μ. Setting $\lambda_t = \mu$ in (12), i.e. considering deterministic loss frequency μ, we can easily obtain that

$$E\left\{e^{-\nu L_t^{(1)}} e^{-\xi L_t^{(2)}}\right\} = \exp\left[-\mu\{1-\hat{c}(\nu,\xi)\}t\right] \tag{17}$$

and using (14) we have

$$E\left\{e^{-\nu L_t^{(1)}} e^{-\xi L_t^{(2)}}\right\} = \exp\left[-\mu\left\{\frac{(\alpha\xi+\beta\nu+\nu\xi)(2\alpha+\nu)(2\beta+\xi)-\theta\alpha\beta\nu\xi}{(\alpha+\nu)(\beta+\xi)(2\alpha+\nu)(2\beta+\xi)}\right\}t\right]. \tag{18}$$

We omit the corresponding expressions for the Laplace transform of the distribution of $L_t^{(i)}, i=1,2,$ which are the Laplace transforms of the distribution of the compound Poisson process with exponential loss sizes, as they can be easily obtained setting $\xi = 0$ and $\nu = 0,$ respectively.

Similar to shot-noise Cox process for N_t, due to the dependence of collateral losses of X and Y with sharing loss frequency rate $\mu,$ it shows that

$$E\left\{e^{-\nu L_t^{(1)}} e^{-\xi L_t^{(2)}}\right\} \neq E\left(e^{-\nu L_t^{(1)}}\right) E\left(e^{-\xi L_t^{(2)}}\right) \tag{19}$$

even if $\theta = 0$. If loss X occurs by arrival process $N_t^{(1)}$ with loss frequency $\mu^{(1)}$ and loss Y occurs by arrival process $N_t^{(2)}$ with loss frequency $\mu^{(2)}$ and everything is independent each other, we can also have the joint Laplace transform of the distribution of aggregate losses $L_t^{(1)}$ and $L_t^{(2)}$ at time t that is the product of the Laplace transforms of the distribution of $L_t^{(i)}, i=1,2$.

3. MOMENTS, COVARIANCE AND LINEAR CORRELATION OF AGGREGATE COLLATERAL LOSSES

In this section, we examine the moments, covariance and linear correlation between $L_t^{(1)}$ and $L_t^{(2)}$ at time t. For the loss arrival process N_t, we use a Cox process with shot noise intensity λ_t and a homogeneous Poisson process with loss frequency μ, respectively.

3.1. Shot-Noise Cox Process

Differentiating (12) w.r.t. ν and ξ and set $\nu = 0$ and $\xi = 0,$ we can derive the joint expectation of $L_t^{(1)}$ and $L_t^{(2)}$ at time t, i.e.

$$E\left\{L_t^{(1)}L_t^{(2)}\right\} = E(\Lambda_t)E(XY)+E\left(\Lambda_t^2\right)E(X)E(Y). \tag{20}$$

Also set $\xi = 0$ and $\nu = 0$ in (12) and differentiate itw.r.t. v and ξ respectively, we can obtain the expectation of $L_t^{(1)}$ and $L_t^{(2)}$ at time t, i.e.

$$E\left\{L_t^{(1)}\right\} = E(\Lambda_t)E(X) \tag{21}$$

and

$$E\left\{L_t^{(2)}\right\} = E(\Lambda_t)E(Y). \tag{22}$$

The higher moments of $L_t^{(1)}$ and $L_t^{(2)}$ at time t can be obtained by differentiating it further, i.e.

$$Var\{L_t^{(1)}\} = E(\Lambda_t)E(X^2) + Var(\Lambda_t)\{E(X)\}^2 \tag{23}$$

and

$$Var\{L_t^{(2)}\} = E(\Lambda_t)E(Y^2) + Var(\Lambda_t)\{E(Y)\}^2. \tag{24}$$

The covariance between $L_t^{(1)}$ and $L_t^{(2)}$ at time t is given by

$$Cov\left(L_t^{(1)}, L_t^{(2)}\right) = E(\Lambda_t)E(XY) + Var(\Lambda_t)E(X)E(Y) \tag{25}$$

and the linear correlation coefficient between $L_t^{(1)}$ and $L_t^{(2)}$ at time t is given by

$$\rho\left(L_t^{(1)}, L_t^{(2)}\right) = \frac{E(\Lambda_t)E(XY) + Var(\Lambda_t)E(X)E(Y)}{\sqrt{\{E(\Lambda_t)\}^2 E(X^2)E(Y^2) + E(\Lambda_t)Var(\Lambda_t)\left[E(X^2)\{E(Y)\}^2 + \{E(X)\}^2 E(Y^2)\right] + \{Var(\Lambda_t)E(X)E(Y)\}^2}} \tag{26}$$

Let us now illustrate the calculations of the covariance and linear correlation between $L_t^{(1)}$ and $L_t^{(2)}$ at time t, where N_t follows a shot-noise Cox process.

Example 1

From [17], we have

$$E(\Lambda_t) = \frac{\rho}{\gamma\delta}t, Var(\Lambda_t) = \frac{2\rho}{\gamma^2\delta^2}t + \frac{2\rho}{\gamma^2\delta^3}e^{-\delta t} - \frac{2\rho}{\gamma^2\delta^3}$$

and using (14) we have

$$E(XY) = \frac{1}{\alpha\beta}\left(1 + \frac{\theta}{4}\right).$$

The parameter values used to calculate the covariance and linear correlation using (25) and (26) are

$$\delta = 0.5, \rho = 2, \gamma = 1, \alpha = 1, \beta = 0.5, t = 1.$$

Hence from (25) and (26), the calculations of covariance and linear correlation between $L_t^{(1)}$ and $L_t^{(2)}$ at time t are shown in Tables 1 and 2 respectively.

3.2. Homogeneous Poisson Process

Similar to a shot-noise Cox process for N_t, differentiating (17) w.r.t. ν and ξ and set $\nu = 0$ and $\xi = 0$ we can easily derive the joint expectation of $L_t^{(1)}$, and $L_t^{(2)}$ at time t, i.e.

$$E\left\{L_t^{(1)} L_t^{(2)}\right\} = \mu t E(XY) + (\mu t)^2 E(X) E(Y). \tag{27}$$

We omit the expressions for the expectation and variance of $L_t^{(1)}$ and $L_t^{(2)}$ at time t as they can be easily derived similar to a shot-noise Cox process for N_t.

The covariance between $L_t^{(1)}$ and $L_t^{(2)}$ at time t is given by

$$Cov\left(L_t^{(1)}, L_t^{(2)}\right) = \mu t E(XY) \tag{28}$$

and the linear correlation coefficient between $L_t^{(1)}$ and $L_t^{(2)}$ at time t is given by

$$\rho\left(L_t^{(1)}, L_t^{(2)}\right) = \frac{E(XY)}{\sqrt{E(X^2)E(Y^2)}}. \tag{29}$$

Let us now illustrate the calculations of the covariance and linear correlation between $L_t^{(1)}$ and $L_t^{(2)}$ at time t, where N_t follows a Poisson process.

Example 2

The parameter values used to calculate the covariance and linear correlation using (28) and (29) are

$$\mu = 4, \alpha = 1, \beta = 0.5, t = 1.$$

From (28) and (29), the calculations of covariance and linear correlation between $L_t^{(1)}$ and $L_t^{(2)}$ at time t are shown in **Table 3** and **Table 4** respectively.

Table 1. Covariance between $L_t^{(1)}$ and $L_t^{(2)}$.

θ	$Cov\left(L_t^{(1)}, L_t^{(2)}\right)$
−1	12.82
−0.5	13.82
0	14.82
0.5	15.82
1	16.82

Table 2. Linear correlation between $L_t^{(1)}$ and $L_t^{(2)}$.

θ	$\rho\left(L_t^{(1)}, L_t^{(2)}\right)$
−1	0.56175
−0.5	0.60557
0	0.64940
0.5	0.69322
1	0.73705

Table 3. Covariance between $L_t^{(1)}$ and $L_t^{(2)}$.

θ	$Cov\left(L_t^{(1)}, L_t^{(2)}\right)$
−1	6
−0.5	7
0	8
0.5	9
1	10

Table 4. Linear correlation between $L_t^{(1)}$ and $L_t^{(2)}$.

θ	$\rho\left(L_t^{(1)}, L_t^{(2)}\right)$
−1	0.375
−0.5	0.4375
0	0.5
0.5	0.5625
1	0.625

3.3. Comparison

The parameter values used in Examples 1-2 provide us with the same means of aggregate collateral losses regardless of the loss arrival process N_t, i.e.

$$\mathbb{E}^{\text{Poisson}}\left\{L_t^{(k)}\right\} = \mathbb{E}^{\text{Cox}}\left\{L_t^{(k)}\right\},$$

where

$$\mathbb{E}^{\text{Poisson}}\left\{L_t^{(1)}\right\} = \mathbb{E}^{\text{Cox}}\left\{L_t^{(1)}\right\} = 4 \tag{30}$$

and

$$\mathbb{E}^{\text{Poisson}}\left\{L_t^{(2)}\right\} = \mathbb{E}^{\text{Cox}}\left\{L_t^{(2)}\right\} = 8. \tag{31}$$

However for each θ, Tables 1 and 3 show that there is an increase in the covariance between $L_t^{(1)}$ and $L_t^{(2)}$ by changing N_t from a homogenous Poisson process to a shot-noise Cox process. Tables 2 and 4 also show that the linear correlation between $L_t^{(1)}$ and $L_t^{(2)}$ increases by changing N_t from a homogenous Poisson process to a shot-noise Cox process for each θ. This implies that the marginal distributions of the aggregate collateral loss with respect to a Cox process have heavier tail than their counterparts with respect to a Poisson process, i.e.

$$Var^{\text{Poisson}}\left\{L_t^{(1)}\right\} = 8 < Var^{\text{Cox}}\left\{L_t^{(1)}\right\} = 11.41 \tag{32}$$

and

$$Var^{\text{Poisson}}\left\{L_t^{(2)}\right\} = 32 < Var^{\text{Cox}}\left\{L_t^{(2)}\right\} = 45.64. \tag{33}$$

It will also become apparent by the joint distributions of aggregate collateral losses and their contours in Section 4 and numerical risk measure values in Examples 3- 6.

4. JOINT DISTRIBUTION OF THE AGGREGATE COLLATERAL LOSSES VIA BIVARIATE FAST FOURIER TRANSFORM

In order to calculate the risk measures of (3)-(6), we invert bivariate Fast Fourier transforms from joint Laplace transforms of the vector $\left(L_t^{(1)}, L_t^{(2)}\right)$ obtained in Section 2. For details on how to use bivariate Fast Fourier transform, we refer to [26-28]. Before we show the calculations of risk measures in Section 5, we present the expressions for the joint probabilities of the aggregate collateral losses and their densities at $L_t^{(1)}=0$ and $L_t^{(2)}=0$. These are required to improve the accuracy of the joint distributions of the aggregate collateral losses inverting bivariate Fast Fourier transforms.

4.1. Shot-Noise Cox Process

If we let $\nu\to\infty$ and $\xi\to\infty$ in (13), we have the expression for the joint probability of aggregate collateral losses at $L_t^{(1)}=0$ and $L_t^{(2)}=0$, i.e.

$$\mathbb{P}\left(L_t^{(1)}=0,L_t^{(2)}=0\right)=\left(\frac{\delta\gamma\mathrm{e}^{-\delta t}}{1-\mathrm{e}^{-\delta t}+\delta\gamma}\right)^{\frac{\rho}{\delta(1+\delta\gamma)}}. \tag{34}$$

Regardless of loss size distributions, we have the same joint probability of aggregate collateral losses at $L_t^{(1)}=0$ and $L_t^{(2)}=0$ when the loss arrival process N_t follows a Cox process with shot noise intensity λ_t. If we set

$$\lim_{\nu\to\infty,\xi\to\infty}\nu\xi\left[\left(\frac{\gamma\mathrm{e}^{-\delta t}}{\gamma+\frac{1-\hat{c}(\nu,\xi)}{\delta}\left(1-\mathrm{e}^{-\delta t}\right)}\right)^{\frac{\rho}{\delta}}\left(\frac{\gamma+\frac{1-\hat{c}(\nu,\xi)}{\delta}\left(1-\mathrm{e}^{-\delta t}\right)}{\gamma\mathrm{e}^{-\delta t}}\right)^{\frac{\gamma\rho}{\delta\gamma+\{1-\hat{c}(\nu,\xi)\}}}-\left(\frac{\delta\gamma\mathrm{e}^{-\delta t}}{1-\mathrm{e}^{-\delta t}+\delta\gamma}\right)^{\frac{\rho}{\delta(1+\delta\gamma)}}\right],$$

the expression for the joint density of aggregate collateral losses at $L_t^{(1)}=0$ and $L_t^{(2)}=0$ is given by

$$\frac{f_{X,Y}(0,0)\times\rho}{1+\delta\gamma}\left(\frac{\delta\gamma\mathrm{e}^{-\delta t}}{1-\mathrm{e}^{-\delta t}+\delta\gamma}\right)^{\frac{\rho}{\delta(1+\delta\gamma)}}\times\left\{\frac{\gamma}{1+\delta\gamma}\ln\left(\frac{1-\mathrm{e}^{-\delta t}+\delta\gamma}{\delta\gamma\mathrm{e}^{-\delta t}}\right)+\frac{1-\mathrm{e}^{-\delta t}}{\delta\left(1-\mathrm{e}^{-\delta t}+\delta\gamma\right)}\right\}. \tag{35}$$

Based on (35), we can easily obtain its expression for exponential loss sizes using (14), i.e.

$$\frac{(1+\theta)\alpha\beta\rho}{1+\delta\gamma}\left(\frac{\delta\gamma\mathrm{e}^{-\delta t}}{1-\mathrm{e}^{-\delta t}+\delta\gamma}\right)^{\frac{\rho}{\delta(1+\delta\gamma)}}\times\left\{\frac{\gamma}{1+\delta\gamma}\ln\left(\frac{1-\mathrm{e}^{-\delta t}+\delta\gamma}{\delta\gamma\mathrm{e}^{-\delta t}}\right)+\frac{1-\mathrm{e}^{-\delta t}}{\delta\left(1-\mathrm{e}^{-\delta t}+\delta\gamma\right)}\right\}. \tag{36}$$

Figures 1-4 are the joint distributions of aggregate collateral losses and their contours at each value of θ with respect to a shot-noise Cox process for N_t.

Figure 2 (or **Figure 1**) shows that joint probabilities of aggregate collateral losses are mainly located between the bottom left corner and the top right corner when $\theta = 1$, which means loss X and Y move in the same direction. On the other hand, compared to when $\theta = 1$, **Figure 4** (or **Figure 3**) shows that joint probabilities of aggregate collateral losses at the bottom left corner and the top right corner moves to its diagonal left and right, respectively when $\theta = -1$ which means loss X and Y move in the opposite direction.

4.2. Homogeneous Poisson Process

If we let $\nu \to \infty$ and $\xi \to \infty$ in (18), we have the expression for the joint probability of aggregate collateral losses at $L_t^{(1)} = 0$ and $L_t^{(2)} = 0$, i.e.

$$\mathbb{P}\left(L_t^{(1)} = 0, L_t^{(2)} = 0\right) = \mathrm{e}^{-\mu t}. \tag{37}$$

Regardless of loss size distributions, we have the same joint probability of aggregate collateral losses at $L_t^{(1)} = 0$ and $L_t^{(2)} = 0$ when the loss arrival process N_t follows a homogeneous Poisson process with loss frequency μ. If we set

$$\lim_{\nu \to \infty, \xi \to \infty} \nu\xi \left[\exp\left[-\mu \left\{ \frac{(\alpha\xi + \beta\nu + \nu\xi)(2\alpha + \nu)(2\beta + \xi) - \theta\alpha\beta\nu\xi}{(\alpha + \nu)(\beta + \xi)(2\alpha + \nu)(2\beta + \xi)} \right\} t \right] - \mathrm{e}^{-\mu t} \right],$$

the expression for the joint density of aggregate collateral losses at $L_t^{(1)} = 0$ and $L_t^{(2)} = 0$ is given by

$$f_{X,Y}(0,0) \times \mu t \mathrm{e}^{-\mu t}. \tag{38}$$

Based on (38), we can easily obtain its expression for exponential loss sizes using (14), i.e.

$$(1+\theta)\alpha\beta \times \mu t e^{-\mu t}. \tag{39}$$

The figures of the joint distributions of aggregate collateral losses and their contours at each value of θ with respect to a Poisson process for N_t are omitted, for which see the early version of this paper in http://ssrn.com/author=383758.

5. Calculating Risk Measures for Collateral Losses Now using the parameter values in Examples 1-2 and Matlab, let us illustrate the calculations of the risk measures of (3)-(6) with respect to a shot-noise Cox process and a Poisson process. They are shown in Tables 5-18.

To reduce the error of used algorithm for inverting bivariate Fast Fourier transforms using Matlab, the following methods have been used:

- The sampling points have been taken as many as possible, i.e. we have used 8192 × 4096 points in the calculation, which was the maximum points we could sampled in 32-bit Matlab.
- If the sampling interval gap is too small, there would be a large amount of truncation error. Also if the sampling interval is too big, there would be a large amount of sampling error. Hence we have chosen carefully the right sampling interval, so that the total errors caused by inverting bivariate Fast Fourier transforms could be negligible.

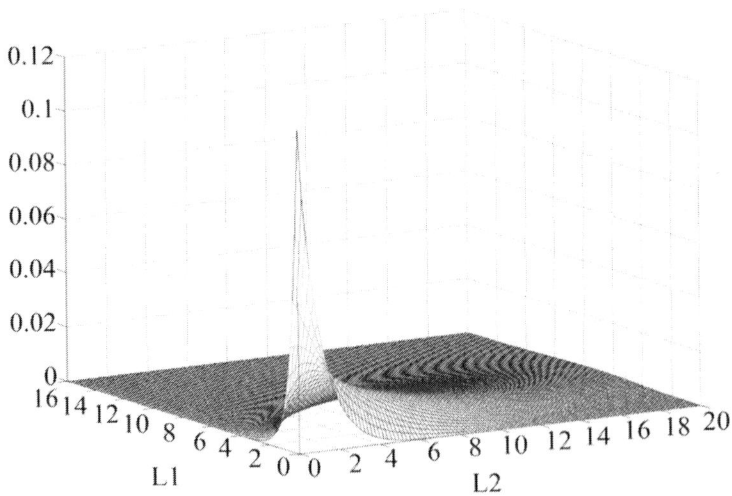

Figure 1. The joint distribution of collateral losses with θ = 1.

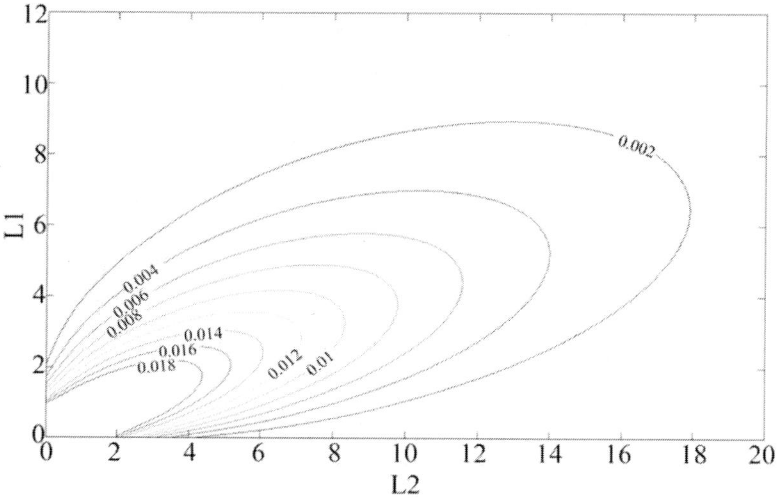

Figure 2. The contour of the joint distribution of collateral losses with θ = 1.

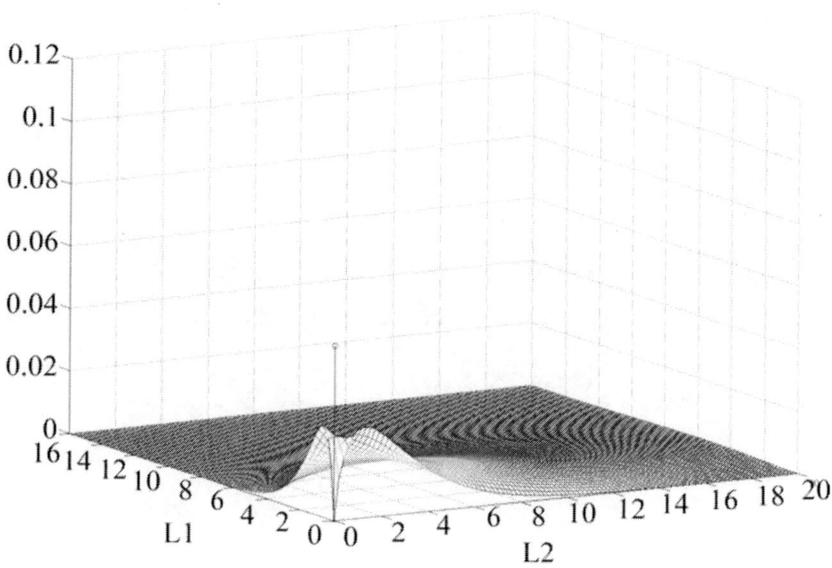

Figure 3. The joint distribution of collateral losses with θ = −1.

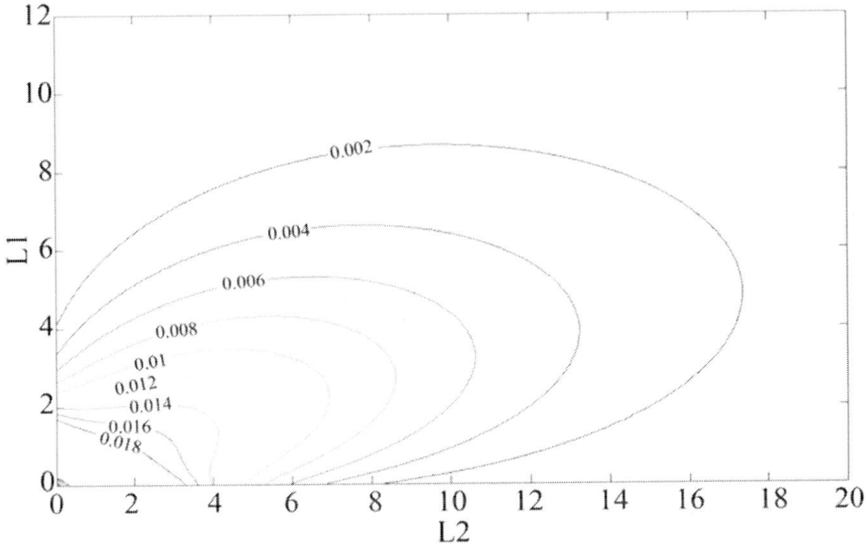

Figure 4. The contour of the joint distribution of collateral losses with θ = −1.

- For the figures in this paper, we have used the 2D low-pass/moving-average filter to filter the jitter noise out. Then we have used the linear interpolation to smooth the borders of the data. The pattern matrix used for this filter was

$$\begin{bmatrix} 0.0625 & 0.125 & 0.0625 \\ 0.125 & 0.25 & 0.125 \\ 0.0625 & 0.125 & 0.0625 \end{bmatrix}.$$

Example 3

Using the different $VaR_{q\%}$ at $q\% = 95\%$ and 99%, the calculations of the risk measure of (3) are shown in Tables 5-8.

Tables 5-8 show that each quantile values are higher with respect to a shot-noise Cox process than their counterparts. They also show that the risk measures of (3) with respect to a shot-noise Cox process are higher than their counterparts. These justify that the marginal/joint distributions of the aggregate collateral losses with respect to a shot-noise Cox process have heavier tail than their counterparts with respect to a Poisson process.

Example 4

Secondly, based on the $VaR_{q\%}$ at 95% and 99% in Example 3, the calculations of the risk measure (4), i.e.

$$\mathbb{E}\left\{L_t^{(1)} \middle| L_t^{(1)} > F^{-1}_{L_t^{(1)}}(u), L_t^{(2)} > F^{-1}_{L_t^{(2)}}(u)\right\}$$ are shown in Tables 9 and 10.

Tables 9 and 10 show that the risk measures of (4) with respect to a shot-noise Cox process are higher than their counterparts regardless of the critical values. Regardless of the loss arrival process N_t, the risk measures of (4) are getting bigger as the critical value goes to 99%. They also show that the differences between the values in **Table 9** and their counterparts in **Table 10** are getting higher as the critical value goes to 99%.

Example 5

Showing the calculations of the $q\%$ quantiles of the sum, $L_t^{(1)} + L_t^{(2)}$, i.e. $F^{-1}_{L_t^{(1)}+L_t^{(2)}}(u)$ in

$$\mathbb{P}\left\{L_t^{(1)} + L_t^{(2)} > F^{-1}_{L_t^{(1)}+L_t^{(2)}}(u)\right\}$$ and the joint probability between $L_t^{(1)} > F^{-1}_{L_t^{(1)}}(u)$ and $L_t^{(1)} + L_t^{(2)} > F^{-1}_{L_t^{(1)}+L_t^{(2)}}(u)$, i.e.

$$\mathbb{P}\left\{L_t^{(1)} > F^{-1}_{L_t^{(1)}}(u), L_t^{(1)} + L_t^{(2)} > F^{-1}_{L_t^{(1)}+L_t^{(2)}}(u)\right\}$$ for each cases in Tables 11-14, the calculations of the risk measure (5) are shown in Tables 15 and 16.

Table 5. A shot-noise Cox process, where $\mathbb{P}\left\{L_t^{(1)} > 10.55\right\} = \mathbb{P}\left\{L_t^{(2)} > 21.10\right\} = 0.05$.

θ	$\mathbb{P}\left\{L_t^{(1)} > 10.55, L_t^{(2)} > 21.10\right\}$	$\mathbb{P}\left\{L_t^{(1)} > 10.55 \middle\| L_t^{(2)} > 21.10\right\} = \mathbb{P}\left\{L_t^{(2)} > 21.10 \middle\| L_t^{(1)} > 10.55\right\}$
1	0.02090	0.41793
0.5	0.01900	0.38002
0	0.01723	0.34449
−0.5	0.01555	0.31105
−1	0.01397	0.27946

Table 6. A Poisson process, where $\mathbb{P}\left\{L_t^{(1)} > 9.37\right\} = \mathbb{P}\left\{L_t^{(2)} > 18.74\right\} = 0.05$.

θ	$\mathbb{P}\left\{L_t^{(1)} > 9.37, L_t^{(2)} > 18.74\right\}$	$\mathbb{P}\left\{L_t^{(1)} > 9.37 \middle\| L_t^{(2)} > 18.74\right\} = \mathbb{P}\left\{L_t^{(2)} > 18.74 \middle\| L_t^{(1)} > 9.37\right\}$
1	0.01459	0.29186
0.5	0.01257	0.25129
0	0.01068	0.21363
−0.5	0.00893	0.17862
−1	0.00730	0.14605

Table 7. A shot-noise Cox process, where $\mathbb{P}\left\{L_t^{(1)} > 14.79\right\} = \mathbb{P}\left\{L_t^{(2)} > 29.58\right\} = 0.01$.

θ	$\mathbb{P}\left\{L_t^{(1)} > 14.79, L_t^{(2)} > 29.58\right\}$	$\mathbb{P}\left\{L_t^{(1)} > 14.79 \middle\| L_t^{(2)} > 29.58\right\} = \mathbb{P}\left\{L_t^{(2)} > 29.58 \middle\| L_t^{(1)} > 14.79\right\}$
1	0.00284	0.28389
0.5	0.00247	0.24727
0	0.00214	0.21401
−0.5	0.00184	0.18373
−1	0.00156	0.15617

Table 8. A Poisson process, where $\mathbb{P}\left\{L_t^{(1)} > 12.61\right\} = \mathbb{P}\left\{L_t^{(2)} > 25.22\right\} = 0.01$.

θ	$\mathbb{P}\left\{L_t^{(1)} > 12.61, L_t^{(2)} > 25.22\right\}$	$\mathbb{P}\left\{L_t^{(1)} > 12.61 \middle\| L_t^{(2)} > 25.22\right\} = \mathbb{P}\left\{L_t^{(2)} > 25.22 \middle\| L_t^{(1)} > 12.61\right\}$
1	0.00152	0.15183
0.5	0.00121	0.12079
0	0.00094	0.09369
−0.5	0.00070	0.07022
−1	0.00050	0.05013

Table 9. A shot-noise Cox process.

θ $q\%$	95%	99%
1	13.99969	18.18178
0.5	13.99910	18.17480
0	13.98970	18.15887
−0.5	13.97155	18.13402
−1	13.94446	18.09996

Table 10. A Poisson process.

θ $q\%$	95%	99%
1	11.89360	14.99842
0.5	11.85723	14.95484
0	11.80964	14.90020
−0.5	11.74893	14.83189
−1	11.67172	14.74505

Table 11. $q\%$ quantiles of the sum, $L_t^{(1)} + L_t^{(2)}$ for a shotnoise Cox process.

θ $q\%$	95%	99%
1	30.27695	41.70510
0.5	30.04685	41.32161
0	29.81675	40.93811
−0.5	29.58665	40.59297
−1	29.35656	40.17112

Table 12. $q\%$ quantiles of the sum, $L_t^{(1)} + L_t^{(2)}$ for a poisson process

θ $q\%$	95%	99%
1	26.44199	34.68714
0.5	26.17355	34.22695
0	25.90510	33.80510
−0.5	25.63665	33.30656
−1	25.32986	32.80801

Table 13. $\mathbb{P}\left\{L_t^{(1)} > F_{L_t^{(1)}}^{-1}(u), L_t^{(1)} + L_t^{(2)} > F_{L_t^{(1)}+L_t^{(2)}}^{-1}(u)\right\}$ for a shot-noise Cox process.

θ $q\%$	95%	99%
1	0.02942	0.00469
0.5	0.02798	0.00439
0	0.02657	0.00410
−0.5	0.02519	0.00380
−1	0.02384	0.00354

Table 14. $\mathbb{P}\left\{L_t^{(1)} > F_{L_t^{(1)}}^{-1}(u), L_t^{(1)} + L_t^{(2)} > F_{L_t^{(1)}+L_t^{(2)}}^{-1}(u)\right\}$ for a Poisson process.

θ $q\%$	95%	99%
1	0.02444	0.00344
0.5	0.02264	0.00311
0	0.02085	0.00275
−0.5	0.01906	0.00243
−1	0.01736	0.00210

Table 15. $\mathbb{P}\left\{L_t^{(1)} > F_{L_t^{(1)}}^{-1}(u) \middle| L_t^{(1)} + L_t^{(2)} > F_{L_t^{(1)}+L_t^{(2)}}^{-1}(u)\right\}$ for a shot-noise Cox process.

θ $q\%$	95%	99%
1	0.58843	0.46900
0.5	0.55950	0.43908
0	0.53137	0.40990
−0.5	0.50387	0.38020
−1	0.47681	0.35389

Table 16. $\mathbb{P}\left\{L_t^{(1)} > F_{L_t^{(1)}}^{-1}(u) \middle| L_t^{(1)} + L_t^{(2)} > F_{L_t^{(1)}+L_t^{(2)}}^{-1}(u)\right\}$ for a Poisson process.

θ q%	95%	99%
1	0.48882	0.34377
0.5	0.45288	0.31048
0	0.41706	0.27504
−0.5	0.38126	0.24282
−1	0.34710	0.21026

Table 17. A shot-noise Cox process.

θ q%	95%	99%
1	11.69317	14.99356
0.5	11.46820	14.64166
0	11.22903	14.25449
−0.5	10.97467	13.75803
−1	10.70397	13.36212

Table 18. A Poisson process.

θ $q\%$	95%	99%
1	9.57781	11.53855
0.5	9.28678	11.14740
0	8.96709	10.60372
−0.5	8.61560	10.13953
−1	8.28519	9.57822

Tables 11 and 12 show that the $q\%$ quantiles of the sum, $L_t^{(1)} + L_t^{(2)}$ with respect to a shot-noise Cox process are higher than their counterparts regardless of the critical values. Regardless of the loss arrival process N_t, they are getting bigger as the critical value goes to 99.9%. They also show that the differences between the values in **Table 11** and their counterparts in **Table 12** are getting higher as the critical value goes to 99.9%.

Tables 13 and 14 show that the joint probability between $L_t^{(1)} > F_{L_t^{(1)}}^{-1}(u)$ and $L_t^{(1)} + L_t^{(2)} > F_{L_t^{(1)}+L_t^{(2)}}^{-1}(u)$ with respect to a shot-noise Cox process are higher than their counterparts regardless of the critical values. Regardless of the loss arrival process N_t, they are getting smaller as the critical value goes to 99.9%. They also show that the differences between the values in **Table 13** and their counterparts in **Table 14** are getting lower as the critical value goes to 99.9%.

Tables 15 and 16 show that the risk measures of (5) with respect to a shot-noise Cox process are higher than their counterparts regardless of the critical values. Regardless of the loss arrival process N_t, the risk measures of (5) are getting smaller as the critical value goes to 99.9%.

Example 6

Lastly, based on the quantile values and the joint probabilities in Example 5, the calculations of the risk measure (6), i.e. $\mathbb{E}\left\{L_t^{(1)} \middle| L_t^{(1)} + L_t^{(2)} > F^{-1}_{L_t^{(1)}+L_t^{(2)}}(u)\right\}$ are shown in Tables 17 and 18.

Tables 17 and 18 show that the risk measures of (6) with respect to a shot-noise Cox process are higher than their counterparts regardless of the quantile values. Regardless of the loss arrival process N_t, the risk measures of (6) are getting bigger as the critical value goes to 99.9%. They also show that the differences between the values in **Table 17** and their counterparts in **Table 18**are getting higher as the critical value goes to 99.9%.

With respect to the FGM copula correlation parameter, θ, the risk measures of (3)-(6) are increasing (decreasing) as it changes to $1(-1)$ regardless of the loss arrival process N_t. It will be interesting to find these risk measure values using other copulas with different marginal distributions.

6. Sensitivity Analysis In this section, we examine the effect on the risk measure of (4), i.e.

$$\mathbb{E}\left\{L_t^{(1)} \middle| L_t^{(1)} > F^{-1}_{L_t^{(1)}}(u), L_t^{(2)} > F^{-1}_{L_t^{(2)}}(u)\right\}$$

caused by changes in the values of the parameters of a shot-noise Cox process from a Poisson process. To do so, we double the means of aggregate collateral losses, (30) and (31) with $\mu = 8$ for a Poisson case and with $\rho = 4$, $\delta = 0.25$ and $\gamma = 0.5$, respectively for shotnoise Cox case, i.e.

$$\mathbb{E}^{\text{Poisson}}\left\{L_t^{(1)}\right\} = \mathbb{E}^{\text{Cox}}\left\{L_t^{(1)}\right\} = 8$$

and

$$\mathbb{E}^{\text{Poisson}}\left\{L_t^{(2)}\right\} = \mathbb{E}^{\text{Cox}}\left\{L_t^{(2)}\right\} = 16,$$

where other parameter values are the same as in Examples 1-2.

Figures 5-6 are the joint distribution of aggregate collateral losses and its contour at $\theta = 1$ with respect to a Poisson process for N_t. Figures 7-12 are the joint distributions of aggregate collateral losses and their con tours at $\theta = 1$ with respect to a shot-noise Cox process for N_t.

Figures 7-12 (shot-noise Cox cases) show how different joint probabilities of aggregate collateral losses are from Figures 5-6 (a Poisson case) by changes in the value of ρ, δ and γ, respectively. Even though they have the same mean, the joint probabilities of aggregate collateral losses for shot-noise Cox cases are located more between the bottom left corner and the top right corner when $\theta = 1$ than its counterpart. Shot-noise Cox cases display that they have heavier tail than their counterparts, which are shown in **Table 19** in terms of the risk measure of (4). Compared to the effect on the risk measure of (4) by changes in the value of ρ, δ and γ, respectively from a Poisson case, **Table 19** shows that γ, which is the parameter in exponential jump size distribution of catastrophic event, is the most sensitive parameter in terms of changes in the value the risk measure of (4).

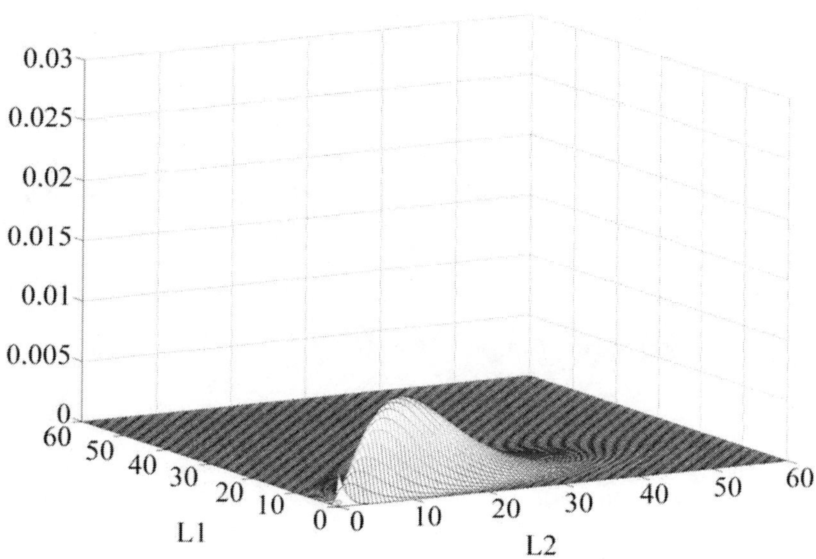

Figure 5. The joint distribution of collateral losses with θ = 1 for a Poisson process.

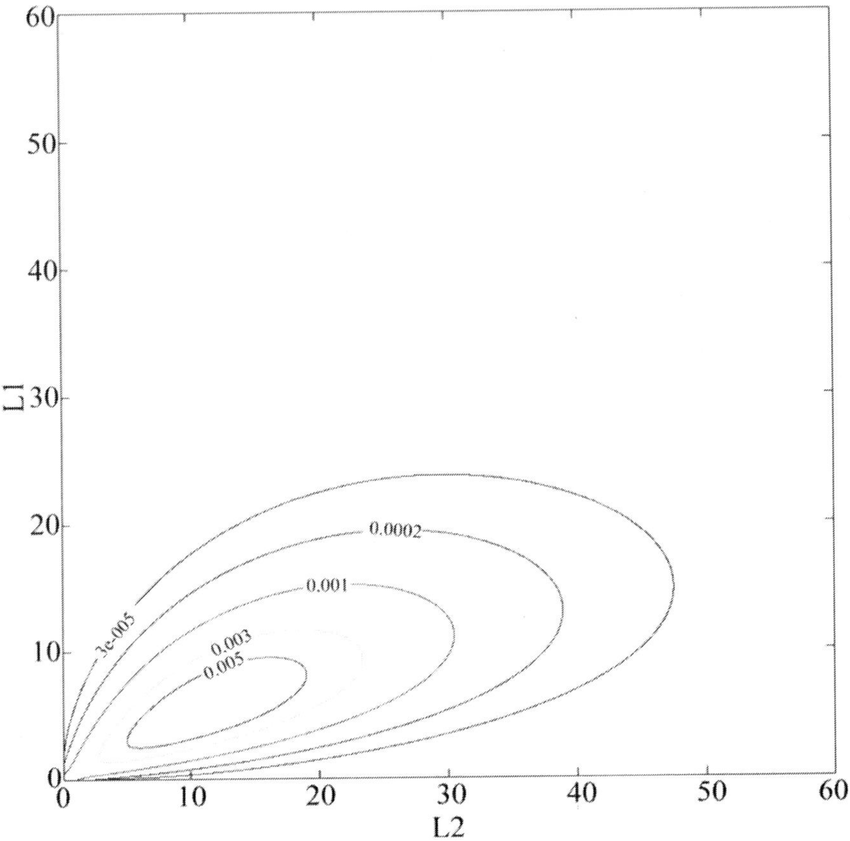

Figure 6. The contour of the joint distribution of collateral losses with θ = 1 for a Poisson process.

Table 19. $\mathbb{E}\left\{L_t^{(1)} \middle| L_t^{(1)} + L_t^{(2)} > F^{-1}_{L_t^{(1)}+L_t^{(2)}}(u)\right\}$.

$\theta = 1$ $q\%$	A Poisson process	A shot-noise Cox process		
		$\rho = 4$	$\delta = 0.25$	$\gamma = 0.5$
95%	18.52079	21.41422	21.37291	23.50179
99%	22.33392	26.15472	26.47892	29.68759

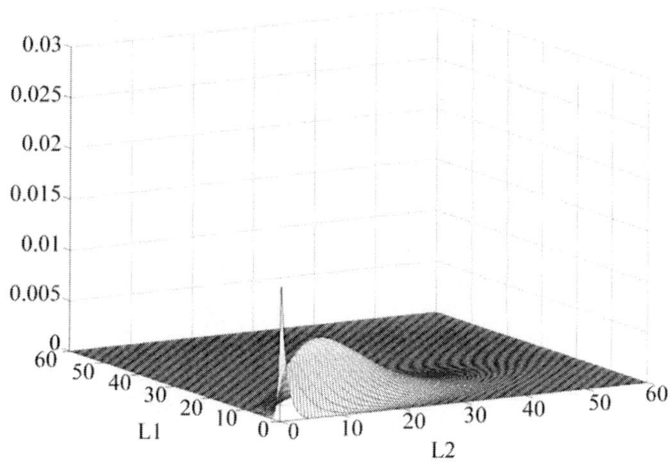

Figure 7. The joint distribution of collateral losses with $\theta = 1$ and $\rho = 4$.

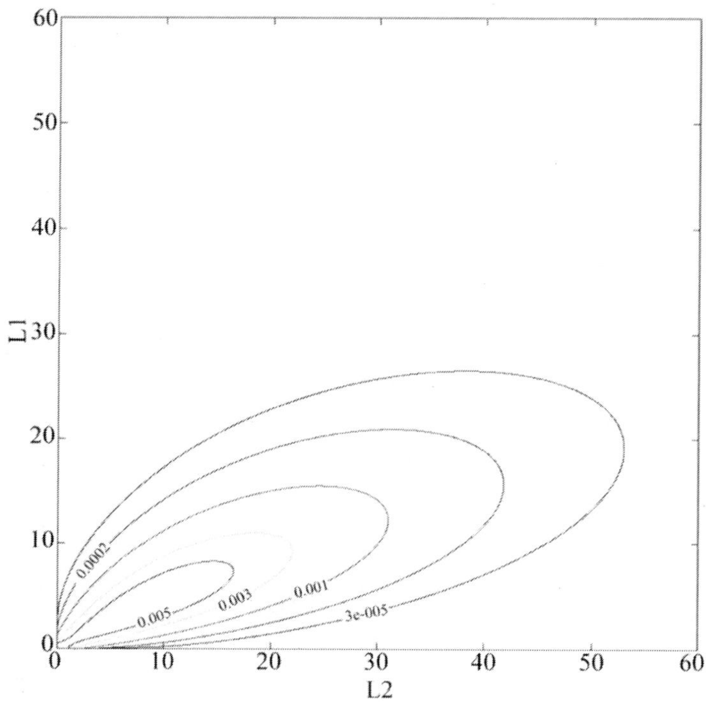

Figure 8. The contour of the joint distribution of collateral losses with $\theta = 1$ and $\rho = 4$.

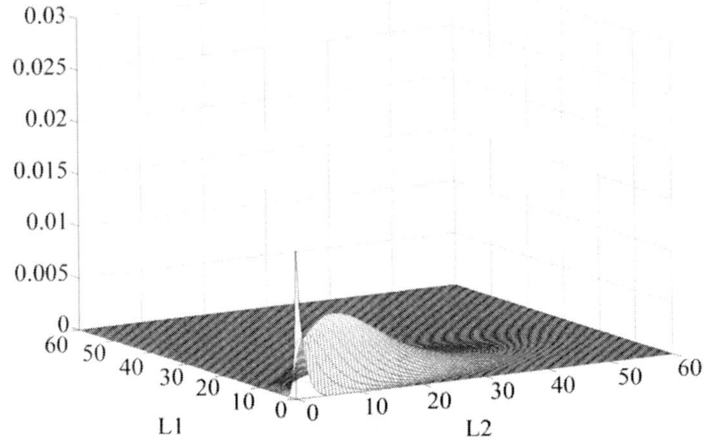

Figure 9. The joint distribution of collateral losses with θ = 1 and δ = 0.25.

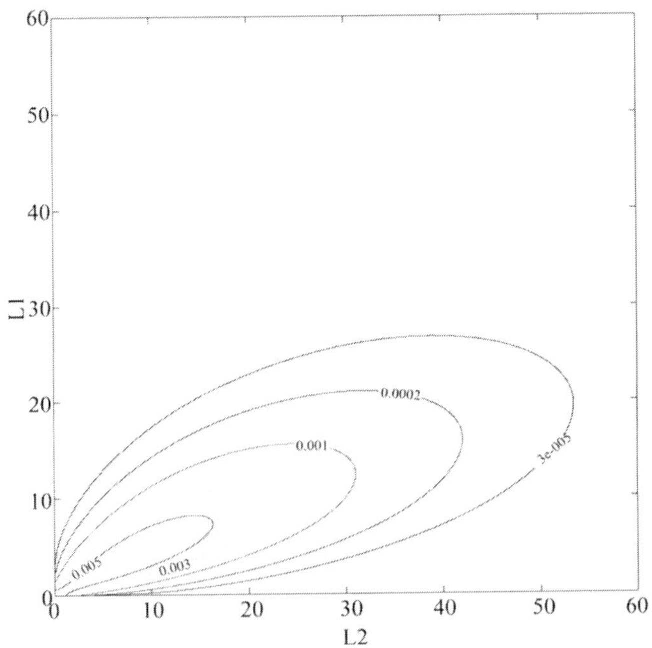

Figure 10. The contour of the joint distribution of collateral losses with θ = 1 and δ = 0.25.

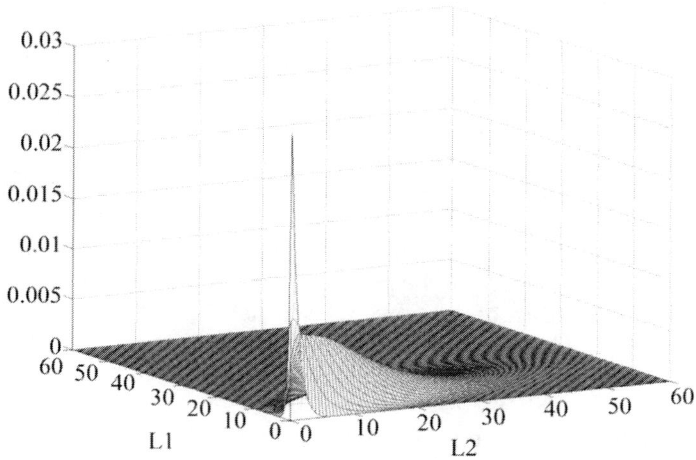

Figure 11. The joint distribution of collateral losses with θ = 1 and γ = 0.5.

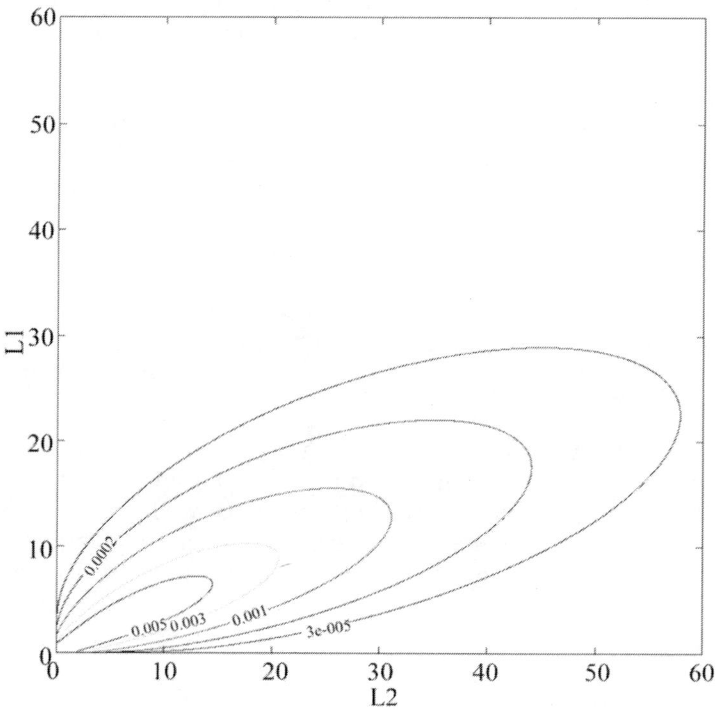

Figure 12. The contour of the joint distribution of collateral losses with θ = 1 and γ = 0.5.

7. CONCLUSIONS

We have used bivariate compound process to model aggregate collateral losses arising from catastrophic events such as flood, storm, hail, bushfire and earthquake. For the number of collateral losses, a Cox process was used to accommodate the stochastic nature of their frequency rate in practice. The shot noise process was used as the intensity of a Cox process as the number of collateral losses arising from catastrophic events depends on the frequency and magnitude of the primary events and the time period needed to determine the effect of these events. We also examined a Poisson process for the number of collateral losses as its counterpart. With the common collateral loss arrival process in the bivariate model, the dependence between the individual losses arising from different risk type has been modelled using the notion of copula where for numerical illustrations, a member of Farlie-Gumbel-Morgenstern copula with exponential margins was used.

As it was difficult to derive the joint distributions the aggregate collateral losses, we derived their Laplace transforms and inverted their Fast Fourier transforms numerically to calculate introduced relevant risk measures. These measures can be used to calculate tail dependences between collateral losses or to calculate insurance premiums. We have presented the expressions for joint probabilities of the aggregate collateral losses and their densities at $L_t^{(1)} = 0$ and $L_t^{(2)} = 0$, which were used to improve the accuracy of joint distributions of the aggregate collateral losses inverting the Fast Fourier transforms. We have compared the simulated risk measure values obtained using a compound Poisson and a compound shot-noise Cox model, respectively. The sensitivity analysis on the parameters of a shot-noise Cox process from a Poisson process also provided using the risk measure of (4).

Jiwook Jang, Genyuan Fu Other counting processes and other copulas with different margins can be considered in the proposed bivariate model, that we leave for further research. We hope that what we have presented in this paper provides practitioners with feasible models to quantify collateral losses that would occur more often due to global warming, climate change and terrorism.

REFERENCES

1. N. Bäuerle and A. Müller, "Modeling and Comparing Dependencies in Multivariate Risk Portfolios," ASTIN Bulletin, Vol. 28, No. 1, 1998, pp. 59-76.

2. H. Cossette, P. Gaillardetz, E. Marceau and J. Rioux, "On Two Dependent Individual Risk Models," Insurance: Mathematics and Economics, Vol. 30, No. 2, 2002, pp. 153-166.
3. C. Genest, E. Marceau and M. Mesfioui, "Compound Poisson Approximations for Individual Models with Dependent Risks," Insurance: Mathematics and Economics, Vol. 32, No. 1, 2003, pp. 73-91.
4. N. Bäuerle and R. Grübel, "Multivariate Counting Processes: Copulas and Beyond," ASTIN Bulletin, Vol. 35, No. 2, 2005, pp. 379-408.
5. M. L. Centeno, "Dependent Risks and Excess of Loss Reinsurance," Insurance: Mathematics and Economics, Vol. 37, No. 2, 2005, pp. 229-238.
6. F. Lindskog and A. J. McNeil, "Common Poisson Shock Models: Applications to Insurance and Credit Risk Modelling," ASTIN Bulletin, Vol. 33, No. 2, 2003, pp. 209-238.
7. D. Pfeifer and J. Nešlehová, "Modeling and Generating Dependent Risk Processes for IRM and DFA," ASTIN Bulletin, Vol. 34, No. 2, 2004, pp. 333-360.
8. V. Chavez-Demoulin, P. Embrechts and J. Nešlehová, "Quantitative Models for Operational Risk: Extremes, Dependence and Aggregation," Journal of Banking and Finance, Vol. 30, No. 10, 2006, pp. 2635-2658.
9. R. Cont and P. Tankov, "Financial Modelling with Jump Processes," Champan & Hall, Boca Raton, 2004.
10. Victorian Bushfires Royal Commission, "Final ReportSummary, Parliament of Victoria," Parliament of Victoria, Melbourne, 2010.
11. M. L. Burton and M. J. Hicks, "Hurricane Katrina: Preliminary Estimates of Commercial and Public Sector Damages," Marshall University, Huntington, 2005.
12. G. Makinen, "The Economic Effects of 9/11: A Retrospective Assessment," Congressional Research Service, Washington DC, 2002.
13. R. B. Nelsen, "An Introduction to Copulas," SpringerVerlag, New York, 1999.
14. B. Schweizer and A. Sklar, "Probabilistic Metric Spaces," Elsevier, New York, 1983.
15. C. Klüppelberg and T. Mikosch, "Explosive Poisson Shot Noise Processes with Applications to Risk Reserves," Bernoulli, Vol. 1, No. 1-2, 1995, pp. 125-147.
16. A. Dassios and J. Jang, "Pricing of Catastrophe Reinsurance & Derivatives Using the Cox Process with Shot Noise Intensity," Finance & Stochastics, Vol. 7, No. 1, 2003, pp. 73-95.
17. A. Dassios and J. Jang, "Kalman-Bucy Filtering for Linear System Driven by the Cox Process with Shot Noise Intensity and Its Application to the Pricing of Reinsurance Contracts," Journal of Applied Probability, Vol. 42, No. 1, 2005, pp. 93-107.
18. A. Dassios and J. Jang, "The Distribution of the Interval between Events of a Cox Process with Shot Noise Intensity," Journal of Applied Mathematics and Stochastic Analysis, Vol. 2008, 2008, Article ID: 367170.
19. J. Jang and Y. Krvavych, "Arbitrage-Free Premium Calculation for Extreme Losses Using the Shot Noise Process and the Esscher Transform," Insurance: Mathematics and Economics, Vol. 35, No. 1, 2004, pp. 97-111.
20. J. Jang and G. Fu, "Transform Approach for Operational Risk Management: VaR and TCE," Journal of Operational Risk, Vol. 3, No. 2, 2008, pp. 45-61.
21. D. R. Cox and V. Isham, "Point Processes," Chapman & Hall, London, 1980.
22. A. McNeil, R. Frey and P. Embrechts, "Quantitative Risk Management: Concepts, Techniques and Tools," Princeton University Press, Princeton, 2005.
23. M. H. A. Davis, "Piecewise Deterministic Markov Processes: A General Class of Non Diffusion Stochastic Models," Journal of the Royal Statistical Society Series B, Vol. 46, No. 3, 1984, pp. 353-388.
24. A. Dassios and P. Embrechts, "Martingales and Insurance Risk," Communications in Statistics. Stochastic Models, Vol. 5, No. 2, 1989, pp. 181-217.
25. T. Rolski H. Schmidli, V. Schmidt and J. L. Teugels, "Stochastic Processes for Insurance and Finance," John Wiley & Sons, Hoboken, 1998.
26. K. R. Castleman, "Digital Image Processing," Prentice Hall, Englewood Cliffs, 1996.

27. R. C. Gonzalez and R. E. Woods, "Digital Image Processing," 2nd Edition, Prentice Hall, Upper Saddle River, 2002.
28. R. C. Gonzalez, R. E. Woods and S. L. Eddins, "Digital Image Processing Using MATLAB," Prentice Hall, Upper Saddle River, 2004.

CHAPTER 11

Lebesgues-Stieltjes Integrals of Fuzzy Stochastic Processes with Respect to Finite Variation Processes

Jinping Zhang, Lingli Luo, Xingmei Li, Xiaoying Wang

Department of Mathematics and Physics, North China Electric Power University, Beijing, China

ABSTRACT

Let $G=\{G_t(\omega),t\in[0,T]\}$ be a fuzzy stochastic process and $\{A_t(\omega),t\in[0,T]\}$ be a real valued finite variation process. We define the Lebesgue-Stieltjes integral denoted by $\int_0^t G_s(\omega)\mathrm{d}A_s(\omega)$ for each $t>0$ by using the selection method, which is direct, nature and different from the indirect definition appearing in some references. We shall show that this kind of integral is also measurable, continuous in time t and bounded a.s. under the Hausdorff metric.

Keywords: Fuzzy Stochastic Process, Finite Variation Process, Fuzzy Stochastic Lebesgue-Stieltjes Integral, Measurability

1. INTRODUCTION

Recently, the theory of fuzzy functions has been developed quickly due to the measurements of various uncertainties arising not only from the randomness but also from the vagueness in some situations. For example, when considering wave height at time t denoted by f_t, due to the influence of random factors and the limitations of the measurement tools and methods, we may not precisely know the height f_t. It is reasonable to consider the wave height as a fuzzy random variable on a probability space $(\Omega, \mathcal{F}, P)$.

Since Puri and Ralescu [1] (1986) defined fuzzy random variable, there had been many further topics such as expectations of fuzzy random variables, fuzzy stochastic processes, integrals of fuzzy stochastic processes, fuzzy stochastic differential equations etc. In order to study a fuzzy function u, it is natural and equivalent to study its α -level set $[u]^{\alpha}$ for any $\alpha \in [0,1]$, where $[u]^{\alpha}$ is a set-valued function. Therefore, as usual, in order to explore the integrals of fuzzy stochastic processes, at first we can study the integrals of set-valued stochastic processes. Kisielewicz (1997) [2] used all selections to define the integral of a set-valued process as a nonempty closed subset of $L^2(\Omega, \mathcal{F}, P; R^n)$, but did not consider its measurability. Based on Kisielewicz's work (1997) [2] , Kim and Kim (1999) [3] studied some properties of this kind of integral. Jung and Kim (2003) [4] modified the definition in 1-dimensional Euclidean space R so that the integral became a set-valued random variable. After the work [4] , there are some references on set-valued integrals and fuzzy integrals. One may refer to papers such as [5] -[13] etc. and references therein. Zhang and Qi [14] (2013) considered the set-valued integral with respect to a finite variation process directly instead of taking the decomposable closure appearing in [4] [6] and other references. As a further work of [14] , here we shall explore the integrals of fuzzy stochastic processes with respect to finite variation processes and prove the measurability and boundedness of this kind of integral, the continuity with respect to t under the Hausdorff metric and its representation theorem.

This paper is organized as follows: in Section 2, we present some notions on set-valued random variables and fuzzy set-valued random variables; in Section 3, we shall give the definition of integral of fuzzy set-valued stochastic processes with respect to finite variation process and prove the measurability and L^2-boundedness which are necessary to our future work on fuzzy stochastic differential equations.

2. PRELIMINARIES

We denote N the set of all natural numbers, R the set of all real numbers, R^d the d-dimensional Euclidean space with the usual norm $\|\cdot\|$, R^+ the set of all nonnegative numbers. Let $(\Omega, \mathcal{F}, P)$ be a complete probability space, $\{\mathcal{F}_t, t \in [0,T]\}$ a σ-field filtration satisfying the usual conditions such that $\mathcal{F}_0$ includes all P-null sets in $\mathcal{F}$. The filtration is non-decreasing and right continuous. Let $\mathcal{B}(E)$ be a Borel field of a topological space E.

Let $(\Omega, \mathcal{F}, P)$ be a complete probability space. $L^p\left(\Omega, \mathcal{F}, P; R^d\right)$ (or brief $L^p\left(\Omega; R^d\right)$) $(p \geq 1)$ the set of all R^d-valued Borel measurable functions $f : \Omega \to R^d$ such that the norm

$$\|f\|_p = \left\{\int_\Omega \|f(\omega)\|^p \, dP\right\}^{\frac{1}{p}}, \text{ if } 1 \leq p < \infty,$$
$$\|f\|_\infty = \text{esssup}_{\omega \in \Omega} \|f(\omega)\|, \text{ if } p = \infty$$

is finite. f is called L^p-integrable if $f \in L^p\left(\Omega; R^d\right)$.

Let $\mathcal{K}\left(R^d\right)$ (resp. $\mathcal{K}_k\left(R^d\right)$, $\mathcal{K}_{kc}\left(R^d\right)$) be the family of all nonempty, closed (resp. nonempty compact, nonempty compact convex) subsets of R^d. For any $x \in R^d$ and $A \in \mathcal{K}\left(R^d\right)$, define the distance between x and A by $d(x, A) = \inf_{y \in A} \|x - y\|$. The Hausdorff metric d_H on $\mathcal{K}\left(R^d\right)$ (cf. [15]) is defined by

$$d_H(A,B) \doteq \max\left\{\sup_{a\in A} d(a,B), \sup_{b\in B} d(b,A)\right\} \quad A,B \in \mathcal{K}(R^d).$$

Denote $\|A\|_K \doteq d_H(\{0\},A) = \sup_{a\in A}\|a\|.$

For $A \subset R^d$, $x^* \in R^d$, the support function of A is defined as follows:

$$S(x^*,A) = \sup\{\langle x^*,x\rangle : x \in A\}.$$

A set-valued function $F:\Omega \to \mathcal{K}(R^d)$ is said to be measurable if for any open set $O \subset R^d$, the inverse $F^{-1}(O) := \{\omega \in \Omega : F(\omega) \cap O \neq \varnothing\}$ belongs to $\mathcal{F}$. Such a function F is called a set-valued random variable.

Let $\mathcal{M}(\Omega,\mathcal{F},P;\mathcal{K}(R^d))$ (resp. $\mathcal{M}(\Omega,\mathcal{F},P;\mathcal{K}_c(R^d))$, $\mathcal{M}(\Omega,\mathcal{F},P;\mathcal{K}_{kc}(R^d))$) be the family of all measurable $\mathcal{K}(R^d)$-valued (resp. $\mathcal{K}_c(R^d)$, $\mathcal{K}_{kc}(R^d)$-valued) functions, briefly by $\mathcal{M}(\Omega,\mathcal{K}(R^d))$ (resp. $\mathcal{M}(\Omega,\mathcal{F},P;\mathcal{K}_c(R^d))$, $\mathcal{M}(\Omega,\mathcal{F},P;\mathcal{K}_{kc}(R^d))$. For $F \in \mathcal{M}(\Omega,K(R^d))$, the family of all L^p-integrable selections is defined by

$$S_F^p(\mathcal{F}) := \{f \in L^p(\Omega,\mathcal{F},P;R^d) : f(\omega) \in F(\omega)\ a.s.\}, p \geq 1 \quad (1)$$

In the following, $S_F^p(\mathcal{F})$ is denoted briefly by S_F^p.

A set-valued random variable F is said to be integrable if S_F^1 is nonempty.

F is called $L^p(p\geq 1)$-integrably bounded if there exits $h \in L^p(\Omega,\mathcal{F},P;R^d)$ s.t. for all $x \in F(\omega)$, $\|x\| \leq h(\omega)$ almost surely.

An R^d-valued stochastic process $f = \{f_t : t \geq 0\}$ (or denoted by $f = \{f(t) : t \geq 0\}$) is defined as a function $f : R_+ \times \Omega \to R^d$ with the $\mathcal{F}$ -measurable section f_t, for $t \geq 0$. We say f is measurable if f is $\mathcal{B}(R_+) \otimes \mathcal{F}$ -measurable. The process $f = \{f_t : t \geq 0\}$ is called $\mathcal{F}_t$ -adapted if f_t is $\mathcal{F}_t$ -measurable for every $t \geq 0$. Let $\Sigma := \bigcap_{t \geq 0} \{Z \in \mathcal{B}(R_+) \otimes \mathcal{F} : Z_t \in \mathcal{F}_t\}$, where $Z_t = \{\omega; (t,\omega) \in Z\}$. We know that Σ is a σ -algebra on $R_+ \times \Omega$. A function $f : R_+ \times \Omega \to R^d$ is measurable and $\mathcal{F}_t$ -adapted if and only if it is Σ-measurable ([9]).

In a fashion similar to the R^d -valued stochastic processes, a set-valued stochastic process $F = \{F_t : t \geq 0\}$ is defined as a set-valued function $F : R_+ \times \Omega \to \mathcal{K}(R^d)$ with $\mathcal{F}$ -measurable section F_t for $t \geq 0$. It is called measurable if it is $\mathcal{B}_+ \otimes \mathcal{F}$ -measurable, and $\mathcal{F}_t$ -adapted if for any fixed t, $F_t(\cdot)$ is $\mathcal{F}_t$ -measurable. $F = \{F_t : t \geq 0\}$ is measurable and $\mathcal{F}_t$ -adapted if and only if it is Σ-measurable. $F = \{F_t : t \geq 0\}$ is called L^p -integrable if every F_t is L^p -integrable.

Let $F(R^d)$ denote the family of all fuzzy sets $v : R^d \to [0,1]$ which satisfy the following two conditions (cf. [3] [6]):

1) The level set $v_1 = \{x \in R^d : v(x) = 1\} \neq \varnothing$;

2) Each v is upper semi-continuous function, i.e. for each $\alpha \in (0,1]$, the level set $[v]^\alpha \doteq \{x \in R^d : v(x) \geq \alpha\}$ is a closed subset of R^d.

Let $F_k(R^d)$ denote the family of all fuzzy sets $v : R^d \to [0,1]$ which satisfy the above conditions 1), 2), and

3) The support set $[v]^0 \doteq cl\{x \in R^d : v(x) > 0\}$ is a compact set.

A fuzzy set v is convex if

$$v(\lambda x + (1-\lambda)y) \geq \min\{v(x), v(y)\} \text{ for any } x, y \in R^d, \lambda \in [0,1].$$

It is know that v is convex if and only if, for any $\alpha \in (0,1]$, the level set $[v]^\alpha$ is a convex subset of R^d. Let $F_c(R^d)$ denote the family of all convex fuzzy sets in $F(R^d)$, and $F_{kc}(R^d)$ be the subset of all convex fuzzy sets in $F_k(R^d)$. Define $d_\infty : F(R^d) \times F(R^d) \to [0,\infty)$ (cf. [1]) by the expression

$$d_\infty(u,v) := \sup_{\alpha \in [0,1]} d_H\left([u]^\alpha, [v]^\alpha\right).$$

We know that d_∞ is a metric in $F(R^d)$ and $(F(R^d), d_\infty)$ a complete metric space (cf. [6] [3]). Moreover, for every $u, v, w, z \in F(R^d), \lambda \in R$, we have

$$d_\infty(u+w, v+w) = d_\infty(u,v);$$

$$d_\infty(u+v, w+z) \leq d_\infty(u,w) + d_\infty(v,z);$$

$$d_\infty(\lambda u, \lambda v) = |\lambda| d_\infty(u,v).$$

Lemma 1. (cf. [16]) Let B be a set and $\{B_\alpha, \alpha \in [0,1]\}$ be a family of subsets of B such that

1) $B_0 = B$;

2) $\alpha \leq \beta$ implies $B_\beta \subseteq B_\alpha$;

3) $\alpha_1 \leq \alpha_2 \leq \cdots$, $\lim_{n\to\infty} \alpha_n = \alpha$ implies $B_\alpha = \bigcap_{i=1}^{\infty} B_{\alpha_i}$.

Then the function $\phi : B \to [0,1]$ defined by $\phi(x) = \sup\{\alpha \in [0,1] : x \in B_\alpha\}$ has the property that $\{x \in B : \phi(x) \geq \alpha\} = B_\alpha$ for every $\alpha \in [0,1]$.

A mapping $G : \Omega \to F(R^d)$ is said to be measurable if $[G(\omega)]^\alpha : \Omega \to \mathcal{K}(R^d)$ is an set-valued random variable for each $\alpha \in (0,1]$. Such a mapping G is called a fuzzy random variable (cf. [17]). Let $\mathcal{M}(\Omega, \mathcal{F}, P; F(R^d))$ (briefly by $\mathcal{M}(\Omega, F(R^d))$) denote the family of all $\mathcal{F}$ -measurable fuzzy random variables. As a similar manner, we have the notations $\mathcal{M}(\Omega, \mathcal{F}, P; F_k(R^d))$, and $\mathcal{M}(\Omega, \mathcal{F}, P; F_{kc}(R^d))$, or briefly by $\mathcal{M}(\Omega, F_k(R^d))$ (resp. $\mathcal{M}(\Omega, F_{kc}(R^d))$).

$G(t,\omega) : \Omega \times [0,T] \to F(R^d)$ is called a fuzzy stochastic process if for any $t \in [0,T]$, $G(t,\omega)$ is a fuzzy random variable. A fuzzy stochastic process $G(t,\omega)$ is said to be $\mathcal{F}_t$ -adapted, if for every $\alpha \in [0,1]$, the set-valued function $[G(t,\omega)]^\alpha : \Omega \to \mathcal{K}(R^d)$ is $\mathcal{F}_t$ -measurable for all $t \in [0,T]$. It is called measurable, if $[G(t,\omega)]^\alpha : \Omega \to \mathcal{K}(R^d)$ is a $\mathcal{B}([0,T]) \otimes \mathcal{F}$ -measurable for all $\alpha \in [0,1]$.

A fuzzy stochastic process G is called L^p -integrably bounded, if there exists a real-valued stochastic process $h \in L^p(\Omega \times [0,T], R^d)$, for any $\alpha \in [0,1]$ such that $\left\| [G(t,\omega)]^\alpha \right\|_K \le h(t,\omega)$ for any $(t,\omega) \in \Omega \times [0,T]$.

It is equivalent to that $\left\| [G]^0 \right\|_K \in L^p\left(\Omega \times [0,T], R^d\right)$.

Let $L^p\left(\Omega, \mathcal{F}, P; F\left(R^d\right)\right)$ denote the family of all measurable $F\left(R^d\right)$-valued L^p-integrably bounded fuzzy functions. Write for brevity by $L^p\left(\Omega, F\left(R^d\right)\right)$, where we consider $F, G \in L^p\left(\Omega, \mathcal{F}, P; F\left(R^d\right)\right)$ as identical if $P\left([F]^\alpha = [G]^\alpha, \forall \alpha \in [0,1]\right) = 1$. Let $L^p\left(\Omega \times [0,T], F\left(R^d\right)\right)$ denote the family of all L^p-integrably bounded $F\left(R^d\right)$-valued $\mathcal{F}_t$-adapted fuzzy stochastic processes.

Let $G : \Omega \to F\left(R^d\right)$ be a fuzzy random variable and $p \ge 1$, The following conditions are equivalent (cf. [15]):

1) $G \in L^p\left(\Omega, \mathcal{F}, P; F\left(R^d\right)\right)$;

2) $[G]^0 \in L^p\left(\Omega, \mathcal{F}, P; \mathcal{K}\left(R^d\right)\right)$;

3) $\left\| [G]^0 \right\|_K \in L^p\left(\Omega, \mathcal{F}, P; R^+\right)$.

We define $\hat{\theta} \in F\left(R^d\right)$ as $\hat{\theta} = I_{\{0\}}$, where for $x \in R^d$, we have $I_{\{x\}}\{y\} = 1$ if $x = y$ and $I_{\{x\}}\{y\} = 0$ if $x \ne y$.

3. LEBESGUE-STIELTJES INTEGRALS WITH RESPECT TO FINITE VARIATION PROCESSES

Let $(\Omega,\mathcal{F},P)$ be a complete probability space equipped with the usual filtration $\{\mathcal{F}_t,t\in[0,T]\}$. Let $A=\{A_t,t\geq 0\}$ be a real valued $\mathcal{F}_t$ -adapted measurable process with finite variation and continuous sample trajectories a.s. from the origin. That is to say, for each compact interval $[s,t]\subset[0,\infty)$ and any partition $\Delta=\{t_1,t_2,\cdots,t_n\}$ of $[s,t]$, the total variation

$$V_A([s,t])=\sup_{\Delta}\sum_{i=1}^{n}\left|A_{t_i}(\omega)-A_{t_{i-1}}(\omega)\right|$$

is finite and $A(0,\cdot)=0$ as. Then for any $T>0$, the process $A=\{A_t,t\geq 0\}$ can generate a random measure denoted by μ_A in the space $([0,T],\mathcal{B}([0,T]))$. For any $(s,t]\subset[0,T]$, let

$$\mu_A((s,t]):=|A(t,\omega)|-|A(s,\omega)|$$

where $A(t,\omega)=A^+(t,\omega)-A^-(t,\omega)$ is the decomposition of A. A^+ and A^- are non-negative and non-de- creasing processes. $|A(t,\omega)|=A^+(t,\omega)+A^-(t,\omega)$.

In the product space $(\Omega\times[0,T],\Sigma)$, Michta (2011) in [7] defined a measure as follows:

$$\nu_A(C):=\int_{\Omega}\int_{[0,T]}I_C(t,\omega)\,\mu_A([0,T])\,\mu_A(\mathrm{d}t)\,P(\mathrm{d}\omega)$$

For $C\in\Sigma$, where I_C is the index function. Then the set function v is a finite measure in the measurable space $(\Omega\times[0,T],\Sigma)$ if and only if $\int_{\Omega}(\mu_A([0,T]))^2 P(\mathrm{d}\omega)<\infty$ (cf. [7]). In the following we always assume $\int_{\Omega}(\mu_A([0,T]))^2 P(\mathrm{d}\omega)<\infty$.

For $1\le p<\infty$, let $L^p(\Omega\times[0,T],\Sigma,\nu_A;R^d)$ be the family of all Σ-measurable R^d-valued stochastic processes f such that

$$\int_{\Omega\times[0,T]}\|f(\omega,t)\|^p \mathrm{d}\nu_A<\infty.$$

For any $f\in L^p(\Omega\times[0,T],\Sigma,\nu_A;R^d)$ and $[s,t]\subset[0,T]$, the stochastic Lebesgue-Stieltjes integral $\int_{[s,t]} f(\tau)\mathrm{d}A_\tau$ is defined by the Bochner integral $\int_{[s,t]} f(\tau)\mu_A(\mathrm{d}\tau)$ path-by-path. One can prove that the integral process $\left\{\int_{[0,t]} f(s)\mathrm{d}A_s, t\in[0,T]\right\}$ is Σ-measurable.

Lemma 2. (cf. [8]) Let $(E,\mathcal{B},\mu)$ be a σ-finite measure space and X a separable Banach space. If $\mathcal{B}$ is separable with respect to μ, (i.e. there exists a countably generated sub-sigma algebra $\mathcal{B}_0\in\mathcal{B}$ such that for every $A\in\mathcal{B}$, there is $B\in\mathcal{B}_0$ satisfying $\mu(A\Delta B)=0$), then space $L^p(E;X)(p\ge 1)$ is separable in norm.

From now on, we always assume the sigma-field $\mathcal{F}$ is separable with respect to P such that the set-valued integral and fuzzy integral can be well defined.

Let $M^p(\Omega\times[0,T],\Sigma,\nu_A;\mathcal{K}(R^d))$ be the family of all Σ-measurable $\mathcal{K}(R^d)$-valued stochastic processes F such that

$$\int_{\Omega\times[0,T]}\|F(\omega,t)\|_K^p \mathrm{d}\nu_A<\infty,$$

where $\left\|F(\omega,t)\right\|_K = \sup_{x\in F(\omega,t)}\|x\|$.

For any $F\in M^p\left(\Omega\times[0,T],\Sigma,\nu_A;\mathcal{K}\left(R^d\right)\right)$, set

$$S^p(F):=\left\{f\in M^p\left(\Omega\times[0,T],\Sigma,\nu_A;R^d\right): f(\omega,t)\in F(\omega,t),\nu_A - a.e\right\}. \quad (2)$$

Definition 1. (cf. [7]) For a set-valued stochastic process $F\in M^p\left(\Omega\times[0,T];\mathcal{K}_{kc}\left(R^d\right)\right)$ the set-valued stochastic Lebesgue-Stieltjes integral (over interval $[s,t]$) of F with respect to the finite variation continuous process A is the set

$$\int_{[s,t]}F(\tau)\mathrm{d}A_\tau := \left\{\int_{[s,t]}f(\tau)\mathrm{d}A_\tau : f\in S^p(F)\right\}.$$

For some fuzzy stochastic process $G\in\mathcal{M}\left(\Omega\times[0,T],F\left(R^d\right)\right)$, it is natural to define the fuzzy integral of G with respect to the finite variation process level-wise.

Let $M^p\left(\Omega\times[0,T],\Sigma,\nu_A;F\left(R^d\right)\right)$ (or abbrev. as $M^p\left(\Omega\times[0,T],F\left(R^d\right)\right)$) be the family of all Σ-measu- rable $F\left(R^d\right)$-valued fuzzy stochastic processes G such that

$$\int_{\Omega\times[0,T]}\left\|\left[G(\omega,t)\right]^0\right\|_K^p \mathrm{d}\nu_A < \infty,$$

where $\left\|\left[G(\omega,t)\right]^0\right\|_K = \sup_{x\in[G(\omega,t)]^0}\|x\|$.

For a fuzzy stochastic process $G\in M^1\left(\Omega\times[0,T],F\left(R^d\right)\right)$, according to Lemma 1 and the properties of set-valued stochastic integrals, the Lebesgue-Stieltjes integral of G (over interval $[s,t]$) can be defined level-wise.

Set

$$\left[\int_s^t G(\tau,\omega)\mathrm{d}A_\tau(\omega)\right]^\alpha := \int_s^t\left[G(\tau,\omega)\right]^\alpha \mathrm{d}A_\tau(\omega) = \left\{\int_s^t g_\tau(\omega)\mathrm{d}A_\tau(\omega), g\in S^1_{[G]^\alpha}\right\} \quad (3)$$

for all $\alpha \in [0,1]$.

Definition 2. For a fuzzy stochastic process $G \in M^1(\Omega \times [0,T], F(R^d))$ and any $0 \le s < t \le T$, the family

$$\left\{ \left[\int_s^t G(\tau,\omega) \mathrm{d}A_\tau(\omega) \right]^\alpha, \alpha \in [0,1] \right\}$$

defined by Equation (3) can determine an $F(R^d)$-valued function denoted by $\int_0^t G(s,\omega) \mathrm{d}A_s$, such a fuzzy function is called the Lebesgue-Stieltjes integral (over interval $[s,t]$) of G with respect to finite variation process $A(t)$.

Theorem 1. ([12]) For $F \in M^1(\Omega \times [0,T], \Sigma, \mathcal{K}_{kc}(R^d); \nu_A)$, $[s,t] \subset [0,T]$ and $\omega \in \Omega$, the Lebesgue- Stieltjes integral $\int_s^t F_\tau(\omega) \mathrm{d}A_\tau(\omega)$ is a compact and convex subset of R^d.

Lemma 3. (cf. [18]) Let $(\Omega, \mathcal{F}, P)$ be a probability space, X a separable Banach space. For random variables F_1, F_2, both the support function $S(x^*, F(\omega))(x^* \in X^*)$ and the metric $d_H(F_1(\omega), F_2(\omega))$ are $\mathcal{F}$-measurable.

Lemma 4. (cf. [14]) Let $A_t(\omega)$ be an R-valued stochastic process with finite variation. For $F \in L^2(\Omega \times [0,T], \Sigma, \nu_A; \mathcal{K}_{kc}(R^d))$ and $[s,t] \subset [0,T]$, we have

1) $\int_s^t \alpha F_\tau \mathrm{d}A_\tau = \alpha \int_s^t F_\tau \mathrm{d}A_\tau, \alpha \in [0,1]$;

2) $S\left(x^*, \int_s^t F_\tau \mathrm{d}A_\tau\right) = \int_s^t S(x^*, F_\tau) \mathrm{d}A_\tau, x^* \in R^d$.

Lemma 5. (cf. [18]) Let $(\Omega,\mathcal{F})$ be a measurable space, X a separable Banach space. Taking $F:\Omega\to\mathcal{K}(X)$ and for any $x^*\in X^*$, assume $S(x^*,F)$ is measurable. Then if one of the following conditions is satisfied:

1) X^* is separable;

2) for any $\omega\in\Omega, F(\Omega)\in\mathcal{K}_{kc}(X)$.

We obtain that F is a set-valued random variable.

From Lemma 3 and Lemma 5, when $X=R^d$, taking $F(\omega)\in\mathcal{K}_{kc}(R^d)$, then for any $x^*\in R^d$, $F(\cdot)$ is measurable if and only if $S(x^*,F(\cdot))$ is $\mathcal{F}_t$ -measurable.

Lemma 6. (cf. [19]) Let $(\Omega,\mathcal{F})$ be a measurable space, X a separable metrizable space, and Y a metrizable space. Then every Caratheodory function $f:\Omega\times X\to Y$ (i.e. for each $x\in X$, the function $f(\cdot,x):\Omega\to Y$ is $\mathcal{F}$ -measurable and for each $\omega\in\Omega$, the function $f(\omega,\cdot):X\to Y$ is continuous) is $(\mathcal{F}\otimes\mathcal{B}(X))$-mea- surable.

Theorem 2. Let $G\in M^1\left(\Omega\times[0,T],\Sigma,v_A;F_{kc}\left(R^d\right)\right)$. Then for each $t\in[0,T]$, the fuzzy stochastic integral $\int_0^t G(s,\omega)\mathrm{d}A_s(\omega)$ is $\mathcal{F}_t$ -measurable. Furthermore, the mapping $\psi(t,\omega)=\int_0^t G(s,\omega)\mathrm{d}A_s(\omega)$ is $\mathcal{B}([0,T])\otimes\mathcal{F}$ -measurable.

Proof. Taking $G\in M^1\left(\Omega\times[0,T],\Sigma,v_A;F_{kc}\left(R^d\right)\right)$, then for each $\alpha\in[0,1]$, the mapping $[G(t,\omega)]^\alpha:\Omega\times[0,T]\to\mathcal{K}(R^d)$ is Σ-measurable. For any $x^*\in R^d$, by Lemma 3, the support function $S\left(x^*,[G(t,\omega)]^\alpha\right)$ is Σ-

measurable too. By Lemma 4, we have $S\left(x^*, \int_0^t [G(s,\omega)]^\alpha \, \mathrm{d}A_s\right) = \int_0^t S\left(x^*, [G(s,\omega)]^\alpha\right) \mathrm{d}A_s$. Since the real-valued Lebesgue-Stieltjes integral $\int_0^t S\left(x^*, [G(s,\omega)]^\alpha\right) \mathrm{d}A_s$ is a Carathedory function, then by Lemma 6, we obtain that $S\left(x^*, \int_0^t [G(s,\omega)]^\alpha \, \mathrm{d}A_s\right)$ is Σ-measurable. Therefore, by Lemma 5, for each $\alpha \in [0,1]$, the mapping $\int_0^t [G(s,\omega)]^\alpha \, \mathrm{d}A_s(\omega) : \Omega \times [0,T] \to \mathcal{K}_{kc}\left(R^d\right)$ is $\mathcal{B}([0,T]) \otimes \mathcal{F}$-measurable and $\mathcal{F}_t$-adapted, which means the integral $\psi(t,\omega) = \int_0^t G(s,\omega) \mathrm{d}A_s(\omega)$ is $\mathcal{B}([0,T]) \otimes \mathcal{F}$-measurable and $\mathcal{F}_t$-adapted.

Theorem 3. Let $G \in M^2\left(\Omega \times [0,T], F_{kc}\left(R^d\right)\right)$. Then for any $t \in [0,T]$, $\int_0^t G(s,\omega) \mathrm{d}A_s(\omega) \in L^2\left(\Omega, F\left(R^d\right)\right)$.

Proof. By Theorem 2, for any $t \in [0,T]$, $\int_0^t G(s,\omega) \mathrm{d}A_s(\omega)$ is $\mathcal{F}_t$-measurable. We will show that for any $\omega \in \Omega$, $t \in [0,T]$, $E\left[\left\| \int_0^t [G(s,\omega)]^0 \, \mathrm{d}A_s(\omega) \right\|_K^2\right] < \infty$.

For any $g \in S^2_{[G]^0}$,

$$\left\| \int_0^t g(s,\omega) \mathrm{d}A_s(\omega) \right\|^2 \le \mu_A([0,T]) \int_0^t \left\| g_s(\omega) \right\|^2 \mathrm{d}A_s(\omega)$$
$$\le \mu_A([0,T]) \int_0^t \left\| [G(s,\omega)]^0 \right\|_K^2 \mathrm{d}A_s(\omega). \quad (4)$$

Then

$$\sup_{g\in S_{[G]^0}}\left\|\int_0^t g_s(\omega)\mathrm{d}A_s(\omega)\right\|^2 \le \mu_A([0,T])\int_0^t\left\|[G(s,\omega)]^0\right\|_K^2 \mathrm{d}A_s.$$

Hence,

$$E\left[\left\|\int_0^t[G(s,\omega)]^0\mathrm{d}A_s\right\|_K^2\right]\le E\left[\mu_A([0,T])\int_0^t\left\|[G(s,\omega)]^0\right\|_K^2\mathrm{d}A_s\right]$$
$$\le\int_{\Omega\times[0,T]}\left\|[G(s,\omega)]^0\right\|_K^2\mathrm{d}\nu_A<\infty, \qquad (5)$$

which means $\int_0^t G(s,\omega)\mathrm{d}A_s(\omega)\in L^2\left(\Omega,F\left(R^d\right)\right)$.

Theorem 4. Let $G\in M^1\left(\Omega\times[0,T],F_{kc}\left(R^d\right)\right)$. Then for any $\omega\in\Omega$, $\int_0^t G(s,\omega)\mathrm{d}A_s(\omega)$ is continuous with respect to t under the metric d_∞.

Proof. Let $0\le r<t\le T$, for any $\omega\in\Omega$, we have

$$\int_0^t G(s,\omega)\mathrm{d}A_s(\omega)=\int_0^r G(s,\omega)\mathrm{d}A_s(\omega)+\int_r^t G(s,\omega)\mathrm{d}A_s(\omega).$$

Then

$$\begin{aligned}
&d_\infty\left(\int_0^r G(s,\omega)\mathrm{d}A_s(\omega),\int_0^t G(s,\omega)\mathrm{d}A_s(\omega)\right)\\
&=d_\infty\left(\int_0^r G(s,\omega)\mathrm{d}A_s(\omega),\int_0^r G(s,\omega)\mathrm{d}A_s(\omega)+\int_r^t G(s,\omega)\mathrm{d}A_s(\omega)\right)\\
&\le d_\infty\left(\int_0^r G(s,\omega)\mathrm{d}A_s(\omega),\int_0^r G(s,\omega)\mathrm{d}A_s(\omega)\right)+d_\infty\left(\hat{\theta},\int_r^t G(s,\omega)\mathrm{d}A_s(\omega)\right)\\
&=d_\infty\left(\hat{\theta},\int_r^t G(s,\omega)\mathrm{d}A_s(\omega)\right)=\sup_{\alpha\in[0,1]}d_H\left(\{0\},\int_r^t[G(s,\omega)]^\alpha\mathrm{d}A_s(\omega)\right)\\
&\le\sup_{\alpha\in[0,1]}\int_r^t d_H\left(\{0\},[G(s,\omega)]^\alpha\right)\mathrm{d}A_s(\omega)\le\int_r^t\sup_{\alpha\in[0,1]}d_H\left(\{0\},[G(s,\omega)]^\alpha\right)\mathrm{d}A_s(\omega)\\
&\le\int_r^t\sup_{\alpha\in[0,1]}\left\|[G(s,\omega)]^\alpha\right\|_K\mathrm{d}A_s(\omega)=\int_r^t\left\|[G(s,\omega)]^0\right\|_K\mathrm{d}A_s(\omega)<\infty.
\end{aligned} \qquad (6)$$

For any $\omega\in\Omega$, we have

$$\lim_{r\to t}\int_r^t\left\|[G(s,\omega)]^0\right\|_K\mathrm{d}A_s(\omega)=0.$$

Then for all $\omega \in \Omega$, $\int_0^t G(s,\omega)\mathrm{d}A_s(\omega)$ is left continuous for $t \in [0,T]$ under the metric d_∞. Similarly, we can prove that $\int_0^t G(s,\omega)\mathrm{d}A_s(\omega)$ is a right continuous for $t \in [0,T]$. Therefore it is continuous in t with respect to d_∞.

Lemma 7. Let fuzzy stochastic process $G \in L^1\left(\Omega \times [0,T], F_{kc}\left(R^d\right)\right)$. Then for each $\alpha \in [0,1]$, there exists a sequence $\left\{f_\alpha^i : i \in N\right\} \subset S^1_{[G]^\alpha}$, such that for every $t \in [0,T]$,

$$S^1_{[G]^\alpha} = cl\left\{f_\alpha^i : i \in N\right\},$$

where the closure is taken in $L^1\left(\Omega \times [0,T], R^d; v_A\right)$.

Proof. Since $\mathcal{F}$ is separable with respect to probability measurable P, we have that $\mathcal{B}([0,T]) \otimes \mathcal{F}$ is separable with respect to product measure $\lambda \times P$. By Lemma 2, $L^1\left(\Omega \times [0,T], R^d; \lambda \times P\right)$ is separable. It can be obtained that $L^1\left(\Omega \times [0,T], \Sigma, R^d; v_A\right)$ is separable under the norm v_A. So that for any $\alpha \in [0,1]$, $S^1_{[G]^\alpha}$ is separable since it is a closed subset of $L^1\left(\Omega \times [0,T], \Sigma, R^d; v_A\right)$. Then there exists a sequence $\left\{f_\alpha^i : i \in N\right\} \subset S^1_{[G]^\alpha}$,

$$S^1_{[G]^\alpha} = cl\left\{f_\alpha^i : i \in N\right\}$$

Theorem 5. For a fuzzy set-valued stochastic process $G \in M^1\left(\Omega \times [0,T], F_{kc}\left(R^d\right)\right)$ and any $\alpha \in [0,1]$, there exists a sequence $\left\{f_\alpha^i : i \in N\right\} \subset S^1_{[G]^\alpha}$ such that

$$\left[G(t,\omega)\right]^{\alpha} = cl\left\{f_{\alpha}^{i}(t,\omega) : i \in N\right\} \ a.e.(t,\omega)$$

and for each t

$$\int_0^t \left[G(s,\omega)\right]^{\alpha} \mathrm{d}A_s(\omega) = cl\left\{\int_0^t f_{\alpha}^{i}(s,\omega)\mathrm{d}A_s(\omega) : i \in N\right\} \ a.s.$$

where "cl" denotes the closure in R^d.

Proof. For each $\alpha \in [0,1]$, by Lemma 7, there exists a sequence $\left\{f_{\alpha}^{n} : n \in N\right\} \subset L^1\left(\Omega \times [0,T], \Sigma, R^d; \nu_A\right)$ such that

$$S^1_{[G]^{\alpha}} = cl\left\{f_{\alpha}^{i} : i \in N\right\},$$

where the closure is taken in $L^1\left(\Omega \times [0,T], \Sigma, R^d; \nu_A\right)$.

For each $\alpha \in [0,1]$, by Castaing represent theorem (cf. [15] [20]), there exists a sequence $\left\{g_{\alpha}^{j} : j \in N\right\} \subset S^1_{[G]^{\alpha}}$ such that

$$\left[G(t,\omega)\right]^{\alpha} = cl\left\{g_{\alpha}^{j}(t,\omega) : j \in N\right\} \ \text{for all } (t,\omega) \in [0,T] \times \Omega.$$

At first we will show that

$$\left[G(t,\omega)\right]^{\alpha} \subset cl\left\{f_{\alpha}^{i}(t,\omega) : i \in N\right\} \ a.e.(t,\omega).$$

In fact, taking $g_{\alpha}^{j} \in S^1_{[G]^{\alpha}}$, there exists a sequence $\left\{f_{\alpha}^{i_k}\right\}$ such that

$$\left\| f_{\alpha}^{i_k} - g_{\alpha}^{j} \right\|_{L^1\left(\Omega \times [0,T], \Sigma, \nu_A, R^d\right)} \to 0,$$

then there exists a subsequence $\left\{f_{\alpha}^{i_{k_j}}\right\}$ such that

$$\left\| f_{\alpha}^{i_{k_j}}(t,\omega) - g_{\alpha}^{j}(t,\omega) \right\|_{R^d} \to 0 \ \ a.e(t,\omega).$$

Therefore

$$\left[G(t,\omega)\right]^{\alpha} \subset cl\left\{f_{\alpha}^{i}(t,\omega): i \in N\right\} \ a.e(t,\omega).$$

On the other hand

$$cl\left\{f_{\alpha}^{i}(t,\omega): i \in N\right\} \subset \left[G(t,\omega)\right]^{\alpha} \ a.e.(t,\omega),$$

since $\left[G(t,\omega)\right]^{\alpha}$ is closed and $f_{\alpha}^{i} \in S_{[G]^{\alpha}}^{1}$, which yields

$$\left[G(t,\omega)\right]^{\alpha} = cl\left\{f_{\alpha}^{i}(t,\omega): i \in N\right\} \ a.e(t,\omega).$$

Since

$$\int_0^t \left[G(s,\omega)\right]^{\alpha} \mathrm{d}A_s(\omega) = \left\{\int_0^t g_{\alpha}(s,\omega)\mathrm{d}A_s(\omega): g_{\alpha} \in S_{[G]^{\alpha}}^{1}\right\}$$

is closed and $f_{\alpha}^{i} \in S_{[G]^{\alpha}}^{1}$, then for each t

$$cl\left\{\int_0^t f_{\alpha}^{i}(s,\omega)\mathrm{d}A_s(\omega): i \in N\right\} \subset \int_0^t \left[G(s,\omega)\right]^{\alpha} \mathrm{d}A_s(\omega) \ a.s. \tag{7}$$

For any $g_{\alpha} \in S_{[G]^{\alpha}}^{1}$, there exists a sequence $\left\{f_{\alpha}^{i_k}: k \in N\right\} \subset \left\{f^{i}: i \in N\right\}$ such that

$$\left\|f_{\alpha}^{i_k} - g_{\alpha}\right\|_{L^1\left(\Omega\times[0,T],\Sigma,\nu_A,R^d\right)} \to 0.$$

Then for each t,

$$\int_0^t g_{\alpha}(s,\omega)\mathrm{d}A_s \in cl\left\{\int_0^t f_{\alpha}^{i}(s,\omega)\mathrm{d}A_s : i \in N\right\} \ a.s.$$

which means

$$\int_0^t \left[G(s,\omega)\right]^{\alpha} \mathrm{d}A_s(\omega) \subset cl\left\{\int_0^t f_{\alpha}^{i}(s,\omega)\mathrm{d}A_s(\omega): i \in N\right\} \ a.s. \tag{8}$$

(7) together with (8) yields

$$cl\left\{\int_0^t f_{\alpha}^{i}(s,\omega)\mathrm{d}A_s(\omega): i \in N\right\} = \int_0^t \left[G(s,\omega)\right]^{\alpha} \mathrm{d}A_s(\omega) \ a.s.$$

Lemma 8. (cf. [15]) Let $F \in \mathcal{M}(\Omega, \mathcal{K}(R^d))$, $\phi : \Omega \times X \to \overline{R} = [-\infty, +\infty]$ satisfy: for fixed $\omega \in \Omega$, $\phi(\omega, \cdot)$ is continuous with respect to x, for fixed $x \in X$, $\phi(\omega, x)$ is measurable with respect to ω, then there exists an $f_0 \in S_F^p$ such that $\int_\Omega \phi(\omega, f(\omega)) \mathrm{d}\mu < \infty$, then we have

$$\inf_{f \in S_F^p} \int_\Omega \phi(\omega, f(\omega)) \mathrm{d}\mu = \int_\Omega \inf_{f \in F(\omega)} \phi(\omega, f(\omega)) \mathrm{d}\mu.$$

Theorem 6. Let $F, G \in M^1(\Omega \times [0,T], F_{kc}(R^d))$. Then for any $t \in [0,T]$,

$$d_\infty\left(\int_0^t F(s,\omega)\mathrm{d}A_s(\omega), \int_0^t G(s,\omega)\mathrm{d}A_s(\omega)\right) \le \int_0^t d_\infty(F(s,\omega), G(s,\omega))\mathrm{d}A_s(\omega)\ a.s.$$

Proof. Let $\phi_t(\omega) = \int_0^t F_s(\omega)\mathrm{d}A_s(\omega), \psi_t(\omega) = \int_0^t G_s(\omega)\mathrm{d}A_s(\omega)$. By Theorem 5, we can obtain that for each $\alpha \in [0,1]$, there exist sequences $\{f_\alpha^i : i \ge 1\} \subseteq S^1_{[F]^\alpha}$ and $\{g_\alpha^j : j \ge 1\} \subseteq S^1_{[G]^\alpha}$ such that $[F(t,\omega)]^\alpha = cl\{f_\alpha^i(t,\omega) : i \ge 1\}\ a.e.(t,\omega)$, $[G(s,\omega)]^\alpha = cl\{g_\alpha^j(t,\omega) : j \ge 1\}\ a.e.(t,\omega)$. For each t,

$$\int_0^t [F(s,\omega)]^\alpha\ \mathrm{d}A_s(\omega) = cl\left\{\int_0^t f_\alpha^i(s,\omega)\mathrm{d}A_s(\omega) : i \in N\right\}\ a.s.$$

and

$$\int_0^t [G(s,\omega)]^\alpha\ \mathrm{d}A_s(\omega) = cl\left\{\int_0^t g_\alpha^j(s,\omega)\mathrm{d}A_s(\omega) : j \in N\right\}\ a.s.$$

Therefore

$$\begin{aligned}
&\inf_{y\in\int_0^t[G(s,\omega)]^\alpha \mathrm{d}A_s(\omega)} \left\|\int_0^t f_\alpha^i(s,\omega)\mathrm{d}A_s(\omega) - y\right\| \\
&= \inf_{j\ge 1}\left\|\int_0^t f_\alpha^i(s,\omega)\mathrm{d}A_s(\omega) - \int_0^t g_\alpha^j(s,\omega)\mathrm{d}A_s(\omega)\right\| \\
&\le \inf_{j\ge 1}\int_0^t \left\|f_\alpha^i(s,\omega) - g_\alpha^j(s,\omega)\right\|\mathrm{d}A_s(\omega)\ a.s.
\end{aligned} \tag{9}$$

By Lemma 8, we have

$$\inf_{j\ge 1}\int_0^t \left\|f_\alpha^i(s,\omega) - g_\alpha^j(s,\omega)\right\|\mathrm{d}A_s(\omega) = \int_0^t \inf_{j\ge 1}\left\|f_\alpha^i(s,\omega) - g_\alpha^j(s,\omega)\right\|\mathrm{d}A_s(\omega)\ a.s. \tag{10}$$

Then

$$\begin{aligned}
&\int_0^t \inf_{j\ge 1}\left\|f_\alpha^i(s,\omega) - g_\alpha^j(s,\omega)\right\|\mathrm{d}A_s(\omega) \\
&\le \int_0^t \sup_{i\ge 1}\inf_{j\ge 1}\left\|f_\alpha^i(s,\omega) - g_\alpha^j(s,\omega)\right\|\mathrm{d}A_s(\omega)\ a.s. \\
&\le \int_0^t d_H\left(\left[F(s,\omega)\right]^\alpha, \left[G(s,\omega)\right]^\alpha\right)\mathrm{d}A_s(\omega)\ a.s. \\
&\le \int_0^t \sup_{\alpha\in[0,1]} d_H\left(\left[F(s,\omega)\right]^\alpha, \left[G(s,\omega)\right]^\alpha\right)\mathrm{d}A_s(\omega)\ a.s. \\
&= \int_0^t d_\infty\left(F(s,\omega), G(s,\omega)\right)\mathrm{d}A_s(\omega).
\end{aligned} \tag{11}$$

Then

$$\sup_{i\ge 1}\inf_{j\ge 1}\left\|\int_0^t f_\alpha^i(s,\omega)\mathrm{d}A_s(\omega) - \int_0^t g_\alpha^j(s,\omega)\mathrm{d}A_s(\omega)\right\| \le \int_0^t d_\infty\left(F(s,\omega), G(s,\omega)\right)\mathrm{d}A_s(\omega)\ a.s.$$

Similarly, we have

$$\sup_{j\ge 1}\inf_{i\ge 1}\left\|\int_0^t g_\alpha^j(s,\omega)\mathrm{d}A_s(\omega) - \int_0^t f_\alpha^i(s,\omega)\mathrm{d}A_s(\omega)\right\| \le \int_0^t d_\infty\left(F(s,\omega), G(s,\omega)\right)\mathrm{d}A_s(\omega)\ a.s.$$

Then for each $\alpha\in[0,1]$,

$$d_H\left(\int_0^t\left[F(s,\omega)\right]^\alpha \mathrm{d}A_s(\omega), \int_0^t\left[G(s,\omega)\right]^\alpha \mathrm{d}A_s(\omega)\right) \le \int_0^t d_\infty\left(F(s,\omega), G(s,\omega)\right)\mathrm{d}A_s(\omega)\ a.s.$$

Therefore

$$\sup_{\alpha\in[0,1]} d_H\left(\int_0^t [F(s,\omega)]^\alpha \, dA_s(\omega), \int_0^t [G(s,\omega)]^\alpha \, dA_s(\omega)\right) \le \int_0^t d_\infty\left(F(s,\omega), G(s,\omega)\right) dA_s(\omega) \; a.s.$$

Hence

$$d_\infty\left(\int_0^t F(s,\omega)\, dA_s(\omega), \int_0^t G(s,\omega)\, dA_s(\omega)\right) \le \int_0^t d_\infty\left(F(s,\omega), G(s,\omega)\right) dA_s(\omega) \; a.s.$$

Theorem 7. Let $F,G \in M^2\left(\Omega\times[0,T], F_{kc}\left(R^d\right)\right)$. Then for each $t\in[0,T]$, we have

$$d_\infty^2\left(\int_0^t F(s,\omega)\, dA_s(\omega), \int_0^t G(s,\omega)\, dA_s(\omega)\right) \le \mu_A([0,T]) \int_0^t d_\infty^2\left(F(s,\omega), G(s,\omega)\right) dA_s(\omega) \; a.s.$$

Proof. For any $\alpha\in[0,1]$, we have

$$\begin{aligned}
&\inf_{j\ge1}\left\|\int_0^t f_\alpha^i(s,\omega)\, dA_s(\omega) - \int_0^t g_\alpha^j(s,\omega)\, dA_s(\omega)\right\|^2 \\
&\le \inf_{j\ge1}\left\{\int_0^t \left\|f_\alpha^i(s,\omega) - g_\alpha^j(s,\omega)\right\| dA_s(\omega)\right\}^2 \\
&\le \inf_{j\ge1}\left\{\mu_A([0,T]) \int_0^t \left\|f_\alpha^i(s,\omega) - g_\alpha^j(s,\omega)\right\|^2 dA_s(\omega)\right\} \; a.s.
\end{aligned} \tag{12}$$

by Lemma 8, we have

$$\begin{aligned}
&\inf_{j\ge1}\left\{\mu_A([0,T]) \int_0^t \left\|f_\alpha^i(s,\omega) - g_\alpha^j(s,\omega)\right\|^2 dA_s(\omega)\right\} \\
&= \mu_A([0,T]) \inf_{j\ge1} \int_0^t \left\|f_\alpha^i(s,\omega) - g_\alpha^j(s,\omega)\right\|^2 dA_s(\omega) \; a.s. \\
&= \mu_A([0,T]) \int_0^t \inf_{j\ge1} \left\|f_\alpha^i(s,\omega) - g_\alpha^j(s,\omega)\right\|^2 dA_s(\omega) \; a.s.
\end{aligned} \tag{13}$$

Then

$$\begin{aligned}
&\mu_A([0,T]) \int_0^t \inf_{j\ge1} \left\|f_\alpha^i(s,\omega) - g_\alpha^j(s,\omega)\right\|^2 dA_s(\omega) \\
&\le \mu_A([0,T]) \int_0^t \sup_{i\ge1}\inf_{j\ge1} \left\|f_\alpha^i(s,\omega) - g_\alpha^j(s,\omega)\right\|^2 dA_s(\omega) \\
&\le \mu_A([0,T]) \int_0^t d_H^2\left([F(s,\omega)]^\alpha, [G(s,\omega)]^\alpha\right) dA_s(\omega) \\
&\le \mu_A([0,T]) \int_0^t d_\infty^2\left(F(s,\omega), G(s,\omega)\right) dA_s(\omega) \; a.s.
\end{aligned} \tag{14}$$

Then

$$\sup_{i\geq 1}\inf_{j\geq 1}\left\|\int_0^t f_\alpha^i(s,\omega)\mathrm{d}A_s(\omega)-\int_0^t g_\alpha^j(s,\omega)\mathrm{d}A_s(\omega)\right\|^2 \leq \mu_A([0,T])\int_0^t d_\infty^2(F(s,\omega),G(s,\omega))\mathrm{d}A_s(\omega)\ a.s.$$

Similarly, we have

$$\sup_{j\geq 1}\inf_{i\geq 1}\left\|\int_0^t g_\alpha^j(s,\omega)\mathrm{d}A_s(\omega)-\int_0^t f_\alpha^i(s,\omega)\mathrm{d}A_s(\omega)\right\|^2 \leq \mu_A([0,T])\int_0^t d_\infty^2(F(s,\omega),G(s,\omega))\mathrm{d}A_s(\omega)\ a.s.$$

Then for each $\alpha\in[0,1]$

$$d_H^2\left(\int_0^t[F(s,\omega)]^\alpha\,\mathrm{d}A_s(\omega),\int_0^t[G(s,\omega)]^\alpha\,\mathrm{d}A_s(\omega)\right)\leq \mu_A([0,T])\int_0^t d_\infty^2(F(s,\omega),G(s,\omega))\mathrm{d}A_s(\omega)\ a.s.$$

Moreover

$$\sup_{\alpha\in[0,1]} d_H^2\left(\int_0^t[F(s,\omega)]^\alpha\,\mathrm{d}A_s(\omega),\int_0^t[G(s,\omega)]^\alpha\,\mathrm{d}A_s(\omega)\right)\leq \mu_A([0,T])\int_0^t d_\infty^2(F(s,\omega),G(s,\omega))\mathrm{d}A_s(\omega)\ a.s.$$

Hence

$$d_\infty^2\left(\int_0^t F(s,\omega)\mathrm{d}A_s(\omega),\int_0^t G(s,\omega)\mathrm{d}A_s(\omega)\right)\leq \mu_A([0,T])\int_0^t d_\infty^2(F(s,\omega),G(s,\omega))\mathrm{d}A_s(\omega)\ a.s.$$

Remark 1. In Theorem 6 and Theorem 7, the inequalities hold too if we take the expectation on both sides.

4. CONCLUSION

In [21] , the author studied the Lebesgue-Stieltjes integral of real stochastic processes with respect to fuzzy valued stochastic processes. In some references such as [5] [6] , the integrals of fuzzy stochastic processes with respect to time t and Brownian motion were studied. In order to guarantee measurability of the integral, Kim (2005) Li and Ren (2007) defined the integral indirectly by taking the decomposable closure. Here, when the integrand taked value in compact and convex subsets of $F(R^d)$, we defined directly the integral of fuzzy stochastic process with respect to real-valued finite variation processes by using selection method, which is different from the above references. Then we proved the measurability (Theorem 2), which was key and guaranteed the reasonability of the definition. Attribute to the good property of finite variation of integrator, the integral was bounded as and L^2-bounded under the metric d_∞ (Theorem 3, Theorem 6 and Theorem 7). This property was much well than the integral with respect to Brownian motion since the latter was of infinite variation.

Thanks to the boundedness of the integral, it was possible to do the further work such as exploring solutions of fuzzy stochastic differential equations.

ACKNOWLEDGEMENTS

We thank the editor and the referees for their comments. This work is partly supported by NSFC (No. 11371135).

CITE THIS PAPER

Jinping Zhang, Lingli Luo, Xingmei Li, Xiaoying Wang (2015) Lebesgues-Stieltjes Integrals of Fuzzy Stochastic Processes with Respect to Finite Variation Processes. *Applied Mathematics*,**06**,2199-2210.

REFERENCES

1. Puri, M.L. and Ralescu, D.A. (1986) Fuzzy Random Variables. Journal of Mathematical Analysis and Applications, 114, 409-422.
2. Kisielewicz, M. (1997) Set-Valued Stochastic Integrals and Stochastic Inclusions. Stochastic Analysis and Applications, 15, 780-800.
3. Kim, B.K. and Kim, J.H. (1999) Stochastic Integrals of Set-Valued Processes and Fuzzy Processes. Journal of Mathematical Analysis and Applications, 236, 480-502.
4. Jung, E.J. and Kim, J.H. (2003) On Set-Valued Stochastic Integrals. Stochastic Analysis and Applications, 21, 401-418.
5. Kim, J.H. (2005) On Fuzzy Stochastic Differentials Equations. Journal of the Korean Mathematical Society, 42, 153-169.
6. Li, S. and Ren, A. (2007) Representation Theorems, Set-Valued and Fuzzy Set-Valued ITO Integral. Fuzzy Sets and Systems, 158, 949-962.
7. Malionwski, M.T. and Michta, M. (2011) On Set-Valued Stochastic Integrals and Fuzzy Stochastic Equations. Fuzzy Sets Systems, 177, 1-19.
8. Zhang, J., Li, S., Mitoma, I. and Okazaki, Y. (2009) On Set-Valued Stochastic Integrals in an M-Type 2 Banach Space. Journal of Mathematical Analysis and Applications, 350, 216-233.
9. Zhang, J., Li, S., Mitoma, I. and Okazaki, Y. (2009) On the Solution of Set-Valued Stochastic Differential Equations in M-Type 2 Banach Space. Tohoku Mathematical Journal, 61, 417-440.
10. Li, J. and Wang, J. (2012) Fuzzy Set-Valued Stochastic Lebesgue Integral. Fuzzy Sets and Systems, 200, 48-64.
11. Mitoma, I., Okazaki, Y. and Zhang, J. (2010) Set-Valued Stochastic Differential Equations in M-Type 2 Banach Space. Communications on Stochastic Analysis, 4, 215-237.
12. Zhang, J., Mitoma, I. and Okazaki, Y. (2013) Set-Valued Stochastic Integral with Respect to Poisson Process in a Banach Space. International Journal of Approximate Reasoning, 54, 404-417.

13. Fei, W. (2013) Existence and Uniquess for Solutions to Fuzzy Stochastic Differential Equations Driven by Local Martingales under the Non-Lipschtiz Condition. Nolinear Analysis, 76, 202-214.
14. Zhang, J. and Qi, J. (2013) Set-Valued Stochastic Integrals with Respect to Finite Variation Processes. Advances in Pure Mathematics, 3, 15-19.
15. Hu, S. and Papageorgiou, N. (1997) Handbook of Multivalued Analysis, Volume I: Theory. Kluwer Academic Publishers, Boston.
16. Negoito, C. and Ralescu, D. (1975) Applications of Fuzzy Sets to Systems Analysis. Wiley, New York.
17. Kwakernaak, H. (1987) Fuzzy Random Variables, Definition and Theoroems. Information Sciences, 15, 1-29.
18. Zhang, W., Li, S. and Wang, Z. (2007) Set-Valued Stochastic Process Introduction. Science Press, Beijing. (In Chinese)
19. Charalambos, D.A. and Kim, C.B. (1994) Infinite Dimensional Analysis. Springer-Verlag, Berlin.
20. Li, S., Ogura, Y. and Kreinovich, Y. (2002) Limit Theorems and Applications of Set-valued and Fuzzy Set-Valued Random Variables. 43rd Edition, Kluwer Academic Publishers, Dordrecht.
21. Feng, Y. (2001) Fuzzy-Valued Mappings with Finite Variation, Fuzzy-Valued Measures and Fuzzy-Valued Lebesgue-Stieltjes Integrals. Fuzzy Sets and Systems, 2, 227-236.

CHAPTER 12

Estimation for Nonnegative First-Order Autoregressive Processes with an Unknown Location Parameter

Andrew Bartlett, William McCormick

University of Georgia, Athens, USA

ABSTRACT

Consider a first-order autoregressive processes $X_t = \phi X_{t-1} + Z_t$, where the innovations are nonnegative random variables with regular variation at both the right endpoint infinity and the unknown left endpoint θ. We propose estimates for the autocorrelation parameter f and the unknown location parameter θ by taking the ratio of two sample values chosen with respect to an extreme value criteria for f and by taking the minimum of $X_t - \hat{\phi}_n X_{t-1}$ over the observed series, where $\hat{\phi}_n$ represents our estimate for f. The joint limit distribution of the proposed estimators is derived using point process techniques. A simulation study is provided to examine the small sample size behavior of these estimates.

Keywords: Nonnegative Time Series; Autoregressive Processes; Extreme Value Estimator; Regular Variation; Point Processes

1. INTRODUCTION

In many applications, the desire to model the phenomena under study by non-negative dependent processes has increased. An excellent presentation of the classical theory concerning these models can be found, for example, in Brockwell and Davis [1]. Recently, advancements in such models have shifted focus to some specialized features of the model, e.g. heavy tail innovations or nonnegativity of the model. In this paper we examine the behavior of traditional estimates under conditions leading to non-Gaussian limits. For example, the standard approach to parameter estimation within the AR(1) process is through the Yule-Walker estimator;

$$\hat{\phi}_{LS} = \frac{\sum_{t=1}^{n-1}(X_t - \bar{X})(X_{t+1} - \bar{X})}{\sum_{t=1}^{n}(X_{t+1} - \bar{X})^2},$$

$$\text{where } \bar{X} = \frac{1}{n}\sum_{t=1}^{n} X_t. \tag{1.1}$$

A slightly different approach presented in Mathew and McCormick [2] used linear programming to obtain estimates for ϕ and θ under certain optimization constraints. While there are many established methods to estimate the autocorrelation coefficient in an AR(1) model, there are just a few approaches on estimating the unknown location parameter in an AR(1) model. Onewas mentioned in McCormick and Mathew [2] where they considered

$$\hat{\theta}_{\text{range}} = X_{j^*} - \hat{\phi}_{\text{range}} X_{j^*-1}, \quad \text{where} \quad \hat{\phi}_{\text{range}} = \frac{X_{t^*+1} - X_{j^*}}{X_{t^*} - X_{j^*-1}},$$

t^* and j^* provides the index of the maximal and minimal X_i respectively for $1 \le i \le n$.

In this paper we examine estimation questions and asymptotic properties of alternative estimates for ϕ and θ respectively, relating to the model

$$X_t = \phi X_{t-1} + Z_t, t \ge 1,$$

where $0<\phi<1, \theta>0$ and $\{Z_t\}$ is an i.i.d. sequence of nonnegative random variables whose innovation distribution F is assumed to be regularly varying at infinity with index $-\beta$ and regularly varying at θ with index α, where θ, denotes the unknown but positive left endpoint. As a result of not restricting the innovations $\{Z_t\}$ to be bounded on a finite range, we can first estimate the autoregressive parameter ϕ through regular variation at infinity and then estimate the positive but unknown location parameter through regular variation at θ, the left endpoint.

While we have mentioned a few established estimation procedures, one notable exception was that of maximum likelihood. Although typically intractable and intricate in the time series setting, when the innovations in the AR(1) model are exponential, the maximum likelihood procedure had a major contribution on the estimation of positive heavy tailed time series. With these considerations in mind, Raftery [3] determined the limiting distribution of the maximum likelihood estimate for the autocorrelation coefficient ϕ. As a result, the estimator

$$\hat{\phi}_n = \bigwedge_{t=1}^{n} \frac{X_t}{X_{t-1}}, \tag{1.2}$$

was considered. The realization of this estimator was the stepping stone for the work done in this paper along with Davis and McCormick [4] which first considered this alternative estimator and used a point process approach to obtain the asymptotic distribution of the natural estimator $\hat{\phi}_n$. This was done in the context that the innovations distribution F varies regularly at 0, the left endpoint, and satisfy some moment condition.

The work presented in this paper is an extension of the work done in Davis and McCormick [4] including the following contributions to dependent time series with heavy-tail innovations. The first contribution involves the development of estimates for the autocorrelation coefficient and unknown location parameter under regular variation at both endpoints, with a rate of convergence $n^{1/\beta}\ell(n)$, where ℓ is slowly varying function. The second contribution involves using an extreme value method, e.g. point processes to establish the asymptotic distribution of the proposed estimators and weak convergence for the asymptotically independent joint distribution. An initial observation is that our estimation procedure is especially easy to implement for both ϕ and θ. That is, the autoregressive coefficient ϕ in the causal AR(1) process is estimated by taking the minimum of the ratio of two sample values while estimation for the unknown location parameter θ was achieved through minimizing $X_t - \hat{\phi}_n X_{t-1}$ over the observed series.

This naturally motivates a comparison between the estimation procedure presented in this paper and the standard linear programming estimates mentioned above, since within a nonnegative AR(1) model the linear programming estimate reduces to the estimate proposed, namely, $\min_{1 \le t \le n}(X_t / X_{t-1})$, where $\{X_t\}$ denotes the AR(1) process. This comparison along with the comparison between Mathew and McCormick's [2] optimization method and Bartlett and McCormick [5] extreme value method was performed through simulation and is presented in Section 3. The results found appear to demonstrate a favorable performance for our extreme value method over the 3 alternative estimators.

The main proofs in this paper rely heavily on point process methods from extreme value theory. The essential idea is to first establish the convergence of a sequence of point processes based on simple quantities and then apply the continuous mapping theorem to obtain convergence of the desired statistics. More background information on point processes, regular variation, and weak convergence can be found in Resnick [6]. Also, a nice survey on linear programming estimation procedures and nonnegative time series can be found in Anděl [7], Anděl [8], and Datta and McCormick [9], whereas more applications on modeling the phenomena with heavy tailed distributions and ensuing estimation issues can be found in Resnick [10].

The rest of the paper is organized as follows: asymptotic limit results for the autocorrelation parameter ϕ, unknown location parameter θ, and joint distribution of (ϕ, θ) are presented in Section 2, while Section 3 is concerned with the small sample size behavior of these estimates through simulation.

2. ASYMPTOTICS

The following point process limit result is fundamental. Since the result makes no use of an ARMA structure, we present it for more general linear models subject to usual summability conditions on the coefficients. In that regard for this result, we assume that $\{X_n, n \geq 0\}$ is the stationary linear process given by

$$X_n = \sum_{j=0}^{\infty} c_j Z_{n-j}$$

with $\sum_{j=0}^{\infty} |c_j|^{\delta} < \infty$ for some $0 < \delta < \beta, \delta \leq 1$. Furthermore for this result we may relax our assumptions on the innovation distribution and we require that $|Z_1|$ has a regularly varying tail distribution, i.e., $P(|Z_1| > x) = x^{-\beta}\ell(x), x > 0$ for a slowly varying function ℓ and the innovation distribution is tail balanced

$$\frac{P(Z_1 > x)}{P(|Z_1| > x)} \to p \text{ and } \frac{P(Z_1 \leq -x)}{P(|Z_1| > x)} \to q \text{ as } x \to \infty.$$

Define point processes

$$\xi_n = \sum_{k=1}^{n} \epsilon_{\left(b_n^{-1} X_{k-1}, Z_k\right)}, n \geq 1$$

and let $\sum_{k \geq 1} \epsilon_{j_k}$ denote PRM (ν) on $\mathbb{R}_0 = \mathbb{R}/\{0\}$ where

ν has Radon-Nikodym derivative with respect to Lebesgue measure

$$\frac{d\nu}{d\lambda}(x) = p\beta x^{-\beta-1} 1_{(0,\infty)}(x) + q\beta |x|^{-\beta-1} 1_{(-\infty,0)}(x).$$

Let $\{Z_{i,k}, i \geq 0, k \geq 1\}$ be an iid array with $Z_{i,k} \overset{d}{=} Z_1$ and independent of $\sum_{k=1}^{\infty} \varepsilon_{j_k}$. Define

$$\xi = \sum_{k=1}^{\infty} \sum_{i=0}^{\infty} \epsilon_{\left(c_i j_k, Z_{i,k}\right)}$$

Our basic result is to show that $M_p(\mathbb{R}_0 \times \mathbb{R})$ equipped with the topology of vague convergence

$$\xi_n \xrightarrow{d} \xi \text{ in}$$

which is close in statement and spirit to Theorem 2.4 in Davis and Resnick [11]. In view of the commonality of the two results, we present only the needed changes to the Davis and Resnick proof to accommodate the current setting. Aside from keeping track of the time when points occur, i.e. large jumps, the difference in the point processes considered here with those in Davis and Resnick [11] is the inclusion of marks, i.e. the second component of the point $\left(b_n^{-1}X_{k-1}, Z_k\right)$. This complication induces an additional weak dependence in the points which is addressed in Lemma 2.2 through a straight forward blocking argument. First, we establish weak convergence of marked point processes of a normalized vector of innovations. For a positive integer m define

$$\boldsymbol{Z}_k^{(m)} = \left(Z_k, Z_{k-1}, \cdots, Z_{k-m+1}\right)$$

and point process

$$\mathcal{I}_n^{(m)} = \sum_{k=1}^{n} \epsilon_{\left(b_n^{-1}\boldsymbol{Z}_k^{(m)}, Z_{k+1}\right)}, n \geq 1.$$

Let $\boldsymbol{e}_i = (0, \cdots, 1, \cdots, 0), 1 \leq i \leq m$ denote the standard basis vectors for $\mathbb{R}^m$. Define an associated marked point process with the first component placed on an axis by

$$\tilde{\mathcal{I}}_n^{(m)} = \sum_{k=1}^{n}\sum_{i=1}^{m} \epsilon_{\left(b_n^{-1}Z_k e_i, Z_{k+i}\right)}$$

In the following Lemma, we show that $\mathcal{I}_n^{(m)}$ and $\tilde{\mathcal{I}}_n^{(m)}$ are asymptotically indistinguishable in the following sense. Let $\mathbb{R}_0^m = \mathbb{R}^m / \{0\}$ and $E = \mathbb{R}_0^m \times \mathbb{R}$. Consider the class of rectangles

$$\mathcal{S} = \left\{ R \times (-\infty, x] : R = \bigtimes_{i=1}^{m} (c_i, d_i], \right.$$

$$\left. x < \infty \text{ and } \bar{R} \times (-\infty, x] \subset E \right\}$$

Lemma 2.1. As n tends to infinity

$$\mathcal{I}_n^{(m)}(B) - \tilde{\mathcal{I}}_n^{(m)}(B) \xrightarrow{p} 0 \text{ for all } B \in \mathcal{S}.$$

Proof. Following the proof presented in Proposition 2.1 of Davis and Resnick [11], suppose that $B \in \mathcal{S}$ is such that for some $1 \le i' \le m,\ B \cap (\mathbb{R}e_{i'} \times \mathbb{R}) \ne \varnothing.$ As noted in Davis and Resnick [11] for all $i \ne i'$, one has $c_i < 0 < d_i$. Observe that

$$\mathcal{I}_n^{(m)}(B) \le \mathcal{I}_n^{(m)}\left(\mathbb{R} \times \cdots \times [c_{i'}, d_{i'}] \times \cdots \times \mathbb{R} \times (-\infty, x]\right)$$

$$= \sum_{k=-i'+2}^{n-i'+1} \epsilon_{\left(b_n^{-1} Z_k e_{i'}, Z_{k+i'}\right)}\left(\mathbb{R} \times \cdots \times [c_{i'}, d_{i'}] \times \cdots \times \mathbb{R} \times (-\infty, x]\right) \quad (2.1)$$

Similarly let $c = \vee_{i \ne i'} c_i < 0$ and $d = \wedge_{i \ne i'} d_i > 0$. Then

$$\begin{aligned} &\mathcal{I}_n^{(m)}(B) \\ &\ge \mathcal{I}_n^{(m)}\left([c,d] \times \cdots \times [c_{i'}, d_{i'}] \times \cdots \times [c,d] \times (-\infty, x]\right) \\ &\ge \mathcal{I}_n^{(m)}\left(\mathbb{R} \times \cdots \times [c_{i'}, d_{i'}] \times \cdots \times \mathbb{R} \times (-\infty, x]\right) \\ &\quad - \sum_{i \ne i'} \mathcal{I}_n^{(m)}\left(E_i \times (-\infty, x]\right) \end{aligned} \quad (2.2)$$

where $y = (y_1, \cdots, y_m) \in E_i$ according to $y_i \notin [c,d]$ and $y_{i'} \in [c_{i'}, d_{i'}]$. Note that

$$E\left[\mathcal{I}_n^{(m)}\left(E_i\times(-\infty,x]\right)\right]$$
$$\le nP\left(\frac{1}{b_n}|Z_1|>|c|\wedge d,\frac{1}{b_n}|Z_1|>|c_{i'}|\wedge|d_{i'}|\right)$$
$$=o(1). \tag{2.3}$$

Thus from (2.1)-(2.3) we obtain

$$E\left|\mathcal{I}_n^{(m)}(B)-\tilde{\mathcal{I}}_n^{(m)}(B)\right|$$
$$\le 2(i'-1)P\left(\frac{1}{b_n}Z_1>|c_{i'}|\wedge|d_{i'}|\right)+o(1)=o(1)$$
$$\text{as } n\to\infty. \tag{2.4}$$

Then the result follows as in Davis and Resnick [11], Proposition 2.1, completing the proof. □

Lemma 2.2. Let $\mathcal{V}_n^{(m)}$ and $\mathcal{V}^{(m)}$ be the point processes on the space $E=\mathbb{R}_0\times\mathbb{R}^m$ defined by

$$\mathcal{V}_n^{(m)}=\sum_{k=1}^{n}\epsilon_{\left(b_n^{-1}Z_k,Z_{k+m}^{(m)}\right)} \text{ and } \mathcal{V}^{(m)}=\sum_{k=1}^{\infty}\epsilon_{\left(j_k,Y_k^{(m)}\right)}$$

where $\left\{\boldsymbol{Y}_k^{(m)},k\ge 1\right\}$ is an iid sequence with

$\boldsymbol{Y}_1^{(m)}\overset{d}{=}(Z_1,\cdots,Z_m)$ and is independent of $\sum_{k=1}^{\infty}\epsilon_{j_k}$. Then in $M_p(E)$,

$\mathcal{V}_n^{(m)}\xrightarrow{d}\mathcal{V}^{(m)}$.

Proof. We employ a blocking argument to establish this result. Let r_n be a sequence of integers such that $r=r_n\to\infty$ as $n\to\infty$ and $r=o(n)$. Let $h=\lfloor n/r\rfloor$ and $l\ge m$. Define blocks

$$I_{n,s} = \left[r(s-1)+1,\cdots,rs-l\right],$$
$$J_{n,s} = \left[rs-l+1,\cdots,rs\right]$$
$$\text{for } 1 \le s \le h \text{ and } J_{n,h+1} = \left[rh+1,\cdots,n\right].$$

Then it is clear that for $s \neq t$

$$\sigma\left(Z_{k+i}, k \in I_{n,s}, 0 \le i \le m\right)$$
$$\perp\!\!\!\perp\, \sigma\left(Z_{k+i}, k \in I_{n,t}, 0 \le i \le m\right).$$

Write

$$\mathcal{V}_{n,s}^{(m)} = \sum_{k \in I_{n,s}} \epsilon_{\left(b_n^{-1} Z_k, \mathbf{z}_{k+m}^{(m)}\right)} \text{ and } \tilde{\mathcal{V}}_{n,t}^{(m)} = \sum_{k \in J_{n,s}} \epsilon_{\left(b_n^{-1} Z_k, \mathbf{z}_{k+m}^{(m)}\right)}$$

Then

$$\mathcal{V}_n^{(m)} = \sum_{s=1}^{h} \mathcal{V}_{n,s}^{(m)} + \sum_{t=1}^{h+1} \tilde{\mathcal{V}}_{n,t}^{(m)}. \tag{2.5}$$

Let $B \subset E$ be a disjoint union of rectangles

$$B = \bigcup_{i-1}^{j} B_i \tag{2.6}$$

where $B_i = \left[c_i, d_i\right] \times R_i$ with $R_i = \times_{l=1}^{m} \left[x_{il}, y_{il}\right]$.

Let $\mu = \nu \times F \times \cdots \times F$ denote the mean measure of $\mathcal{V}^{(m)}$ which is PRM(μ) on E. To complete the proof we first show that for all sets B of the form given in (2.6) that

$$\lim_{n \to \infty} P\left(\mathcal{V}_n^{(m)}(B) = 0\right) = \exp\left(-\mu(B)\right). \tag{2.7}$$

The above limit result follows from the easily verifiable relations:

$$P\left(\mathcal{V}_{n,1}^{(m)}(B) = \cdots = \mathcal{V}_{n,h}^{(m)}(B) = 0\right)$$
$$= P^h\left(\mathcal{V}_{n,1}^{(m)}(B) = 0\right); \tag{2.8}$$

$$P\left(\mathcal{V}_{n,s}^{(m)}(B_k)\wedge\mathcal{V}_{n,s}^{(m)}(B_l)\geq 1\right)=O\left(\frac{r^2}{n^2}\right)$$

for $1\leq k\neq l\leq j$; (2.9)

$$P\left(\mathcal{V}_{n,s}^{(m)}\left(\bigcup_{i=1}^{j}B_i\right)\geq 1\right)=\sum_{i=1}^{j}P\left(\mathcal{V}_{n,s}^{(m)}(B_i)\geq 1\right)+O\left(\frac{r^2}{n^2}\right); \tag{2.10}$$

$$P\left(\mathcal{V}_{n,1}^{(m)}(B_i)\geq 1\right)=\frac{r}{n}\mu(B_i)(1+o(1)); \tag{2.11}$$

and

$$\sum_{t=1}^{h+1}\tilde{\mathcal{V}}_{n,t}^{(m)}(B)\xrightarrow{p}0 \text{ as } n\to\infty. \tag{2.12}$$

Indeed, in view of (2.5) and (2.12), (2.7) is equivalent to showing

$$\lim_{n\to\infty}P\left(\sum_{s=1}^{h}\mathcal{V}_{n,s}^{(m)}(B)=0\right)=\exp(-\mu(B)) \tag{2.13}$$

and the above relation holds by (2.8), (2.10), and (2.11), viz.

$$P\left(\sum_{s=1}^{h}\mathcal{V}_{n,s}^{(m)}(B)=0\right)=\left(1-P\left(\mathcal{V}_{n,1}^{(m)}(B)\geq 1\right)\right)^h$$
$$=\left(1-\sum_{i=1}^{j}\frac{r}{n}\mu(B_i)+o(r/n)\right)^h$$
$$\to\exp\left(-\sum_{i=1}^{j}\mu(B_i)\right).$$

It is immediate that for a rectangle

$B=[c,d]\times\times_{i=1}^{m}[x_i,y_i]\subset E$ we have

$$\lim_{n\to\infty}E\left[\mathcal{V}_n^{(m)}(B)\right]=\mu(B). \tag{2.14}$$

Therefore the result is seen to hold by (2.7) and (2.14) by application of Theorem 4.7 in Kallenberg [12]. □

Lemma 2.3. Let $\mathcal{I}_n^{(m)}$ and $\mathcal{I}^{(m)}$ be point processes on the space $E = \mathbb{R}_0^{(m)} \times \mathbb{R}$

$$\mathcal{I}_n^{(m)} = \sum_{k=1}^{n} \epsilon_{\left(b_n^{-1} \mathbf{Z}_k^{(m)}, Z_{k+1}\right)}, n \geq 1 \text{ and } \mathcal{I}^{(m)} = \sum_{k=1}^{\infty} \sum_{i=1}^{m} \epsilon_{\left(j_k e_i, Z_{i,k}\right)}$$

Then in $M_p(E)$,

$$\mathcal{I}_n^{(m)} \xrightarrow{d} \mathcal{I}^{(m)}.$$

Proof. We begin by applying the argument used in Theorem 2.2 of Davis and Resnick [11] with the modification that the relevant composition of maps of point processes is given by

$$\sum_{k\geq 1} \epsilon_{(u_k, v_{mk}, \cdots, v_{1k})} \mapsto \left(\sum_{k\geq 1} \epsilon_{(u_k, v_{mk})}, \cdots, \sum_{k\geq 1} \epsilon_{(u_k, v_{1k})} \right)$$

$$\mapsto \left(\sum_{k\geq 1} \epsilon_{(u_k e_1, v_{1k})}, \cdots, \sum_{k\geq 1} \epsilon_{(u_k e_m, v_{mk})} \right)$$

$$\mapsto \sum_{k\geq 1} \sum_{i=1}^{m} \epsilon_{(u_k e_i, v_{ik} \cdot)}$$

Each map being continuous, the composition is a continuous map from $M_p\left(\mathbb{R}_0 \times \mathbb{R}^m\right)$ to $M_p\left(\mathbb{R}_0^{(m)} \times \mathbb{R}\right)$ with each space being equipped with the topology of vague convergence. Therefore by the continuous mapping theorem and Lemma 2.2 we obtain

$$\tilde{\mathcal{I}}_n^{(m)} = \sum_{k=1}^{\infty} \sum_{i=1}^{m} \epsilon_{\left(b_n^{-1} Z_k e_i, Z_{k+i}\right)} \xrightarrow{d} \sum_{k=1}^{\infty} \sum_{i=1}^{m} \epsilon_{\left(j_k \mathbf{e}_i, Z_{i,k}\right)}$$

$$= \mathcal{I}^{(m)}. \tag{2.15}$$

Finally we complete the proof by Lemma 2.1 and (2.15) arguing as in Davis and Resnick [11].

We are now ready to present our fundamental result.

Theorem 2.1. Let ξ_n and ξ be the point processes on the space $E = \mathbb{R}_0 \times \mathbb{R}$ defined by

$$\xi_n = \sum_{k=1}^{n} \epsilon_{\left(b_n^{-1} X_{k-1}, Z_k\right)}, n \geq 1 \text{ and } \xi = \sum_{k=1}^{\infty} \sum_{i=0}^{\infty} \epsilon_{\left(c_i j_k, Z_{i,k}\right)}$$

where $\sum_{k=1}^{\infty} \epsilon_{j_k}$ is $PRM(v)$ and $\{Z_{i,k}, i \geq 0, k \geq 1\}$ is an iid array with $Z_{i,k} \overset{d}{=} Z_1$ and independent of $\sum_{k=1}^{\infty} \epsilon_{j_k}$.

Then in $M_p(E)$,

$$\xi_n \xrightarrow{d} \xi.$$

Remark. Apart from considering a time coordinate and restricting the process to an AR(1) process, the above Theorem 2.1 and Theorem 3.1 in Mathew and McCormick [2] consider essentially the same point process limit result. However, their result gave a wrong limit point process. This error is corrected in the current paper.

Proof. Observe that the map

$$\left(z_k, z_{k-1}, \cdots, z_{k-m+1}\right) \mapsto \sum_{i=0}^{m-1} c_i z_{k-i}$$

induces a continuous map on point processes given by

$$\sum_{k=1}^{\infty} \epsilon_{\left(z_k, z_{k-1}, \cdots, z_{k-m+1}, z_{k+1}\right)} \mapsto \sum_{k=1}^{\infty} \epsilon_{\left(\sum_{j=0}^{m-1} c_j z_{k-j}, z_{k+1}\right)}.$$

Thus we obtain from Lemma 2.3 that

$$\sum_{k=1}^{n} \epsilon_{\left(b_n^{-1} \sum_{i=0}^{m-1} c_i Z_{k-i}, Z_{k+1}\right)} \xrightarrow{d} \sum_{k=1}^{\infty} \sum_{i=0}^{m-1} \epsilon_{\left(c_i j_k, Z_{i,k}\right)}. \tag{2.16}$$

The result now follows from (2.16) by the same argument in Davis and Resnick [11] to finish their Theorem 2.4.

Returning to the AR(1) model under discussion in this paper and the estimate $\hat{\phi}_n$ given in (1.2), we obtain the following asymptotic limit result.

Theorem 2.2. Let $\{X_t, t \geq 0\}$ be the stationary solution to the AR(1) recursion $X_t = \phi X_{t-1} + Z_t$ and

$\hat{\phi}_n = \bigwedge_{t=1}^{n} \frac{X_t}{X_{t-1}}$. Under the assumptions that $0 < \phi < 1$

and the innovation distribution F has regularly varying right tail with index $-\beta$ and finite positive left endpoint θ,

$$\lim_{n \to \infty} P\left(b_n\left(\hat{\phi}_n - \phi\right) > x\right) = \mathrm{e}^{-x^{\beta} E W^{-\beta}} \quad \text{for all } x > 0,$$

where $W = \bigwedge_{i=0}^{\infty} \frac{Z_i}{\phi^i}$ and $b_n = F^{\leftarrow}(1 - 1/n)$.

Proof. For $x > 0$ define a subset

$$Q_x = \left\{(x_1, x_2) : \frac{x_2}{x_1} \leq x, x_1 > 0, x_2 \geq \theta\right\}.$$

Then note that for the point processes

$\xi_n = \sum_{k=1}^{n} \epsilon_{\left(b_n^{-1} X_{k-1}, Z_k\right)}$, we have

$$\{\xi_n(Q_x) = 0\} = \left\{\bigwedge_{k=1}^{n} \frac{Z_k}{b_n^{-1} X_{k-1}} > x\right\} = \left\{b_n\left(\hat{\phi}_n - \phi\right) > x\right\}.$$

Applying Theorem 2.1 in the case of an AR(1) process so that $c_i = \phi^i, i \geq 0$, we have

$$\xi_n \xrightarrow{d} \xi = \sum_{k=1}^{\infty} \sum_{i=0}^{\infty} \epsilon_{\left(\phi^i j_k, Z_{i,k}\right)}.$$

Note that as a subset of $E = (0, \infty] \times [\theta, \infty)$, the set Q_x is a bounded continuity set with respect to the limit point process ξ so that

$$\lim_{n\to\infty} P\left(b_n\left(\hat{\phi}_n - \phi\right) > x\right) = P\left(\xi\left(Q_x\right) = 0\right)$$
$$= P\left(\bigwedge_{k=1}^{\infty}\bigwedge_{i=0}^{\infty}\frac{Z_{i,k}}{\phi^i j_k} > x\right) = P\left(\bigwedge_{k=1}^{\infty}\frac{W_k}{j_k} > x\right)$$

where $\left(W_k = \bigwedge_{i=0}^{\infty}\phi^{-i}Z_{i,k}, k \geq 1\right)$ is an i.i.d. sequence independent of $\left(j_k, k \geq 1\right)$. Let

$$\tilde{\xi} = \sum_{k\geq 1}\epsilon_{(j_k, W_k)}. \tag{2.17}$$

Then by Proposition 5.6 in Resnick [10], we have that if G denotes the distribution of W_1, then $\tilde{\xi}$ is a Poisson random measure on E with mean measure $\mu = \nu \times G$, where $\nu(\mathrm{d}x) = \beta x^{-\beta-1}1_{[0,\infty)}(x)\mathrm{d}x$. Using (2.17) we can write

$$P\left(\bigwedge_{k=1}^{\infty}\frac{W_k}{j_k} > x\right) = P\left(\tilde{\xi}\left(Q_x\right) = 0\right) = \exp\left(-\mu\left(Q_x\right)\right).$$

Since $\mu\left(Q_x\right) = x^{\beta}EW^{-\beta}$, the result follows.

Corollary 2.1. Under conditions given in Theorem 2.2,

$$\hat{\phi}_n \xrightarrow{a.s} \phi.$$

Proof. Since $b_n = F^{\leftarrow}\left(1 - 1/n\right) \to \infty$ we have

$\hat{\phi}_n \xrightarrow{p} \phi$. But this implies $\hat{\phi}_n \xrightarrow{a.s} \phi$. since $\hat{\phi}_n \geq \phi$

and is non-increasing.

Let us now define our estimate of θ:

$$\hat{\theta}_n = \bigwedge_{t\in I_n}\left(X_t - \hat{\phi}_n X_{t-1}\right),$$

where we define the index set

$I_n = \{t : 1 \le t \le n \text{ and } X_{t-1} \le (a_n b_n)^\rho\}$ where $0 < \rho < 1$

is a fixed value.

Lemma 2.4. Under the assumptions that F is regularly varying with index α at its positive left endpoint θ and $\overline{F}$ is regularly varying with index $-\beta$ at infinity, its right endpoint, and $\alpha > \beta$, then

$$a_n^{-1}\left(\hat{\theta}_n - \bigwedge_{t \in I_n} Z_t\right) \xrightarrow{p} 0, \text{as } n \to \infty,$$

where $a_n = F^{\leftarrow}(1/n) - \theta$. Furthermore, for any $y \ge 0$

$$\lim_{n\to\infty} P\left(a_n^{-1}\left(\hat{\theta}_n - \theta\right) > y\right)$$

$$= \lim_{n\to\infty} P\left(a_n^{-1}\left(\bigwedge_{t=1}^{n} Z_t - \theta\right) > y\right) = \mathrm{e}^{-y^\alpha}.$$

Proof. Since $\alpha > \beta$, we have $\lim_{n\to\infty} a_n b_n = \infty$.

Therefore since $\left(b_n\left(\hat{\phi}_n - \phi\right), n \ge 1\right)$ is a tight sequence by Theorem 2.2 and since $\max_{t \in I_n} X_{t-1} / (a_n b_n)^\rho \le 1$ with

$0 < \rho < 1$, we have

$$a_n^{-1}\left(\hat{\phi}_n - \phi\right) \bigvee_{t \in I_n} X_{t-1} \xrightarrow{p} 0.$$

The first statement now follows since

$$a_n^{-1}\left|\hat{\theta}_n - \bigwedge_{t \in I_n} Z_t\right| \le a_n^{-1}\left|\hat{\phi}_n - \phi\right| \bigvee_{t \in I_n} X_{t-1}.$$

For the second statement observe

$$0 \le P\left(a_n^{-1}\left(\bigwedge_{t\in I_n} Z_t - \theta\right) > y\right) - P\left(a_n^{-1}\left(\bigwedge_{t=1}^{n} Z_t - \theta\right) > y\right)$$

$$= P\left(a_n^{-1}\left(\bigwedge_{t=1}^{n} Z_t - \theta\right) \le y < a_n^{-1}\left(\bigwedge_{t\in I_n} Z_t - \theta\right)\right)$$

$$\le P\left(\bigcup_{1\le t\le n}\left(X_{t-1} > (a_n b_n)^{\rho} \text{ and } a_n^{-1}(Z_t - \theta) \le y\right)\right)$$

$$\le nP\left(X_0 > (a_n b_n)^{\rho}\right)P\left(Z_1 \le a_n y + \theta\right) = o(1). \tag{2.18}$$

The result for the second statement now follows from (2.18) and the first part of the lemma. Finally, the identification of the limit distribution is well known. □

A useful observation follows from this lemma which we state as a corollary.

Corollary 2.2. Under the assumptions of Lemma 2.4 for any $x, y > 0$

$$\lim_{n\to\infty}\left\{Pb_n\left(\left(\hat{\phi}_n - \phi\right) > x, a_n^{-1}\left(\hat{\theta}_n - \theta\right) > y\right)\right.$$
$$\left. - P\left(b_n\left(\hat{\phi}_n - \phi\right) > x, a_n^{-1}\left(\bigwedge_{t=1}^{n}(Z_t - \theta) > y\right)\right)\right\} = 0$$

Proof. By Lemma 2.4 we have

$$\lim_{n\to\infty}\left\{P\left(b_n\left(\hat{\phi}_n - \phi\right) > x, a_n^{-1}\left(\hat{\theta}_n - \theta\right) > y\right)\right.$$
$$\left. - P\left(b_n\left(\hat{\phi}_n - \phi\right) > x, a_n^{-1}\left(\bigwedge_{t\in I_n} Z_t - \theta\right) > y\right)\right\} = 0$$

The result then follows from

$$0 \le E1\left\{b_n\left(\hat{\phi}_n - \phi\right) > x\right\}$$
$$\cdot\left(1\left\{a_n^{-1}\left(\bigwedge_{t\in I_n} Z_t - \theta\right) > y\right\} - 1\left\{a_n^{-1}\left(\bigwedge_{t=1}^{n} Z_t - \theta\right) > y\right\}\right)$$
$$\le P\left(a_n^{-1}\left(\bigwedge_{t\in I_n} Z_t - \theta\right) > y\right) - P\left(a_n^{-1}\left(\bigwedge_{t=1}^{n} Z_t - \theta\right) > y\right).$$

□

Corollary 2.2 allows a simplification in determining the joint asymptotic behavior of $\left(\hat{\phi}_n, \hat{\theta}_n\right)$ by allowing us to replace $\hat{\theta}_n$ with $\min_{1\le t\le n} Z_t$. The next lemma will provide another useful simplification—this time on $\hat{\phi}_n$.

For a positive integer m define

$$X_t^{(m)} = \sum_{i=0}^{m-1} \phi^i Z_{t-i}.$$

Lemma 2.5. Let $U_n^{(m)}$ and U_n be defined as

$$U_n^{(m)} = \bigwedge_{t=1}^{n} \frac{Z_t}{b_n^{-1} X_{t-1}^{(m)}} \quad \text{and} \quad U_n = \bigwedge_{t=1}^{n} \frac{Z_t}{b_n^{-1} X_{t-1}}.$$

Then for any $\epsilon > 0$

$$\lim_{m\to\infty} \lim_{n\to\infty} P\left(\left|U_n^{(m)} - U_n\right| > \epsilon\right) = 0.$$

Proof. We first note for any positive M that

$$P\left(\left|U_n^{(m)} - U_n\right| > \epsilon\right) = P\left(U_n^{(m)} U_n \left|\frac{1}{U_n^{(m)}} - \frac{1}{U_n}\right| > \epsilon\right)$$

$$\le P\left(\left|\frac{1}{U_n^{(m)}} - \frac{1}{U_n}\right| > \epsilon / M^2\right) + P\left(U_n^{(m)} > M\right).$$

In order to calculate $P\left(\left|\frac{1}{U_n^{(m)}} - \frac{1}{U_n}\right| > \epsilon\right)$ we partition X_t. That is, we write

$$X_t = X_t^{(m)} + X_t'^{(m)}$$

where $X_t'^{(m)} = \sum_{j=m}^{\infty} \phi^j Z_{t-j}$, so that

$$0 \le \bigvee_{t=1}^{n} \frac{X_{t-1}}{Z_t} - \bigvee_{t=1}^{n} \frac{X_{t-1}^{(m)}}{Z_t} \le \bigvee_{t=1}^{n} \frac{X_{t-1}'^{(m)}}{Z_t}.$$

Define point processes

$$\xi_n^{(m)} = \sum_{t=1}^{n} \varepsilon_{\left(b_n^{-1} X_{t-1}'^{(m)}, Z_t\right)} \text{ and } \xi^{(m)} = \sum_{k=1}^{\infty} \sum_{i=m}^{\infty} \epsilon_{\left(\phi^i j_k, Z_{i,k}\right)}$$

where $\{j_k, k \ge 1\}$ and $\{Z_{i,k}, k \ge 1, i \ge 0\}$ have the distributional properties given in Theorem 2.2. Applying Theorem 2.2 with $c_i = \phi^i$ for $i \ge m$ and c_i equal to 0 otherwise, we obtain

$\xi_n^{(m)} \xrightarrow{d} \xi^{(m)}$.

Then letting for $x > 0$

$$\mathcal{R}_x = \left\{ (u, v) : u > 0, v > \theta, \text{and } \frac{u}{v} > x \right\},$$

we have

$$\begin{aligned} &\lim_{n\to\infty} P\left(\bigvee_{t=1}^{n} b_n^{-1} X_{t-1}'^{(m)} / Z_t \le x \right) \\ &= \lim_{n\to\infty} P\left(\xi_n^{(m)} (\mathcal{R}_x) = 0 \right) = P\left(\xi^{(m)} (\mathcal{R}_x) = 0 \right). \end{aligned}$$

Setting

$$V_k^{(m)} = \bigvee_{i=m}^{\infty} \frac{\phi^i}{Z_{i,k}} \text{ and } \zeta^{(m)} = \sum_{k=1}^{\infty} \epsilon_{\left(j_k, V_k^{(m)}\right)}$$

we have

$$\begin{aligned} P\left(\xi^{(m)} (\mathcal{R}_x) = 0 \right) &= P\left(\bigvee_{k=1}^{\infty} j_k V_k^{(m)} \le x \right) \\ &= P\left(\zeta^{(m)} (\mathcal{R}_x) = 0 \right), \end{aligned}$$

where

$$\mathcal{S}_x = \{(u,v) : u > 0, v > \theta, \text{and } uv > x\}.$$

Since $\zeta^{(m)}$ is Poisson random measure with mean measure $\mu_m = \nu \times H_m$ where $V_1^{(m)} \sim H_m$ and

$$\mu_m(\mathcal{S}_x) = x^{-\beta} E\left(V_1^{(m)}\right)^{\beta},$$

we obtain

$$\lim_{n\to\infty} P\left(\bigvee_{t=1}^{n} b_n^{-1} X_{t-1}'^{(m)} / Z_t \le x\right) = \exp\left(-x^{-\beta} E\left(V_1^{(m)}\right)^{\beta}\right).$$

Next since

$$\left|\frac{1}{U_n^{(m)}} - \frac{1}{U_n}\right| \le \bigvee_{t=1}^{n} \frac{b_n^{-1} X_{t-1}'^{(m)}}{Z_t},$$

we have for large n that

$$P\left(\left|\frac{1}{U_n^{(m)}} - \frac{1}{U_n}\right| > \epsilon/M^2\right) \le P\left(\bigvee_{t=1}^{n} \frac{b_n^{-1} X_{t-1}'^{(m)}}{Z_t} > \epsilon/M^2\right)$$
$$\le 2\left(1 - \exp\left(-\epsilon^{-\beta} M^{2\beta} E\left(V_1^{(m)}\right)^{\beta}\right)\right)$$
$$\le 2\epsilon^{-\beta} M^{2\beta} E\left(V_1^{(m)}\right)^{\beta}.$$

Therefore, since $V_1^{(m)} \le \phi^m / \theta$,

$$\lim_{m\to\infty} \lim_{n\to\infty} P\left(\left|\frac{1}{U_n^{(m)}} - \frac{1}{U_n}\right| > \epsilon/M^2\right) = 0.$$

Next, note that from the limit law for the maximum obtained above, by replacing $X_{t-1}'^{(m)}$ with $X_{t-1}^{(m)}$ and by taking reciprocals, we derive the limit law for minimum,

$$\lim_{n\to\infty} P\left(U_n^{(m)} = \bigwedge_{t=1}^{n} \frac{Z_t}{b_n^{-1} X_{t-1}^{(m)}} \le x\right)$$
$$= 1 - \exp\left(-x^{\beta} E\tilde{W}_m^{-\beta}\right) \tag{2.19}$$

where $\tilde{W}_m$ has the distribution of $\bigwedge_{i=0}^{m-1} Z_{i,k}/\phi^i$. Thus, for any integer $m \ge 1$,

$$\lim_{n\to\infty} P\left(U_n^{(m)} \le x\right) \ge 1 - \exp\left(-(x/\theta)^{\beta}\right).$$

Thus for any $\epsilon > 0$, we have for M large enough that

$$\limsup_{n\to\infty} \sup_{m\ge 1} P\left(U_n^{(m)} > M\right) < \epsilon$$

completing the proof. □

With Lemma 2.5 in hand we can now focus our attention to the limiting joint distribution of $\left(U_n^{(m)}, \bigwedge_{t=1}^{n} Z_t\right)$.

This will be accomplished by a blocking argument. To that end for a fixed positive integer k, let $r_n = \lfloor n/k \rfloor$ and define blocks for $i = 1, \cdots, \lfloor n/r_n \rfloor$ by

$$J_i = \left[(i-1)r_n + 1, \cdots, ir_n - q\right],$$
$$\text{and } J_i' = \left[ir_n - q + 1, \cdots, ir_n\right]$$

where q is a positive integer greater than m. Furthermore, let

$$J_0' = \left[r_n \lfloor n/r_n \rfloor + 1, \cdots, n\right].$$

Now we define the events

$$\chi_i = \left\{\exists l \in J_i : \left(\frac{Z_l}{b_n^{-1} X_{l-1}^{(m)}}\right) \le x \text{ or } a_n^{-1}(Z_l - \theta) \le y\right\},$$
$$i = 1, \cdots, l_n$$

and

$$\chi_i' = \left\{ \exists l \in J_i' : \left(\frac{Z_l}{b_n^{-1} X_{l-1}^{(m)}} \right) \le x \text{ or } a_n^{-1}(Z_l - \theta) \le y \right\},$$
$$i = 0, \cdots, l_n$$

where $l_n = \lfloor n/r_n \rfloor$. We begin by showing that the events χ_i' are negligible.

Lemma 2.6. For any $x, y > 0$

$$\lim_{n \to \infty} P\left(\bigcup_{i=0}^{l_n} \chi_i' \right) = 0.$$

Proof. Observe that

$$nP\left(\frac{Z_l}{b_n^{-1} X_{l-1}^{(m)}} \le x \right) \le nP\left(\frac{\theta}{b_n^{-1} X_{l-1}^{(m)}} \le x \right)$$
$$\sim x^{\beta} \theta^{-\beta} \sum_{i=0}^{m-1} \phi^{i\beta} \le \frac{x^{\beta}}{\theta^{\beta}\left(1 - \phi^{\beta}\right)} \tag{2.20}$$

and

$$nP\left(a_n^{-1}(Z_l - \theta) \le y \right) \sim y^{\alpha}. \tag{2.21}$$

Thus for some constant c and any $n \ge 1$,

$$P\left(\bigcup_{i=0}^{l_n} \chi_i' \right) \le ck/n$$

establishing the lemma.

Define events A_i and B_i by

$$A_i = \left\{ \exists l \in J_i : \left(\frac{Z_l}{b_n^{-1} X_{l-1}^{(m)}} \right) \le x \right\}$$
$$\text{and } B_i = \left\{ \exists l \in J_i : a_n^{-1}(Z_l - \theta) \le y \right\}.$$

The following result provides the asymptotic behavior of the probability of these events.

Lemma 2.7. For any $x > 0, y > 0$, we have as $k \to \infty$

$$\lim_{n\to\infty} P(A_i) \sim \frac{1}{k} x^{\beta} E\tilde{W}_m^{-\beta} \text{ and } \lim_{n\to\infty} P(B_i) \sim \frac{y^{\alpha}}{k}.$$

Proof. Since the events A_i are independent, we have

$$P\left(\bigcap_{i=1}^{k}\left(\frac{Z_l}{b_n^{-1} X_{l-1}^{(m)}} > x, \forall l \in J_i\right)\right)$$

$$= \left(P\left(\frac{Z_l}{b_n^{-1} X_{l-1}^{(m)}} > x, \forall l \in J_1\right)\right)^k.$$

Using Lemma 2.6 we have that

$$P\left(\bigwedge_{l=1}^{n} \frac{Z_l}{b_n^{-1} X_{l-1}^{(m)}} > x\right)$$

$$= P\left(\bigcap_{i=1}^{k}\left(\left(\frac{Z_l}{b_n^{-1} X_{l-1}^{(m)}}\right) > x, \forall l \in J_i\right)\right) + O\left(\frac{1}{n}\right)$$

$$= \left(P\left(\frac{Z_l}{b_n^{-1} X_{l-1}^{(m)}} > x, \forall l \in J_1\right)\right)^k + O\left(\frac{1}{n}\right).$$

From (2.19) we showed that

$$\lim_{n\to\infty} P\left(U_n^{(m)} = \bigwedge_{l=1}^{n} \frac{Z_l}{b_n^{-1} X_{l-1}^{(m)}} \le x\right)$$

$$= 1 - \exp\left(-x^{\beta} E\tilde{W}_m^{-\beta}\right).$$

Hence using this limit law on $\bigwedge_{l=1}^{n}\left(Z_l / b_n^{-1} X_{l-1}^{(m)}\right)$, we obtain

$$\lim_{k\to\infty}\lim_{n\to\infty}\left(P\left[\frac{Z_l}{b_n^{-1}X_{l-1}^{(m)}}>x,\forall l\in J_1\right]\right)^k$$
$$=\exp\left(-x^{\beta}E\tilde{W}_m^{-\beta}\right).$$

Thus,

$$\lim_{n\to\infty}P\left[A_i\right]\sim\frac{x^{\beta}}{k}E\tilde{W}_m^{-\beta},\text{as }k\to\infty. \tag{2.22}$$

Similarly using the result of Lemma 2.4, we obtain

$$\lim_{n\to\infty}P\left[B_i\right]\sim\frac{y^{\alpha}}{k},\text{as }k\to\infty. \tag{2.23}$$

Hence the lemma holds. □

Lemma 2.8. For some constant c

$$P\left(A_i\cap B_i\right)\le cP\left(A_i\right)P\left(B_i\right).$$

Remark. Since the cardinality of J_i depends on r_n which depends on k and the events A_i and B_i depend on n, $P(A_i)$ and $P(B_i)$ depend on k and n. The conclusion of this lemma provides that for all k and n, there is a constant dependent on no parameters for which the inequality stated there holds.

Proof. To calculate the intersection we define the following sets

$$K_1=K_{1,i}=\left\{\left(l_1,l_2\right):l_1,l_2\in J_i\text{ and }l_2\notin\left[l_1-1-m,l_1\right]\right\}$$

and

$$K_2=K_{2,i}=\left\{\left(l_1,l_2\right):l_1,l_2\in J_i\text{ and }l_2\in\left[l_1-1-m,l_1\right]\right\}.$$

Now for $l_1-1-m\le l_2\le l_1-1$ and n sufficiently large,

$$\left\{\left(\frac{Z_{l_1}}{b_n^{-1}X_{l_1-1}^{(m)}}\right)\le x\right\}\cap\left\{a_n^{-1}\left(Z_{l_2}-\theta\right)\le y\right\}$$
$$\subset\left\{\sum_{i\ne l_1-l_2-1,0\le i\le m-1}\phi^i Z_{l_1-1-i}\ge b_n\frac{Z_{l_1}}{x}-\theta-1\right\}$$
$$\cap\left\{Z_{l_2}\le\theta+a_n y\right\}.$$

It then follows from (2.20), (2.21), and independence that

$$P\left(\left\{\sum_{i\ne l_1-l_2-1,0\le i\le m-1}\phi^i Z_{l_1-1-i}\ge b_n\frac{Z_{l_1}}{x}-\theta-1\right\}\right.$$
$$\left.\cap\left\{Z_{l_2}\le\theta+a_n y\right\}\right)$$
$$=O\left(\frac{1}{n^2}\right).$$

If $l_1=l_2=l$, we have

$$P\left(\left\{\frac{Z_l}{b_n^{-1}X_{l-1}^{(m)}}\le x\right\}\cap\left\{a_n^{-1}(Z_l-\theta)\le y\right\}\right)$$
$$=\int_{\theta}^{\theta+a_n y}P\left(X_{l-1}^{(m)}\ge\frac{z}{b_n^{-1}x}\right)\mathrm{d}F(z)=O\left(\frac{1}{n^2}\right).$$

Therefore, for some constant c

$$P\left(\bigcup_{(l_1,l_2)\in K_2}\left\{\frac{Z_{l_1}}{b_n^{-1}X_{l_1-1}^{(m)}}\le x\right\}\cap\left\{a_n^{-1}\left(Z_{l_2}-\theta\right)\le y\right\}\right)$$
$$=c/nk.$$

In order to handle set K_1, observe from construction of the blocks J_i and set K_1 that if $(l_1,l_2)\in K_1$ then the events

$$\left\{\frac{Z_{l_1}}{b_n^{-1}X_{l_1-1}^{(m)}} \le x\right\} \text{ and } \left\{a_n^{-1}\left(Z_{l_2}-\theta\right) \le y\right\}$$

are independent. Thus, if we define $\{Z_i', i \in \mathbb{Z}\}$ as an independent copy of $\{Z_i', i \in \mathbb{Z}\}$, then

$$P\left(\bigcup_{(l_1,l_2)\in K_1}\left\{\frac{Z_{l_1}}{b_n^{-1}X_{l_1-1}^{(m)}} \le x\right\}\cap\left\{a_n^{-1}\left(Z_{l_2}-\theta\right)\le y\right\}\right)$$

$$= P\left(\bigcup_{(l_1,l_2)\in K_1}\left\{\frac{Z_{l_1}}{b_n^{-1}X_{l_1-1}^{(m)}} \le x\right\}\cap\left\{a_n^{-1}\left(Z_{l_2}'-\theta\right)\le y\right\}\right)$$

$$\le P\left(A_i \cap B_i'\right) = P\left(A_i\right)P\left(B_i\right) \le c/k^2$$

where $B_i' = \left\{\exists l \in J_i : a_n^{-1}\left(Z_l'-\theta\right) \le y\right\}$ and where we used Lemma 2.7 in the last step. Thus, we have that for some constant c

$$P\left(A_i \cap B_i\right)$$

$$= P\left(\bigcup_{(l_1,l_2)\in K_1}\left\{\frac{Z_{l_1}}{b_n^{-1}X_{l_1-1}^{(m)}} \le x\right\}\cap\left\{a_n^{-1}\left(Z_{l_2}-\theta\right)\le y\right\}\right.$$

$$\left.\cup \bigcup_{(l_1,l_2)\in K_2}\left\{\frac{Z_{l_1}}{b_n^{-1}X_{l_1-1}^{(m)}} \le x\right\}\cap\left\{a_n^{-1}\left(Z_{l_2}-\theta\right)\le y\right\}\right)$$

$$\le c/k^2 = O\left(P\left(A_i\right)P\left(B_i\right)\right),$$

which completes the proof in view of Lemma 2.7. □

Lemma 2.9. For any $x > 0, y > 0$,

$$\lim_{n\to\infty} P\left(\bigwedge_{t=1}^{n}\left(\frac{Z_t}{b_n^{-1}X_{t-1}^{(m)}}\right) > x, a_n^{-1}\bigwedge_{t=1}^{n}\left(Z_t-\theta\right) > y\right)$$

$$= \exp\left(-x^{\beta}E\tilde{W}_m^{-\beta} - y^{\alpha}\right).$$

Proof. First from Lemma 2.6, we have as $n \to \infty$ that

$$P\left(\bigwedge_{t=1}^{n}\left(\frac{Z_t}{b_n^{-1} X_{t-1}^{(m)}}\right) > x, a_n^{-1} \bigwedge_{t=1}^{n}(Z_t - \theta) > y\right)$$
$$= P\left(\bigcap_{i=1}^{k} \chi_i^c\right) + o(1).$$

Next by (2.22), (2.23), and Lemma 2.8 we obtain that as k tends to infinity

$$\lim_{n\to\infty} P(\chi_i) \sim \frac{1}{k}\left(x^{\beta} E\tilde{W}_m^{-\beta} + y^{\alpha}\right).$$

Therefore, we obtain

$$\lim_{k\to\infty}\lim_{n\to\infty} P\left(\bigcap_{i=1}^{k} \chi_i^c\right) = \lim_{k\to\infty}\left(1 - \frac{1}{k}\left(x^{\beta} E\tilde{W}_m^{-\beta} + y^{\alpha}\right)\right)^k$$
$$= \exp\left(-\left(x^{\beta} E\tilde{W}_m^{-\beta} + y^{\alpha}\right)\right).$$

Hence

$$\lim_{n\to\infty} P\left(\bigwedge_{t=1}^{n}\left(\frac{Z_t}{b_n^{-1} X_{t-1}^{(m)}}\right) > x, a_n^{-1} \bigwedge_{t=1}^{n}(Z_t - \theta) > y\right)$$
$$= \exp\left(-x^{\beta} E\tilde{W}_m^{-\beta} - y^{\alpha}\right).$$

Theorem 2.3. Let $\{X_t, t \geq 1\}$ denote the stationary AR(1) process such that the innovation distribution F satisfies $\bar{F}$ is $RV_{-\beta}$ atinfinity and F is RV_{α} at θ.

If $\alpha > \beta$ then for any $x > 0, y > 0$ we have

$$\lim_{n\to\infty} P\left(b_n\left(\hat{\phi}_n - \phi\right) > x, a_n^{-1}\left(\hat{\theta}_n - \theta\right) > y\right)$$
$$= e^{-x^{\beta} EW^{-\beta} - y^{\alpha}},$$

where $W = \wedge_{j=0}^{\infty} Z_j / \phi^j$.

Proof. Let us first observe that for $\epsilon > 0$

$$P\left(U_n^{(m)} > x+\epsilon, a_n^{-1} \bigwedge_{t=1}^{n} (Z_t - \theta) > y\right) - P\left(\left|U_n^{(m)} - U_n\right| > \epsilon\right)$$

$$\leq P\left(U_n > x, a_n^{-1} \bigwedge_{t=1}^{n} (Z_t - \theta) > y\right)$$

$$\leq P\left(U_n^{(m)} > x, a_n^{-1} \bigwedge_{t=1}^{n} (Z_t - \theta) > y\right).$$

Thus by Lemma 2.9 we obtain

$$\exp\left\{-(x+\epsilon)^{\beta} E\tilde{W}_m^{-\beta} - y^{\alpha}\right\}$$

$$-\limsup_{n\to\infty} P\left(\left|U_n - U_n^{(m)}\right| > \epsilon\right)$$

$$\leq \liminf_{n\to\infty} P\left(U_n > x, a_n^{-1} \bigwedge_{t=1}^{n} (Z_t - \theta) > y\right)$$

$$\leq \limsup_{n\to\infty} P\left(U_n > x, a_n^{-1} \bigwedge_{t=1}^{n} (Z_t - \theta) > y\right)$$

$$\leq \exp\left(-x^{\beta} E\tilde{W}_m^{-\beta} - y^{\alpha}\right).$$

Letting m tend to infinity in the above and then ϵ tend to 0, we obtain from Lemma 2.5 and $\lim_{m\to\infty} E\tilde{W}_m^{-\beta} = EW^{-\beta}$ that

$$\lim_{n\to\infty} P\left(U_n > x, a_n^{-1} \bigwedge_{t=1}^{n} (Z_t - \theta) > y\right)$$

$$= \exp\left(-x^{\beta} EW^{-\beta} - y^{\alpha}\right).$$

The theorem now follows from this and Corollary 2.2.

3. SIMULATION STUDY

In this section we assess the reliability of our extreme value estimation method through a simulation study. This included a comparison between our estimation procedure and that of three alternative estimation procedures for both the autocorrelation coefficient ϕ and the unknown location parameter θ under two different innovation distributions. Additionally, the degree of approximation for the empirical probabilities of $\hat{\phi}_{\min}$ and $\hat{\theta}_{\min}$ to its respective limiting distribution was reported.

To study the performance of the estimators

$\hat{\phi}_{\min} = \bigwedge_{t=1}^{n} \frac{X_t}{X_{t-1}}$ and $\hat{\theta}_{\min} = \bigwedge_{t\in I_n} \left(X_t - \hat{\phi}_{\min} X_{t-1}\right)$ respectively, we generated 5000 replications for the nonnegative time series $\left(X_0, X_1, \cdots, X_n\right)$ for two different sample sizes (500,1000), where $\{X_t\}$ is an AR(1) process satisfying the difference equation

$$X_t = \phi X_{t-1} + Z_t, \text{ for } \left(1 \le t \le n\right) \text{ and } Z_t \ge \theta.$$

The autoregressive parameter ϕ is taken to be in the range from 0 to 1 guaranteeing a nonnegative time series and the unknown location parameter θ is positive when the innovations Z_t are taken to be

$$F(z) = \begin{cases} c\left(z-\theta\right)^{\alpha}, \text{if } \theta < z < \theta + 1, \\ 1 - d\left(z-\theta\right)^{-\beta}, \text{if } \theta + 1 < z < \infty. \end{cases}$$

For this innovation distribution let c and d be nonnegative constants such that $c + d = 1$, then this distribution is regularly varying at both endpoints with index of regular variation $-\beta$ at infinity and index of regular variation α at θ. For this simulation study two distributions were considered: $i)\ F_1, c = 0, d = 1,\ ii)\ F_2, c = 0.5, d = 0.5.$

Now observe in case i) the innovation distribution F_1 is a Pareto distribution with a regular varying tail distribution at ∞ with index of regular variation $-\beta$ and regular varying at $\theta+1$ with a fixed index $\alpha=1$, whereas in case ii) the innovation distribution F_2 is regular varying at ∞ and θ with no restriction on α or β.

First we examine the simulation results for $\phi=0.9$ under F_1 for each of the six different β values considered by computing 5000 estimates using

$\hat{\phi}_{\min}=\wedge_{t=1}^{n}\frac{X_t}{X_{t-1}},\hat{\phi}_{\max}=\frac{X_{t^*+1}}{X_{t^*}},\hat{\phi}_{\text{range}}=\frac{X_{t^*+1}-X_{j^*}}{X_{t^*}-X_{j^*-1}}$ where t^* and j^* provides the index of the maximal and minimal X_i respectively for $1\leq i\leq n$, and

$$\hat{\phi}_{LS}=\begin{cases}\sum_{t=1}^{n-1}X_tX_{t+1}\Big/\sum_{t=1}^{n}X_t^2,\text{if } 0<\beta<1\\ \sum_{t=1}^{n}\left(X_t-\bar{X}\right)\left(X_{t+1}-\bar{X}\right)\Big/\sum_{t=1}^{n}\left(X_t-\bar{X}\right)^2,\text{if } 1<\beta<3\end{cases}$$

where $\bar{X}=\sum_{t=1}^{n}X_t/n$. The means and standard deviations (written below in parentheses), of these estimates are reported in **Table 1** along with the average length for a 95 percent empirical confidence intervals with exact coverage. Since the main purpose of this section is to compare our estimator $\hat{\phi}_{\min}$ to Bartlett and McCormick [5] estimator $\hat{\phi}_{\max}$, McCormick and Mathew [2] estimator $\hat{\phi}_{\text{range}}$, and Davis and Resnick's [13] estimator $\hat{\phi}_{LS}$, the confidence intervals were directly constructed from the empirical distributions of

Table 1. Comparison of estimators for f = 0.9 under F1.

β	n	$\hat{\phi}_{min}$	$\hat{\phi}_{max}$	$\hat{\phi}_{range}$	$\hat{\phi}_{LS}$	95% C.I. Avg. Length			
						Min est.	Max est.	Range est.	*LS* est.
0.2	500	0.9000	0.9002	0.9002	0.8988	<0.0001	<0.0001	<0.0001	0.0288
		(<0.0001)	(0.0015)	(0.0016)	(0.0219)	-	-	-	-
	1000	0.9000	0.9001	0.9002	0.8997	<0.0001	<0.0001	<0.0001	0.0091
		(<0.0001)	(0.0009)	(0.0009)	(0.0112)	-	-	-	-
0.8	500	0.9004	0.9026	0.9032	0.9231	0.0046	0.0139	0.0228	0.0657
		(0.0016)	(0.0083)	(0.0083)	(0.0251)	-	-	-	-
	1000	0.9002	0.9020	0.9014	0.9158	0.0023	0.0072	0.0133	0.0529
		(0.0008)	(0.0064)	(0.0064)	(0.0186)	-	-	-	-
1.2	500	0.9049	0.9072	0.9067	0.8923	0.0271	0.0449	0.0548	0.0623
		(0.0078)	(0.0146)	(0.0139)	(0.0176)	-	-	-	-
	1000	0.9029	0.9117	0.9047	0.8964	0.0187	0.0325	0.0389	0.0439
		(0.0056)	(0.0117)	(0.0109)	(0.0117)	-	-	-	-
1.8	500	0.9191	0.9202	0.9136	0.8947	0.0562	0.0763	0.0846	0.0695
		(0.0622)	(0.0673)	(0.0612)	(0.0381)	-	-	-	-
	1000	0.9113	0.9208	0.9116	0.8989	0.0501	0.0690	0.0743	0.0485
		(0.0313)	(0.0367)	(0.0384)	(0.0223)	-	-	-	-
2.2	500	0.9236	0.9313	0.9176	0.8917	0.0670	0.0847	0.0944	0.0709
		(0.0909)	(0.1050)	(0.0244)	(0.0281)	-	-	-	-
	1000	0.9207	0.9293	0.9151	0.8961	0.0623	0.0821	0.0874	0.0494
		(0.0412)	(0.0456)	(0.0217)	(0.0228)	-	-	-	-
2.8	500	0.9475	0.9505	0.9205	0.8873	0.0767	0.0905	0.1050	0.0737
		(0.1378)	(0.1471)	(0.0723)	(0.0691)	-	-	-	-
	1000	0.9437	0.9484	0.9187	0.8906	0.0735	0.0885	0.0956	0.0516
		(0.0834)	(0.1163)	(0.0317)	(0.0331)	-	-	-	-

$$n^{1/\beta}\left(\hat{\phi}_{\min}-\phi\right), n^{1/\beta}\left(\hat{\phi}_{\max}-\phi\right), n^{1/\beta}\left(\hat{\phi}_{\text{range}}-\phi\right)$$
$$\text{and } \left(n/\log n\right)^{1/\beta}\left(\hat{\phi}_{LS}-\phi\right)$$
respectively.

To evaluate and compare the performance of four location estimators, six different scenarios for α and β are presented in **Table 2** under F_2. When $\theta = 2$, 5000 estimates for each estimator;

$$\hat{\theta}_{\min} = \bigwedge_{t\in I_n}\left(X_t - \hat{\phi}_{\min} X_{t-1}\right),\ \hat{\theta}_{\text{range}} = X_{j^*} - \hat{\phi}_{\text{range}} X_{j^*-1},\ \hat{\theta}_{e1} = \left(1-\hat{\phi}_{\text{range}}\right)\bigwedge_{t=1}^{n} X_t,$$

and $\hat{\theta}_{e2} = \left(1-\hat{\phi}_{LS}\right)\bigwedge_{t=1}^{n} X_t$ were obtained. The exponent ρ inside the index set

$$I_n = \left\{t : 1 \le t \le n \text{ and } X_{t-1} \le \left(a_n b_n\right)^{\rho}\right\}$$, was set to 0.9.

The means and standard deviations (written below in parentheses), of these estimates are reported in **Table 2** along with the average length for a 95 percent empirical confidence intervals. For convenience, the empirical distributions of $n^{-1/\alpha}\left(\hat{\theta}_{\min}-\theta\right)$, $n^{-1/\alpha}\left(\hat{\theta}_{\text{range}}-\theta\right)$, $q_n\left(\hat{\theta}_{e1}-\theta-\left(1-\hat{\phi}_{\text{range}}\right)w_n\right)$, and $q_n\left(\hat{\theta}_{e2}-\theta-\left(1-\hat{\phi}_{LS}\right)w_n\right)$ were respectively used, where the normalizing constants q_n and w_n are obtained through Equations (3.12)-(3.16) of McCormick and Mathew [2].

Table 2. Comparison of estimators for f = 2 under F_2.

							95% C.I. Avg. Length			
β	α	n	$\hat{\theta}_{\min}$	$\hat{\theta}_{\text{range}}$	$\hat{\theta}_{e1}$	$\hat{\theta}_{e2}$	Min est.	Range est.	e1 (range) est.	e2 (*LS*) est.
0.5	0.6	500	2.00	2.44	3.16	3.98	<0.0001	2.34	2.09	3.02
			(0.0087)	(0.4068)	(0.6775)	(0.7657)	-	-	-	-
		1000	2.00	2.39	3.77	3.67	<0.0001	1.89	1.69	2.60
			(0.0002)	(0.3605)	(0.4381)	(0.4960)	-	-	-	-
	1.6	500	1.94	2.68	3.34	4.10	0.0273	2.37	3.18	3.66
			(0.0879)	(0.3566)	(0.6656)	(0.7503)	-	-	-	-
		1000	1.99	2.63	3.94	3.81	0.0152	1.87	2.23	2.79
			(0.0503)	(0.3199)	(0.4311)	(0.4780)	-	-	-	-
	2.6	500	1.87	2.77	3.41	4.16	0.1422	2.19	2.85	2.38
			(0.4697)	(0.7553)	(0.6733)	(0.7632)	-	-	-	-
		1000	1.89	2.73	3.99	3.87	0.0988	1.74	2.21	2.03
			(0.4045)	(0.3143)	(0.4276)	(0.4822)	-	-	-	-
1.5	0.6	500	NA	1.72	2.35	2.15	NA	1.57	1.28	2.21
			(-)	(0.9410)	(0.5327)	(0.3706)	-	-	-	-
		1000	NA	1.80	2.19	2.65	NA	1.20	1.25	1.88
			(-)	(0.9141)	(0.4381)	(0.2565)	-	-	-	-
	1.6	500	2.80	1.95	2.54	2.26	1.72	2.23	2.18	1.94
			(0.3863)	(0.9569)	(0.5298)	(0.3815)	-	-	-	-
		1000	2.38	2.03	2.37	2.84	1.28	2.10	1.78	1.31
			(0.2263)	(0.9308)	(0.4461)	(0.2701)	-	-	-	-
	2.6	500	2.99	2.16	2.79	2.35	1.83	2.47	2.21	2.09
			(0.3604)	(0.9443)	(0.5271)	(0.3982)	-	-	-	-
		1000	2.68	2.12	2.27	2.93	1.26	2.10	1.81	1.42
			(0.2956)	(0.9010)	(0.4398)	(0.2837)	-	-	-	-

Remark. In the case that $0<\beta<1$, $\hat{\theta}_{e1}$ converge at a faster rate than $\hat{\theta}_{e2}$ and in the case that $1<\beta<3$, $\hat{\theta}_{e2}$ converges at a faster rate than $\hat{\theta}_{e1}$. Lastly, since the McCormick and Mathew [2] paper has the restriction that $Var\left(Z_1\right)<\infty$, only when $\alpha>2$ can the estimators $\hat{\theta}_{e1}$ and $\hat{\theta}_{e2}$ be fairly compared, whereas only when $\alpha>\beta$ is our estimator applicable.

Now observe for the selected β values being considered, **Table 1** shows that our estimator performs at least as well as the three other alternative estimators. This is particularly true under the heavier tail models, i.e. when $0 < \beta < 2$. In this regime our estimate shows little bias and the average lengths of the confidence intervals are smaller than the other three estimates, sometimes by a wide margin. In particular, when $\beta = 0.8$ and n = 1000 the 95% confidence interval average length for our method is 3.13, 5.78 and 23 times smaller than the three alternative estimators respectively. This is in part due to the use of one-sided confidence intervals since $\hat{\phi}_{\min} \geq \phi$, for all $t \geq 1$. Naturally, when $1 < \beta < 3$, Davis and Resnick least square estimator is more efficient than all three extreme value estimators. While our estimator $\hat{\phi}_{\min}$ will always perform slightly better than the $\hat{\phi}_{\max}$ estimator, Bartlett's and McCormick [5] estimator $\hat{\phi}_{\max}$ main advantage lies with its versatility to perform well for various nonnegative time series, including but not restricted to higher order autoregressive models, along with ARMA models.

Table 2 reveals that our estimator for θ generally performs better than the three alternative estimators for $\alpha = 0.6, 1.6, 2.6$ when $0 < \beta < 1$. This is particularly true when comparing average confidence interval lengths. Although all three estimators $\hat{\theta}_{\min}$, $\hat{\theta}_{\text{range}}$, and $\hat{\theta}_{e1}$ converge to the true value of the parameter θ as n tends to infinity respectively, in this setting they may not compete asymptotically with, say, a conditional least square estimator $\hat{\theta}_{e2}$ when $\beta > 1$. Nonetheless for small sample sizes our simulation study favors $\hat{\theta}_{\text{range}}$ over the other three estimators. The difficulty for a least square estimate is that a small negative bias for the estimate of the autocorrelation parameter ϕ gives rise to a much larger positive bias in the estimate of $\hat{\theta}_{e2}$. While the affect is not as great, the positive bias found in our estimator $\hat{\phi}_{\min}$ and the others for ϕ has a significant effect on the estimate for θ.

Figures 1-4 show a comparison be tween the probability that estimators $\hat{\phi}_{\min}$, $\hat{\phi}_{\max}$, $\hat{\phi}_{\text{range}}$, and $\hat{\phi}_{LS}$ are within 0.01 of the true autocorrelation parameter value, respectively. With a sample size of 500, these figures plotted the sample fraction of estimates which fell within a bound of $\varepsilon = 0.01$ of the true value. Good performance with respect to this measure is reflected in curves near to 1.0 with diminishing good behavior as curves approach 0.0. When $0 < \beta < 1$, the figures seem to show that our estimator compared to the other three produced a higher fraction of precise estimates, especially compared to Davis and Resnick estimator. When the regular variation index value is closer to 2, we see a higher fraction of the Davis-Resnick estimates showing better accuracy by this measure. The figures also indicate that McCormick and Mathew's range estimator produced a consistent high fraction of precise estimates when $\beta > 2$.

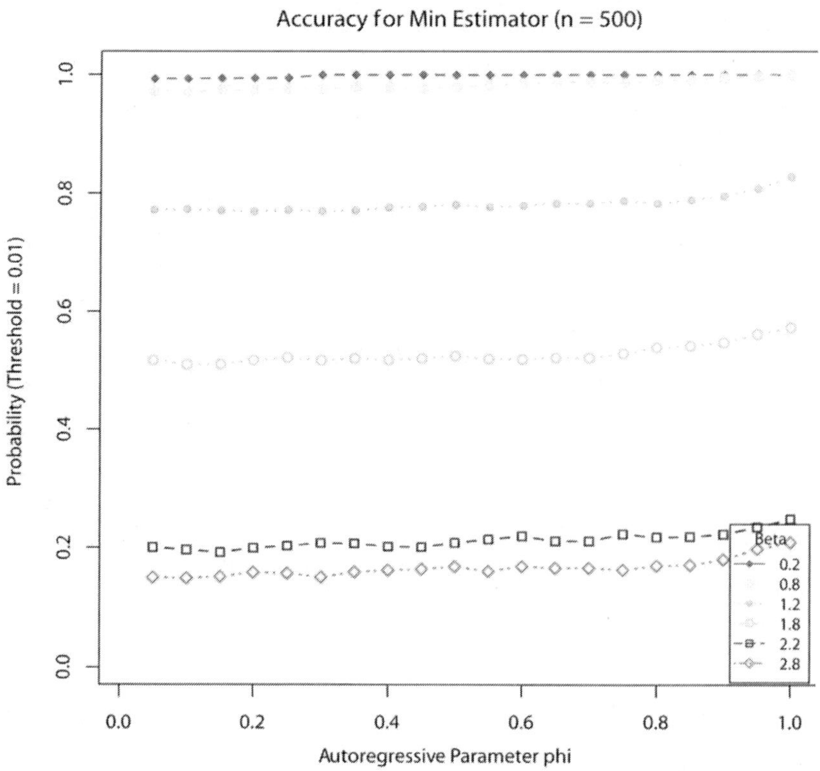

Figure 1. $P\left[\left|\hat{\phi}_{\min} - \phi\right| < 0.01\right]$ for $RV_{-\beta}$.

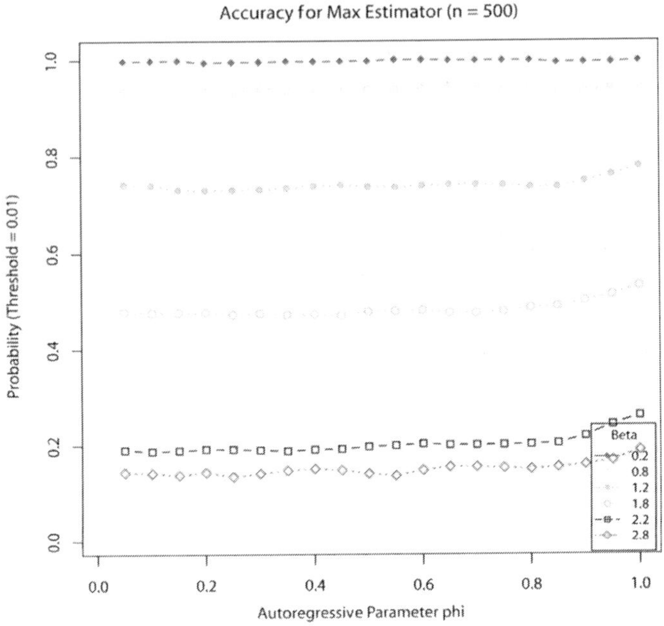

Figure 2. $P\left[\left|\hat{\phi}_{\max}-\phi\right|<0.01\right]$ for $RV_{-\beta}$.

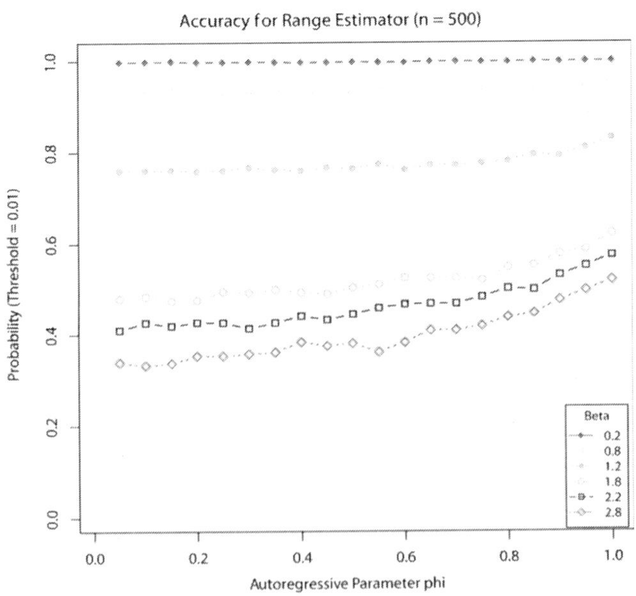

Figure 3. $P\left[\left|\hat{\phi}_{\text{range}}-\phi\right|<0.01\right]$ for $RV_{-\beta}$.

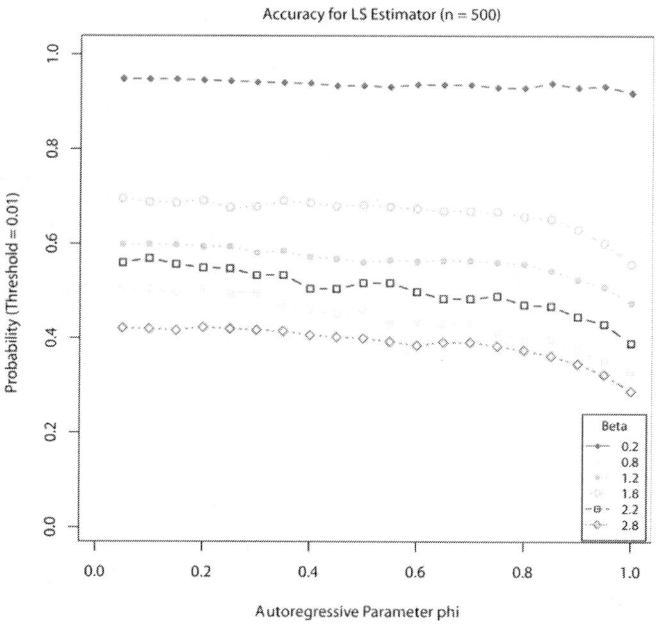

Figure 4. $P\left[\left|\hat{\phi}_{LS}-\phi\right|<0.01\right]$ for $RV_{-\beta}$.

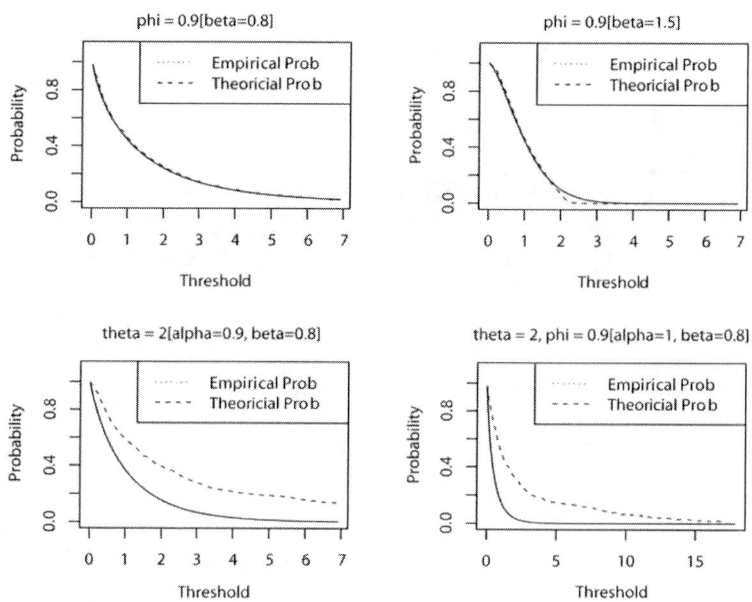

Figure 5. Empirical vs. theoretical probability.

Lastly, we performed a Monte Carlo simulation to study the degree of approximation for the empirical probability $P\left[b_n\left(\hat{\phi}_{\min}-\phi\right)>x\right]$, $P\left[a_n^{-1}\left(\hat{\theta}_{\min}-\theta\right)>y\right]$, and $P\left[b_n\left(\hat{\phi}_{\min}-\phi\right)>x, a_n^{-1}\left(\hat{\theta}_{\min}-\theta\right)>x\right]$ to its limiting values $e^{-x^{\beta}EW^{-\beta}}$, $e^{-y^{\alpha}}$, and $e^{-x^{\beta}EW^{-\beta}-y^{\alpha}}$ respectively. The empirical distributions were calculated from 5000 replications of the nonnegative time series $\left(X_0, X_1, \cdots, X_n\right)$ for a sample size of 5000, where

$EW^{-\beta}=1-\sum_{i=0}^{M}\frac{\phi^{(\beta i(i+3)/2)}}{i+2}\left(1-\phi^{\beta(i+2)}\right)$, and M was set to 500. Additionally, we restricted $\alpha>\beta$. The top two plots in **Figure 5** below shows the performance when $Z_t \sim F_1$ and the autocorrelation coefficient ϕ is 0.9 for $\alpha=1$ and β equal to 0.8, 1.5 respectively. Observe for $0<x<7$ that the empirical tail probability

$b_n\left(\hat{\phi}_{\min}-\phi\right)>x$ mirrors the theoretical probability quite nicely. The lower left plot in **Figure 5**displays the asymptotic performance when $Z_t \sim F_2$ and the location parameter θ is 2 for $\alpha=0.9$, $\beta=0.8$. Notice that the convergence rate of the empirical probability to the theoretical probability is extremely slow. This is not surprising since on average our estimate falls more than 0.1 from the true value when $\beta=0.8$. The lower right plot in **Figure 5** displays the asymptotic performance when

Andrew Bartlett, William McCormick $Z_t \sim F_1$ for the joint distribution of $\left(\hat{\phi}_{\min}, \hat{\theta}_{\min}\right)$. Observe that this plot solidifies the asymptotic independence between $b_n\left(\hat{\phi}_{\min}-\phi\right)$ and $a_n^{-1}\left(\hat{\theta}_{\min}-\theta\right)$.

REFERENCES

1. P. J. Brockwell and R. A Davis, "Time Series: Theory and Methods," 2nd Edition, Springer, New York, 1987.
2. W. P. McCormick and G. Mathew, "Estimation for Nonnegative Autoregressive Processes with an Unknown Location Parameter," Journal of Time Series Analysis, Vol. 14, No. 1, 1993, pp. 71-92.

3. A. E. Raftery, "Estimation Eficace Pour un Processus Autoregressif Exponentiel a Densite Discontinue," Publications de l'Institut de Statistique de l'Université de Paris, Vol. 25, No. 1, 1980, pp. 65-91.
4. R. A. Davis and W. P. McCormick, "Estimation for FirstOrder Autoregressive Processes with Positive or Bounded Innovations," Stochastic Processes and Their Applications, Vol. 31, No. 1, 1989, pp. 237-250.
5. A. Bartlett and W. P. McCormick, "Estimation for NonNegative Time Series with Heavy-Tail Innovations," Journal of Time Series Analysis, 2012. http://onlinelibrary.wiley.com/journal/10.1111/%28ISSN%291467-9892/earlyview
6. S. I. Resnick, "Point Processes, Regular Variation and Weak Convergence," Advances in Applied Probability, Vol. 18, No. 1, 1986, pp. 66-138.
7. Jiří Anděl, "Non-Negative Autoregressive Processes," Journal of Time Series Analysis, Vol. 10, No. 1, 1989, pp. 1-11.
8. Jiří Anděl, "Non-Negative Linear Processes," Applications of Mathematics, Vol. 36, No. 1, 1991, pp. 277-283.
9. S. Datta and W. P. McCormick, "Bootstrap Inference for a First-Order Autoregression with Positive Innovations," Journal of American Statistical Association, Vol. 90, No. 1, 1995, pp. 1289-1300.
10. S. I. Resnick, "Heavy-Tail Phenomena: Probabilistic and Statistical Modeling," Springer, New York, 2007.
11. R. A. Davis and S. Resnick, "Limit Theory for Moving Averages of Random Variables with Regularly Varying Tail Probabilities," Annals of Probability, Vol. 13, No. 1, 1985, pp. 179-195.
12. O. Kallenberg, "Random Measures," Akademie-Verlag, Berlin, 1976.
13. R. A. Davis and S. Resnick, "Limit Theory for the Sample Covariance and Correlation Functions of Moving Averages," Annals of Statistics, Vol. 14, No. 1, 1986, pp. 533-558.

CHAPTER 13

Parameter Dependence in Stochastic Modeling—Multivariate Distributions

Jerzy K. Filus[1], Lidia Z. Filus[2]

[1]*Department of Mathematics and Computer Science, Oakton Community College, Des Plaines, USA*

[2]*Department of Mathematics, Northeastern Illinois University, Chicago, USA*

ABSTRACT

We start with analyzing stochastic dependence in a classic bivariate normal density framework. We focus on the way the conditional density of one of the random variables depends on realizations of the other. In the bivariate normal case this dependence takes the form of a parameter (here the "expected value") of one probability density depending continuously (here linearly) on realizations of the other random variable. The point is, that such a pattern does not need to be restricted to that classical case of the bivariate normal. We show that this paradigm can be generalized and viewed in ways that allows one to extend it far beyond the bivariate or multivariate normal probability distributions class.

Keywords: Multivariate Probability Distributions, Stochastic Dependence Paradigms, Multivariate Gaussian Distributions, Parameter Dependence Method of Construction, Conditioning, Stress, Biomedical Applications

1. INTRODUCTION

This paper can be viewed as an extension of our previous work (Filus and Filus [1]) on the bivariate Gaussian pdf structure's genesis of wide classes of newly constructed bivariate probability distributions. These distributions we constructed in our papers since 2000 up to recently (see, Filus and Filus [1] -[6] , also see Kotz, Balakrishnan and Johnson [7] pp. 217-218). Here, our explanations concerning the relation of our models to the bivariate normal were modified and we extended the topic to higher dimensional models and to the relation between our "parameter dependence method" of models construction and the multivariate normal paradigm.

It is a well-known fact that among existing multivariate probability distributions, there are no more than a few classes that are widely and successfully applied in practical stochastic modeling procedures. Typically, the underlying random variables are assumed to be independent or having an "approximately Gaussian" bivariate or multivariate distribution. The normality often is assumed even when corresponding data hardly agree with that mathematical model (showing asymmetry, for example). On the other hand, from all the multivariate distributions used in applications, the normal seems to be "the best". The reason for this is that the Gaussian models catch the stochastic relationship (mainly by a regression function) between its marginal random variables in the most natural way. We first analyze and interpret the specific way the multivariate normal density of the random vector $(X_1,\cdots,X_m)$ relates to the marginal quantities $X_1,\cdots,X_m$. Next we extend the "Gaussian pattern" to more general classes of bivariate and multivariate distributions including cases with non-Gaussian marginals. First of all we show that the common "mechanism" of the stochastic dependences both in the Gaussian distributions structure and the structure of the distributions we define, relies on the same way of conditioning. Namely, in all the considered cases, the conditional density of one random variable, say, $X_j\,(j=2,3,\cdots,m)$, given realizations, say, $x_1,\cdots,x_{j-1}$, of the marginal random variables $X_1,\cdots,X_{j-1}$ can be obtained by setting (often arbitrarily) an "initially constant" parameter, say, q_j of the density of X_j to be any continuous function $\theta_j\left(x_1,\cdots,x_{j-1}\right)$ of the realizations. In this way the conditional pdf of X_j, given $x_1,\cdots,x_{j-1}$ is defined, which stands for the "source" of the stochastic dependences.

Pursuing this method successively for $j = 2, 3, \cdots, m$ we always arrive at a unique model. This manner, however, is characteristic for the m-variate normal where we turn an "original" normal $N(m_j, \sigma_j)$ density of X_j into the conditional density by setting a new value μ_j^* of the "affected old parameter m_j" to be the following (linear regression) function: $\mu_j^* = \mu_j + c_1 x_1 + \cdots + c_{j-1} x_{j-1}$. Our main contribution is to generalize the latter function to any, not necessarily linear, function and consider not only the parameter m_j but also any other parameter of any probability density to define the corresponding conditional distributions. This is the essence of the socalled parameter dependence method. Specifically in this paper, our task will be showing more closely relation of this method to the multivariate Gaussian model construction.

In Section 2, we analyze the stochastic dependences between marginal random variables of the bivariate normal in order to point out the original version of the parameter dependence pattern next extending to other constructed bivariate probability densities. The explanation as well as the example of applications of the bivariate normal is different from that in Filus and Filus [1] . In Section 3 we present the extension of the bivariate normal pdf to the bivariate FF-normal (formerly called "pseudonormal"). In Section 4 we apply the parameter dependence method to construct the bivariate FF-Weibull (formerly "pseudo-Weibull") density in reliability framework of joint density of parallel system components life times. Comparison with other, similar, methods in the literature is presented in Section 5. In Section 6 we extend the constructions from bivariate to multivariate probability densities, first showing their relation with the multivariate normal dependences structure. Examples of the construction of multivariate FF-normal and multivariate FF-exponential ("pseudo-exponential") densities are given.

In Section 7, we point out that the "method of parameter dependence" is used in some more areas of reliability theory for different situations than we are considering. This is a part of the accelerated life testing theory where the dependence of life time distribution's parameter from a given (high) stress is investigated.

Another (fairly new) area is the "load optimization theory" sometimes associated with the load sharing phenomena analysis that we sketch in Subsection 7.2. The differences between these approaches and our theory are pointed out in 7.2.

2. THE BIVARIATE NORMAL CASE

We start with the following situation. Suppose the normally distributed random variable X_2describes an attribute of a physical or biological object, say u. Consider the (stochastic) behavior of the object u in two distinct "physical" situations. In the first situation, u is exposed to some random stress whose magnitude is described by a normally distributed random variable X_1. In the second situation we assume there is no such a stress present or the stress takes on a fixed predetermined value. The usual task here is to determine the joint distribution of X_1, X_2. Let the densities of X_1, X_2be normal, i.e., $g_1(x_1)=N(\mu_1,\sigma_1)$, $g_2(x_2)=N(\mu_2,\sigma_2)$. [It is clear that we must assume that the value (m_2 – $k\sigma_2$) is positive for at least k = 3, in order to assure approximate positivity of the normal life-time X_2].

2.1. Simplified Biomedical Example

Imagine the following fictitious experiment whose goal is to establish the possible stochastic impact of a medication's dose change on some cancer treatment results. Suppose a person of a certain fixed age, was diagnosed with a kind of cancer. Assume that one of the significant characteristics of that kind of cancer is a tumor with a size X_2. During a given time period T after the patient was diagnosed, a specific medication was administered. Also suppose that this medication was routinely administered in the past, and that the average dose is estimated (or fixed) to be m_1 milligrams per kilo of weight daily. Assume that, originally, the known (either measured or estimated) average size of the tumor is, say, m_2 millimeters and after the period T of treatment the tumor size X_2 is measured again and its negative or positive increment $X_2-\mu_2$ is statistcally confronted with the dose X_1 of the medication administered.

We assume that the goal of the underlying experiment is to make a prediction on effect $X_2-\mu_2$ of the treatment when the dose is changed from its "original level" $X_1=\mu_1$ to a level $X_1\neq\mu_1$. Randomness of the doze X_1 may be justified when only "historical data" are analyzed and then extrapolated for a larger population of cases not yet recorded. In the case of extrapolation of historical data for a larger population we assume that the only information one possess on the applied in the past dose X_1 is its probability distribution, which is the Gaussian $N(\mu_1,\sigma_1)$ with given values for both the parameters m_1, σ_1. Also, for $X_1=\mu_1$ the tumor size X_2 (after the treatment) is assumed to be a random variable having a normal $N(\mu_2,\sigma_2)$ density, where $X_2-\mu_2$ is the value of the tumor "increment" under the treatment characterized by the dose level m_1. For any other applied doze $X_1 = x_1$ the, associated with a single patient, value $(x_1 - m_1)$ statistically affects the change in the tumor size $X_2 - m_2$ i.e., the treatment result. The word "statistically" here means that the impact of a nonzero quantity $(x_1 - m_1)$ ("the dose is not the standard one") on the (former) probability density $N(m_2, \sigma_2)$ of the tumor size X_2 realizes through affecting the value of the mean m_2 rather than directly affecting the numerical value x_2 of X_2.

If we were interested in finding the joint probability distribution of X_1, X_2 it is enough to determine the conditional density $g_2(x_2|x_1)$ of $X_2|X_1$, since the marginal density of X_1 is not changing.

In accordance with the "linear regression rule", the dependence of the (new) expected value μ_2^* (so new probability density $N(\mu_2^*,\sigma_2^*)$) of the tumor size X_2 on the event $X_1 = x_1$, is determined by the familiar functional relationship:

$$E\left(X_2|X_1=x_1\right)=\mu_2^*=\mu_2+a\left(x_1-\mu_1\right) \tag{1}$$

where $a = r(\sigma_2/\sigma_1)$ and r is the (linear) correlation coefficient of the variables X_1, X_2.

This approach directly leads to the determination of the conditional density of the random variable X_2 given any realization $X_1 = x_1$. It is a well-known fact that the conditional density $g_2(x_2|x_1)$ is, again, normal and

$$g_2\left(x_2|x_1\right)=\frac{1}{\sigma_2\sqrt{2\pi\left(1-\rho^2\right)}}\exp\frac{-\left(x_2-\mu_2-a\left(x_1-\mu_1\right)\right)^2}{2\sigma_2^2\left(1-\rho^2\right)} \quad ,(2)$$

i.e., the $N\left(\mu_2+a\left(x_1-\mu_1\right),\sigma_2\sqrt{1-\rho^2}\right)$ density in x_2.

The joint density $g(x_1,x_2)$ of the random variables X_1, X_2 is given by the usual arithmetic product $g_2(x_2|x_1)g_1(x_1)$.

2.2. Remark

In the example above, one can reinterpret the "response random variable" X_2 to be for example the patient's "residual life-time", or blood pressure, or level of some important chemical in the blood (such as cholesterol). In such cases the mathematics of the problem would remain the same.

Note the obvious fact that the tumor size X_2 does not have a physical influence on the medication dose X_1 so that the original marginal pdf $g_1(x_1)$ remains the same. However, the stochastic dependence between X_1, X_2 is mutual, since, in general, $g_1(x_1|x_2) \neq g_1(x_1)$.

It is well known that the actual problem with the bivariate normal density construction is to get to the conditional density (2), which fully represents the underlying stochastic dependence of random variable X_2 on X_1.

Our claim is that the above paradigm for the stochastic dependence (characteristic for the bivariate Gaussians) can be extended to other classes of bivariate and multivariate distributions (see Filus and Filus [3] [4]).

3. THE FF-NORMAL (PSEUDONORMAL) EXTENSION

Historically, people relied on the nice symmetry in the stochastic dependence of X_1 and X_2 when using their joint bivariate normal distribution. This kind of symmetry (i.e., both marginal and both conditional distributions are normal and both sides regression functions are linear) can only be achieved with the linear regression functions as described above (1). However, are linear regression functions really the only functions that one can successfully apply within this framework? Assuming that the function $m_2(X_1)$ is any continuous function in X_1, one obtains a wide and interesting extension of the class of bivariate normal densities. We called this class FF-normal (previously named "pseudonormal", see [2] and also [7]). In this case the parameter σ_2 can as well become a continuous function of the stress X_1 (X_1 may have a "stress" interpretation in a very wide sense). This stress may change the parameter σ_2 of the (normal) density of, say, the "life-time" X_2 into another value $\sigma_2^* = \sigma_2^*(x_1)$ that depends on the particular realization x_1 of the random variable X_1. The price for such a wide generalization is loss of the, mentioned above, symmetry (the marginal of X_2 ceases to be normal) but the gains are considerable. Anyway, the bivariate normal remains to be a special cases of the FF-normals.

With the bivariate FF-normal densities of (X_1, X_2) we can use general continuous $m_2(x_1)$ and $\sigma_2(x_1)$ functions, and, performing similar calculations as above, we find, rather surprisingly, that $g_2(x_2|x_1)$ is once more a regular normal density in x_2.

Consider now the following situation with a bivariate FF-normal distribution in which the "physical" interpretation of the underlying random variables can now be more general than above. Let u_1, $u_2$be two objects (or phenomena) which are characterized by the random variables X_1, X_2respectively. If the objects are physically separated then the random variables X_1, X_2 are assumed to be independent, having normal pdfs $g_1(x_1)=N(\mu_1,\sigma_1)$ and $g_2(x_2)=N(\mu_2,\sigma_2)$ respectively. When the objects physically interact (or rather only $u_1$physically impacts u_2), then the corresponding joint FF-normal density $g(x_1, x_2)$ of the random vector (X_1, X_2) is given by the usual product formula:

$g(x_1,x_2)=g_1(x_1)g_2(x_2|x_1)$ with the invariant marginal density $g_1(x_1)=N(\mu_1,\sigma_1)$.

For the conditional density of $X_2|x_1$ we have: $g_2(x_2|x_1)=N(\mu_2^*(x_1),\sigma_2^*(x_1))$.

The functions $\mu_2^*(x_1),\sigma_2^*(x_1)$ are "formed" from the "no-stress" parameters m_2, σ_2 of X_2's density.

More explicitly, one obtains the bivariate FF-normal pdf in the form:

$$g(x_1,x_2)=\frac{1}{\sqrt{2\pi}\sigma_1}\exp\frac{-(x_1-\mu_1)^2}{2\sigma_1^2}\frac{1}{\sqrt{2\pi}\sigma_2^*(x_1)}\exp\frac{-(x_2-\mu_2^*(x_1))^2}{2\sigma_2^{*2}(x_1)}$$

where $\mu_2(x_1)=E[X_2|x_1]$ is the (in general, nonlinear) "regression function", and $[\sigma_2^*(x_1)]^2=\text{Var}[X_2|x_1]$ is the conditional variance obtained from the "previous" variance σ_2^2.

In particular, one may consider the "nonlinear regression function"

$$E\left[X_2 \middle| x_1\right] = \mu_2(x_1) = \mu_2 + a^*(x_1 - \mu_1) + A(x_1 - \mu_1)^n$$

with arbitrary real parameters a^* and A, $n = 2,3,\cdots$.

Realize that in the case A = 0 and $\sigma_2(x_1) = \sigma_2\sqrt{1-\rho^2}$ we obtain the regular bivariate Gaussian density. The coefficient A of the term $A(x_1 - \mu_1)^n$ may be considered as a nonlinear "correction" of the regular Gaussian (linear) regression. The main purpose of that correction is to enhance the accuracy in various modeling procedures. For some type of asymmetric data, especially interesting may be the "non-symmetric" "quadratic" case n = 2.

Equally important is the "symmetric" "cubic" case when n = 3. Also interesting is the case where $a^*= 0$ and A ≠ 0. Here, the linear correlation coefficient is zero, while the (nonlinear) dependence may be quite "heavy". Another interesting case (which may be combined with those above) is when we choose the standard deviation function as belonging to the "parametric class" of "quadratic functions" defined by $\sigma_2^*(x_1) = k\sigma_2(1 + r^2 x_1^2)$ with k positive real. It is rather obvious that the idea of the construction of bivariate normal (and FF-normal as well) can be extended to other probability distributions such as exponential, gamma, Weibull, lognormal, etc.

Generally speaking, the essence of the construction method is that for any pair of ("initially independent") random variables X_1 and X_2 with given probability densities $g_1(x_1, \omega)$ and $g_2(x_2; \Theta)$ respectively, one can simply "declare" some parameter (or a vector parameter), say Θ, of the density $g_2(x_2; \Theta)$ to be dependent on the values of the other random variable X_1. This means that when $X_1 = x_1$ we may assume that $g_2(x_2, \Theta(x_1)) = g_2(x_2 | x_1)$ is the "affected by x_1" distribution of the random variable X_2 (given the event $X_1 = x_1$ occurred, with probability density $g_1(x_1)$).

Then the joint density of the pair (X_1, X_2) is always $g(x_1, x_2) = g_2(x_2, \Theta(x_1))g_1(x_1)$.

This situation is especially natural if we consider X_2 to be the life-time of an object and X_1 is the stress put on it.

Roughly, one can say that the construction method of bivariate distributions, presented above is an extension of the method used in the construction of the bivariate normal.

4. RELIABILITY EXAMPLE

Consider a 2-component (say u_1, u_2) parallel system reliability setting in which X_1, X_2 represent the components' life-times (see Barlow and Proschan [8]). We start with the situation where the system's components act separately. We call that pattern the "laboratory conditions". In this latter case the components are physically separated and consequently their life-times (represented by the, statistically estimated, "baseline probability densities" $g_1(x_1)$ and $g_2(x_2)$ respectively), are stochastically independent. When the two components are installed into the system, they start to interact. Assume that during that interaction some irregularities in the work of component u_1 cause corresponding changes in u_2's inner physical structure. This increases the hazard rate of that physically affected component u_2. Such physical phenomena are then "responsible" for the occurrence of stochastic dependence in the "in-system" component life-times X_1 and X_2.

One can also imagine this situation as follows. During the two components' "in-system" performance, component u_1 creates a situation in which component u_2 is "constantly bombarded" by a string of harmful "micro-shocks" (see Filus and Filus [5]). Each such micro-shock causes a corresponding "micro-damage" in the affected component u_2's physical constitution. We also assume that these micro-damages in component u_2's inner physical structure "cause" some corresponding "micro-changes in the original (baseline) failure rate" (and, in parallel, in the corresponding probability distribution) of its life-time X_2. After a, possibly long, time period X_1 of such interaction all these micro-damages cumulate their effects. As a result of this accumulation, the overall change in the corresponding "hazard rate function" will become significant. To describe formally the change in the hazard rate function we have chosen to consider corresponding changes in its parameter(s). In what follows we present a particular bivariate model for a 2-component system reliability which we called FF-Weibullian (formerly "pseudo-Weibullian" in Filus and Filus [4]).

Suppose the lifetimes of the components u_1 and u_2 in "laboratory conditions" are independent and distributed according to the Weibull density random variables X_1 and X_2.

Let $g_k(x_k) = \lambda_k \alpha_k x_k^{\alpha_k - 1} \exp - \lambda_k x_k^{\alpha_k}$ be the pdf of X_k (k = 1, 2).

Here, for k = 1, 2, we have the "vector parameter" $\theta_k = (\lambda_k, a_k)$.

Next consider the components u_1 and u_2 as acting within the system. Let the resulting (changed) values λ_2^*, σ_2^* of the parameters of the (original) pdf $g_2(x_2; l_2, a_2)$ be determined by the following continuous functions of x_1:

$\lambda_2^* = \lambda_2^*(x_1)$ and $\alpha_2^* = \alpha_2^*(x_1)$.

One then obtains the wide class of bivariate FF-Weibullian densities:

$$g(x_1, x_2) = g(x_1) g_2(x_2 | x_1) = \lambda_1 \alpha_1 x_1^{\alpha_1 - 1} \exp(-\lambda_1 x_1^{\alpha_1}) \lambda_2^*(x_1) \alpha_2^*(x_1) x_2^{\alpha_2^*(x_1) - 1} \exp\left(-\lambda_2^*(x_1) x_2^{\alpha_2^*(x_1)}\right) \quad (3)$$

where, for ease of computation, we recommend to apply as "sub-model" the following family of "parameter functions":

$\lambda_2^*(x_1) = \lambda_2(1 + A x_1^r)$ and $\alpha_2^*(x_1) = s\alpha_2$ with parameters A, r, s positive reals.

In particular, s may depend on x_1.

Another analytically interesting "sub-model" is given by:

$\lambda_2^*(x_1) = \lambda_2 \exp[A x_1^r]$ with A and r real, and $\alpha_2^*(x_1) = s\alpha_2(1 + c x_1)$ with s > 0, c^{3} 0.

Note that both factors $g_1(x_1)$ and $g_2(x_2|x_1)$ of the joint density $g(x_1, x_2)$ given by (3) are Weibullian densities. In particular, $g_2(x_2|x_1)$ is Weibullian with respect to the argument x_2 alone.

For the simpler FF-exponential example, see [1] .

5. NOTICE ON SIMILAR INVESTIGATIONS IN THE LITERATURE

A parallel and basically independent path of investigation, which also has its roots in the bivariate normal distribution's dependence paradigm, is present in the literature under the key word "conditioning".

This method, used in the construction of numerous multivariate probability distributions, was extensively developed mostly since around 1987. See, for example, Arnold, Castillo and Sarabia [9] with citations. Also consider Castillo and Galambos [10] .

The underlying method (by numerous authors called the "conditioning method") relies on imposing conditional structure X|Y and Y|X on, given in advance, "baseline" probability densities f(x; A) and g(y; B) of some ("initially independent") random variables X and Y respectively, where A and B are scalar or vector parameters. The two conditional densities are defined as we did above, i.e.

g(y|x) = g(y; B(x)) and f(x|y) = f(x; A(y))

where A(y) and B(x) are continuous functions of realizations of the random variables Y, X respectively.

In this case the task is to find two proper (unknown) marginal densities for the bivariate probability distribution of (X, Y) which are, as a rule, not unique and sometimes do not exist.

Despite similarities this method essentially differs from ours. In our case, instead of the two conditional densities g(y|x) and f(x|y), we define only one, say, g(y|x), but together with the marginal f(x).

Pursuing this way we always directly obtain a unique model simply as the product of the two (known) densities.

In such a way, we have obtained a wide class of bivariate densities which is essentially disjoint from the class obtained by that alternative method. Also, the physical interpretation of the, so defined, conditional densities differs in the two approaches. However, both approaches are devoted to the same purpose which is to extend of the paradigm of the bivariate normal in stochastic modeling. Nevertheless, using the conditioning method it is very difficult to construct the multivariate distributions of any higher than two dimensions.

Practically that method reduces to the bivariate cases while the method we present has a remarkable easiness of construction of probability distributions of, actually, arbitrary finite dimension. There is, namely, a recurrence procedure which allows to construct any j-th dimensional pdf based on corresponding (j – 1)-th dimensional pdf $(j = 3, 4, \cdots)$ constructed "at an earlier stage". That procedure was also used in Filus and Filus [11] for the construction of discrete time stochastic processes.

The next section is devoted to the construction of multivariate distributions for any arbitrary finite dimension.

6. METHOD OF PARAMETER DEPENDENCE FOR MULTIVARIATE PROBABILITY DISTRIBUTIONS CONSTRUCTION

For the construction, mentioned in the title, we successively use the simple recurrence method that yields the j-th dimensional probability density, given the (j – 1)-th one. Realize, that (for j = 3) we have already defined the 2- dimensional densities $g^2(x_1, x_2)$ by means of the products $g_1(x_1)g_2(x_2|x_1)$, where each underlying conditional density was given by $g_2(x_2|x_1) = g_2(x_2, q_2(x_1))$.

So the "first step" is already done. Suppose now that we have at our disposal the (j – 1)-th dimensional (j ³ 3) pdf, say $g^{j-1}(x_1, x_2, \cdots, x_{j-1})$ of the random vector $(X_1, X_2, \cdots, X_{j-1})$. The task of obtaining the joint density $g^j(x_1, x_2, \cdots, x_j)$ of the random vector $(X_1, X_2, \cdots, X_{j-1}, X_j)$ always reduces to defining the conditional density $g_j(x_j|x_1, x_2, \cdots, x_{j-1})$, given any univariate baseline pdf $h_j(x_j, \theta_j)$ by the method of parameter dependence. Assuming that originally q_j is a constant parameter, we define the conditional pdf $g_j(x_j|x_1, x_2, \cdots, x_{j-1})$ by setting (according to the new physical situation of "being in the system" together with the other j – 1 objects):

$$g_j(x_j|x_1, x_2, \cdots, x_{j-1}) = h_j(x_j, \theta_j^*(x_1, x_2, \cdots, x_{j-1}))$$

Now, the "new" value θ_j^* of the parameter is a continuous function of the realizations ("multi-stresses") $(x_1, x_2, \cdots, x_{j-1})$ of the random vector $(X_1, X_2, \cdots, X_{j-1})$.

The j-dimensional pdf of the random vector $(X_1, X_2, \cdots, X_{j-1}, X_j)$ one obtains simply as the product:

$$g^j(x_1, x_2, \cdots, x_j) = g_j(x_j | x_1, x_2, \cdots, x_{j-1}) g^{j-1}(x_1, x_2, \cdots, x_{j-1})$$

The latter pdf becomes the basis for identical construction of the (j + 1)-dimensional pdf and so on.

We then stop the procedure once j + 1 = m, where m is the total dimension of the considered (maximal) random vector, say, $(X_1, \cdots, X_m)$.

Since the analogy with the construction of each j-dimensional normal pdf $(j = 3, 4, \cdots)$ is not as straightforward as in the bivariate case, we found it beneficial to show this analogy closer, by reducing the normal's construction to the "diagonal case". Let us start with recalling that any normally distributed random vector, say, $X = (X_1, \cdots, X_j)$, $j = 2, 3, \cdots, m$ is obtainable from the random vector $Z = (Z_1, \cdots, Z_j)$ by an affine transformation

$$X^{\mathrm{T}} = AZ^{\mathrm{T}} + \mu^{\mathrm{T}}, \tag{4}$$

where the random variables $Z_1, \cdots, Z_j$ are independent and each having the standard normal N(0, 1) pdf. A is an arbitrary j × j matrix with real entries (here, without losing generality, we restrict ourselves to nonsingular matrices A, only) and $\mu = (\mu_1, \cdots, \mu_j)$ is an arbitrary fixed vector in R^j. The symbol^{T} denotes the usual matrix transpose. Recall that every matrix A can be decomposed as

$$A = MB, \tag{5}$$

where B is a lower triangular and M is an orthogonal matrix. From (5) we obtain that any nonsingular lower triangular matrix B can be represented as the product:

$$B = M^{\mathrm{T}} A, \tag{6}$$

where A is an arbitrary nonsingular matrix. If we replace representation (4) of the random normal vector X by the following representation

$$Y^{\mathrm{T}} = BZ^{\mathrm{T}} + \mu^{\mathrm{T}}, \tag{7}$$

then we replace the arbitrary random vector X by an arbitrary "triangular" random vector Y related to X by:

$$(Y-\mu)^{\mathrm{T}} = M^{\mathrm{T}}(X-\mu)^{\mathrm{T}}, \tag{8}$$

where M^T is an arbitrary orthogonal transformation.

Since the two zero-expectation random vectors X – μ, Y – μ are obtained one from another by an isometry (here, rotation) M^T in the Euclidean space R^j, they may be considered as representing the same "stochastic data" expressed in two different (but still rectangular) coordinate systems. So from a stochastic viewpoint the "difference" between the random vectors X and Y is inessential and we can consider the random vector Y as an "arbitrary normal" ("with accuracy to the rotation" M^T).

Collecting all the above, we will consider the normal random vector Y, given by (7), where matrix B is any lower triangular matrix and Z is the standard normal j-vector. Write (7) in the form:

$$\begin{aligned}
Y_1 &= b_{11}Z_1 + \mu_1 \\
Y_2 &= b_{21}Z_1 + b_{22}Z_2 + \mu_2 \\
&\vdots \\
Y_{j-1} &= b_{j-1,1}Z_1 + \cdots + b_{j-1,j-1}Z_{j-1} + \mu_{j-1} \\
Y_j &= b_{j,1}Z_1 + \cdots + b_{j,j-1}Z_{j-1} + b_{j,j}Z_j + \mu_j \quad j \to m
\end{aligned} \tag{7*}$$

where m ^{3} j is the actual dimension of the constructed (final) random vector, say $(X_1, \cdots, X_m)$ (if m = ⍰, one defines, in effect, a stochastic process with time j).

Considering the first –1 lines in (7*) as a system of linear equations, one obtains all $Z_1, \cdots, Z_{j-1}$ as linear combinations of $Y_1, \cdots, Y_{j-1}$. Substituting these solutions back into (7*) one obtains the following form:

$$\begin{aligned}
Y_1 &= c_{11}Z_1 + \mu_1 \\
Y_2 &= c_{22}Z_2 + \left(c_{21}Y_1 + \mu_2\right) \\
Y_3 &= c_{33}Z_3 + \left(c_{31}Y_1 + c_{32}Y_2 + \mu_3\right) \\
&\vdots \\
Y_{j-1} &= c_{j-1,j-1}Z_{j-1} + \left(c_{j-1,1}Y_1 + \cdots + c_{j-1,j-2}Y_{j-2} + \mu_{j-1}\right)
\end{aligned} \tag{9}$$

Realize that transformation (9) is easily reversible.

Assuming that realizations $y_1, \cdots, y_{k-1}$ of the random variables $Y_1, \cdots, Y_{k-1}$ are known, we obtain for each $k = 1, \cdots, j$:

$$Z_k = \frac{Y_k - \left(c_{k,1}y_1 + \cdots + c_{k,k-1}y_{k-1} + \mu_k\right)}{c_{kk}} \tag{10}$$

where from the above assumed nonsingularity we have $c_{kk} \neq 0$. From (10) it follows that the conditional density of each Y_k, given the values $y_1, \cdots, y_{k-1}$, is normal and for the corresponding conditional expectation we have

$$E\left[Y_k \middle| y_1, \cdots, y_{k-1}\right] = \mu_k + c_{k,1}y_1 + \cdots + c_{k,k-1}y_{k-1}$$

while for the (constant) conditional variance we obtain

$$\mathrm{Var}\left[Y_k \middle| y_1, \cdots, y_{k-1}\right] = c_{kk}^2$$

To adopt the above procedure to our concept of "baseline" T_j versus "in system" Y_j random variables, replace in (9) the independent standard random variables $Z_1, Z_2, \cdots$ by independent random variables, say, $T_1, T_2, \cdots$, where each T_k has the ("baseline") normal $N(\mu_k, s_k)$ pdf $(k = 1, 2, \cdots)$.

Replace transformation (7) by

$$Y = BT^{\mathrm{T}}, \tag{11}$$

where $T = (T_1, \cdots, T_j)$. Using this change, (10) will be replaced by the following inverse transformation:

$$T_k = \frac{Y_k - (c_{k,1} y_1 + \cdots + c_{k,\,k-1} y_{k-1})}{c_{kk}} \tag{10*}$$

This yields the conditional pdf of $Y_k | y_1, \cdots, y_{k-1}$ to be the normal

$$N(\mu_k + c_{k,1} y_1 + \cdots + c_{k,\,k-1} y_{k-1}, |c_{kk}| \sigma_k)$$

Finally, the general pattern of "creation" of any successive j-variate normal pdf can be explained as follows.

Given are the first j – 1 lines of transformation (11) in the form:

$$\begin{aligned} Y_1 &= c_{11} T_1 \\ Y_2 &= c_{22} T_2 + c_{21} Y_1 \\ Y_3 &= c_{33} T_3 + (c_{31} Y_1 + c_{32} Y_2) \\ &\vdots \\ Y_{j-1} &= c_{j-1,\,j-1} T_{j-1} + (c_{j-1,1} Y_1 + \cdots + c_{j-1,\,j-2} Y_{j-2}) \end{aligned} \tag{12}$$

for some $j - 1 = 1, 2, \cdots$. (Realize that the joint normal pdf $g^{j-1}(y_1, y_2, \cdots, y_{j-1})$ of the random vector $(Y_1, Y_2, \cdots, Y_{j-1})$ was defined in j – 1 "previous" steps. In particular for j – 1 = 1, it is the univariate normal $N(\mu_1, |c_{11}| s_1)$ density of the variable Y_1.)

We may assume that the next baseline random variable T_j, originally having the $N(\mu_j, s_j)$ pdf, is incorporated to the "system" by transforming

$$T_j \rightarrow Y_j | Y_1, \cdots, Y_{j-1}$$

This transformation is thought of as adding to (12) the following j-th line:

$$Y_j = c_{j,j} T_j + (c_{j,1} Y_1 + \cdots + c_{j,\,j-1} Y_{j-1}) \tag{13}$$

["Physically" this could mean that the variables $Y_1,\cdots,Y_{j-1}$ "become" explanatory ("stresses") for the "new" variable Y_j obtained from T_j that (originally) was independent from these stresses].

From (13) one can determine the conditional pdf of Y_j, given any realization $\left(y_1,\cdots,y_{j-1}\right)$ of the random vector $\left(Y_1,\cdots,Y_{j-1}\right)$, as the following normal pdf in y_j:

$$N\left(\mu_j+c_{j,1}y_1+\cdots+c_{j,j-1}y_{j-1},\left|c_{jj}\right|\sigma_j\right).$$

Thus, as the j-th "object" (originally independent from the "system" and characterized by the random quantity T_j) was "put into the system" the quantity T_j turns to the quantity Y_j and, in parallel, the parameters μ_j and s_j of its normal density are turned into $\mu_j+c_{j,1}y_1+\cdots+c_{j,j-1}y_{j-1}$ and $\sigma_j\left|c_{jj}\right|$, respectively, while normality is preserved.

Clearly, the new value μ_j^* of the (conditional) expectation became the continuous (here linear) function

$$\mu_j^*\left(y_1,\cdots,y_{j-1}\right)=\mu_j+c_{j,1}y_1+\cdots+c_{j,j-1}y_{j-1}$$

of realizations $y_1,y_2,\cdots,y_{j-1}$ while unfortunately the new value of the standard deviation does not depend on $y_1,y_2,\cdots,y_{j-1}$ but remains constant even if multiplied by the specific, determined by the "system", number c_{jj}. This can be "made up" if we allow the number c_{jj} in (13) to be dependent on $y_1,y_2,\cdots,y_{j-1}$, but then the so obtained multivariate FF-normal distribution ceases to be normal since (13) ceases to be linear.

As we have shown also in multivariate cases, the origin of the "parameter dependence method for the construction", lies in the construction of the multivariate normal distributions. Recall that having defined the conditional pdf $g_j\left(y_j\middle|y_1,\cdots,y_{j-1}\right)$ and the joint pdf $g^{j-1}\left(y_1,\cdots,y_{j-1}\right)$ we automatically have the joint pdf $g^{j}\left(y_1,\cdots,y_{j}\right)$ as the simple arithmetic product of the two. In the case just considered, all the densities $g^{j}\left(y_1,\cdots,y_{j}\right), j=1,2,\cdots,$ are (arbitrary with the accuracy to the rotations in R^j) multivariate normal. Preserving the general spirit of the multivariate normal pdf derivation, let us extend all the Equations (13) for $j=2,3,\cdots,m$ by allowing the translations $c_{j,1}Y_1+\cdots+c_{j,j-1}Y_{j-1}$ to be any nonlinear continuous function of $Y_1,\cdots,Y_{j-1}$ and replacing the constant $c_{j,j}$ by any continuous function of the same variables. Now, for any $j=2,3,\cdots,m$ (13) may be rewritten into the following "triangular" (see Filus, Filus and Arnold [12]) form:

$$Y_j=\Phi_j\left(Y_1,\cdots,Y_{j-1}\right)T_j+\Psi_j\left(Y_1,\cdots,Y_{j-1}\right) \tag{13*}$$

where $F_j()$ and $Y_j()$ are arbitrary continuous functions and $P\left(\Phi_j\left(Y_1,\cdots,Y_{j-1}\right)\neq 0\right)=1$.

From (13*) we obtain its inverse:

$$T_j=\frac{Y_j-\Psi_j\left(Y_1,\cdots,Y_{j-1}\right)}{\Phi_j\left(Y_1,\cdots,Y_{j-1}\right)} \tag{13**}$$

and then for each observation $\left(y_1,\cdots,y_{j-1}\right)$ of $\left(Y_1,\cdots,Y_{j-1}\right)$ the conditional pdf of $Y_j|y_1,\cdots,y_{j-1}$ as follows:

$$g_j\left(y_j|y_1,\ \ldots\ ,y_{j-1}\right)=N\left(\mu_j+\Psi_j\left(y_1,\cdots,y_{j-1}\right);\sigma_j\left|\Phi_j\left(y_1,\cdots,y_{j-1}\right)\right|\right) \tag{14}$$

It is clear that the sequence of the densities (14) $\left(j=2,3,\cdots,m\right)$ together with the normal initial density $g_1(y_1)$ of Y_1 uniquely determines the m-variate FF-normal pdf $g(y_1,\cdots,y_m)$ of the random vector $(Y_1,\cdots,Y_m)$. Remarkably, this non-normal density has its natural representation as the product of m normal densities:

$$g(y_1,\cdots,y_m)=g_1(y_1)\prod_{j=2}^{m}g_j(y_j|y_1,\cdots,y_{j-1}) \tag{15}$$

However, the marginal pdfs of $Y_2,\cdots,Y_m$ are not normal anymore.

The main conclusion which follows the considerations in Sections 6.2 - 6.4 may be stated as: There is a generic relationship that associates the construction method of the parameter dependence with the stochastic dependence structure present within the multivariate normal distribution of any dimension.

As an example of this relationship realize that the transformations (13) and (13*), when applied to the independent normal random variables $T_j\ (j=1,2,\cdots,m)$, define multivariate normal and FF-normal pdfs respectively. They produce other m-variate probability distributions if the normality assumption for T_j is dropped.

Let now $T_j\ (j=1,2,\cdots,m)$ be independent random variables all having the standard exponential pdf:

$f_j(t_j)=\exp[-t_j]$.

Applying to the random vector $(T_1,\cdots,T_m)$ transformation (13*) and then (13**) for $j=1,2,\cdots,m$, one obtains the joint density $g(y_1,\cdots,y_m)$ of the resulting random vector $(Y_1,\cdots,Y_m)$ to be given by the product of m factors (15), where according to first row of (12) we have $g_1(y_1)=(1/|c_{11}|)\exp[-y_1/|c_{11}|]$ and according to (13**)

$$g_j(y_j|y_1,\cdots,y_{j-1})=\frac{1}{|\Phi_j(y_1,\cdots,y_{j-1})|}\exp\frac{-(y_j-\Psi_j(y_1,\cdots,y_{j-1}))}{|\Phi_j(y_1,\cdots,y_{j-1})|}$$

where the latter is the two parameter exponential density with respect to y_j for $j = 2,3,\cdots,m$.

Another interesting case of the m-variate FF-Weibullian pdf can be obtained by applying transformations (13*) to m independent Weibullian random variables. An even more general class of FF-Weibullians one obtains using the pseudopower transformations (see Filus and Filus [4]) instead of the pseudoaffine (13*) which actually is a special case of the pseudopower.

All these distributions (including the m-variate normal) can as well be obtained by direct use of the "parameter dependence pattern" which produces more m-variate models than the considered above transformations. On the other hand existence of the defining transformations facilitates an underlying statistical analysis and simulations.

7. OTHER PARAMETER DEPENDENCE PARADIGMS IN THE LITERATURE

Some paradigms, applied in the reliability literature, are exactly those of the "parameter dependence" that we describe in this paper. However, in most of the cases they are not directly related to the problem of construction of multivariate probability distributions (so, also are different from the "conditioning" procedures in [9] [10] ; see above, Section 5). There are two such subjects that we discuss in the following.

7.1. The Accelerated Life Testing

When testing the life times of some high reliability products, the stresses usually encountered such as temperature, humidity, voltage sometimes are kept on significantly higher than usual levels in order to make the life times shorter than they are in normal conditions. The so obtained data (a "sample") is then extrapolated into those (hypothetical) life times that would, possibly, be obtained under the regular values of the stresses. Existence of rules, that associate the products' life times with values of the stresses applied, is necessary for performing proper extrapolations. Several such rules, typically known as the Arrhenius or Eyring (see, Meeker and Escobar [13] , Nelson [14] and [15] an internet source) relationships, are based on physical and chemical considerations on the rates of some chemical reactions that give rise to a given unit's failure. The obtained models, in general, allow determining the ratio of life times of the same product under higher and under normal temperature or, in the case of the Eyring model, some other physical quantities that play the role of the stresses (see, [13] , formula (18.5) page 476). Methods like that (i.e., the so called SAFT models [13]) directly relate the (life) times by means of a simple coefficient called the "acceleration factor".

Unfortunately, with this method the simplicity often comes along with inaccuracy of the predictions. Other methods apply the "Proportional Hazards Relationships" known also as Cox Model (see Cox [16]) which instead of times relate hazard rates.

More recently ([15]), the relationships between the life times under different stresses are related indirectly through their probability distributions via distribution's parameter in that way that considers distribution's parameter as a function of a given stress (mostly temperature, humidity, pressure, voltage). Those relationships, even if considered in a different context, obey the same paradigm (of the parameter dependence) as that considered in this paper in association with the construction of the multivariate probability distributions.

In what follows we discuss the differences.

1) The generality of the "parameter dependence theory" we built in this paper, is significantly higher than the very special case applied to the accelerated life testing theory. There are three reasons for that.

Firstly, in our approach the subject of constructing conditional probability distributions is not limited to the life testing, and not even to the "stress-life time" pattern only. The range of applications of our theory is very wide, including many biomedical (see, Collett [17]) and econometric relationships, (see, Filus and Filus [11] also Filus, Filus and Krysiak [18]).

Secondly, in the paradigm we consider, the relation between a parameter and stress (or any other random quantity) is given by an arbitrary continuous function, while the number of such functions applied in association with the accelerated life time testing is very limited. Actually, the functions are restricted to few "models" such as the Arrhenius, Eyrie, inverse power law, log-linear, and not many more (see, for example, the Eyring-Weibull model in [15] , Section 5). Those models were obtained from physical and chemical considerations that are only valid for some simple failure (degradation) mechanisms while very often real failure mechanism is too complicated to be analyzed that way.

Our idea is to omit the complicated physical or chemical phenomena that often are poorly understood and to apply two steps purely empirical approach.

Speaking roughly, the first step is an "educated guess" (for choice of a proper function) and the second is statistical verification of this guess.

Thirdly, in our theory we may consider an arbitrary parameter of an arbitrary probability distribution as a stress dependent, while, according to our knowledge (see, for example, [15]), the only life time distributions so far considered in the accelerated life testing are exponential, Weibullian, and lognormal (normal), and for each the distribution only one parameter is taken under consideration.

2) Besides the generality (of the constructed conditional distributions) our concept also differs with regard to the purpose. Namely, independently of the conditional distributions construction, we also have the construction of bivariate and multivariate probability distributions such as the FF-normal, FF-exponential, FF-Weibullian, FF-gamma and other (for comparison with similar "conditioning methods" of construction present in the literature, see Section 5).

The construction of high dimension multivariate distributions based on parameter dependence can easily be extended to Markovian and non-Markovian (still simple!) stochastic processes (see Filus and Filus [11]). The latter constructions seem to be rather unique in the literature.

7.2. Load Control and Load Sharing

1) Other than the accelerated life testing subject, where the "parameter dependence paradigm" is applied, is a set of problems centered around the notion of "load optimization" (see Filus [19] [20] , Levitin and Amari [21] , Nourelfath and Yalaoui [22] and others). This relatively new topic can be described as follows. Some working systems, such as cargo transporting trucks, trains, electric power lines, highways, computer processors, or other systems supporting varying amounts of load, require control of that load. On the one hand more load yields more gain, but, on the other hand, load (as stress) increment yields a corresponding increment of the system's failure rate which, in turn, depends on certain parameters [20] . The optimization problem can be formulated in several ways ([20] [21]) but, in all cases, the main idea is to balance between an expected gain that follows a good (load) transportation and the loss in reliability of the transporting medium which decreases the overall gain. Thus, for some (optimal) value of the load to be found, the pure expected gain earned by the system is maximal. In this framework the relationship between stress (the load) and some parameters of the system life time probability distribution (failure rate) is vital for finding a proper model. As examples of such relations, the power and exponential functions of the load were chosen in [20] .

2) Similar application of the parameter dependence pattern also occurs when "load sharing phenomena" takes place. Suppose that we have a parallel system supporting a load such as several engines aircraft or two electric power lines. Failure of any system's component may cause the total load to be redistributed among fewer components, so that the load on each of them increases by some predictable value. Now we may encounter either the load optimization problem [21] or simply the task of determination of system's life time probability distribution (see Freund [23] for the exponential case as well as Lu [24] and Filus [25] for Weibullian and lognormal cases). In all these cases the parameter dependence pattern is involved or it is desirable to apply it to get a deeper insight into the underlying stochastic phenomena.

Remark. As a final remark, let me mention the relationship between the parameter dependence presented in this paper, and the stochastic dependence based on models initiated in 1961 by Freund [23] . Besides some similarity the most basic difference lies in the fact that in the Freund scheme the system components act independently until the first failure. In general, the components successive failures cause the total load to be shared by fewer and fewer remaining components, affecting their failure rates (via the parameters).

Quite opposite to that, in the models we introduce, the component interactions take place only when the components work. Any failure of a system component stops its influence on the remaining components' life times. Therefore, the two paradigms, the "Freund's load sharing" and our "parameter dependence", are "disjoint" and in a sense "complementary". In reality, both (physical) phenomena may take place at the same time and it seems to be quite possible in the future to construct stochastic models (i.e., multivariate probability distributions) that would obey both paradigms.

Jerzy K. Filus, Lidia Z. Filus Nevertheless, we stress the generic relation of all the multivariate probability distributions based on the parameter dependence with the multivariate Gaussians.

REFERENCES

1. Filus, J.K. and Filus, L.Z. (2013) A Method for Multivariate Probability Distributions Construction via Parameter Dependence. Communications in Statistics: Theory and Methods, 42, 716-721.
2. Filus, J.K. and Filus, L.Z. (2000) A Class of Generalized Multivariate Normal Densities. Pakistan Journal of Statistics, 16, 11-32.
3. Filus, J.K. and Filus, L.Z. (2007) On New Multivariate Probability Distributions and Stochastic Processes with Systems Reliability and Maintenance Applications. Methodology and Computing in Applied Probability, 9, 426-446,

4. Filus, J.K. and Filus, L.Z. (2006) On Some New Classes of Multivariate Probability Distributions. Pakistan Journal of Statistics, 1, 21-42.
5. Filus, J.K. and Filus, L.Z. (2008) On Multicomponent System Reliability with Microshocks-Microdamages Type of Components' Interaction. Proceedings of the International Multiconference of Engineers and Computer Scientists, Lecture Notes in Engineering and Computer Science, Hong Kong, 19-21 March 2008, 1945-1951.
6. Filus, J.K. and Filus, L.Z. (2010) Weak Stochastic Dependence in Biomedical Applications. American Institute of Physics Conference Proceedings 1281, Numerical Analysis and Applied Mathematics, III, 1873-1876.
7. Kotz, S., Balakrishnan, N. and Johnson, N.L. (2000) Continuous Multivariate Distributions (Vol. 1). 2nd Edition, J. Wiley & Sons, Inc, New York, 217-218.
8. Barlow, R.E. and Proschan, F. (1975) Statistical Theory of Reliability and Life Testing. Holt, Rinehart and Winston, New York.
9. Arnold, B.C., Castillo, E. and Sarabia, J.M. (1999) Conditional Specification of Statistical Models. Springer Series in Statistics, Springer Verlag, New York.
10. Castillo, E. and Galambos, J. (1990) Bivariate Distributions with Weibull Conditionals. Analysis Mathematica, 16, 3-9.
11. Filus, J.K., Filus, L.Z. (2008) Construction of New Continuous Stochastic Processes. Pakistan Journal of Statistics, 24, 227-251.
12. Filus, J.K., Filus, L.Z. and Arnold, B.C. (2010) Families of Multivariate Distributions Involving "Triangular" Transformations. Communications in Statistics—Theory and Methods, 39, 107-116.
13. Meeker, W.Q. and Escobar, L.A. (1998) Statistical Methods for Reliability Data. John Wiley & Sons, Inc., New York.
14. Nelson, W. (1990) Accelerated Testing: Statistical Models, Test Plans and Data Analysis. Wiley, New York.
15. (2013) www.ReliaWiki.org/index.php/Accelerated_Life_Testing_Data_Analysis_Reference
16. Cox, D.R. (1972) Regression Models and Life Tables (with Discussion). Journal of the Royal Statistical Society, B74, 187-220.
17. Collett, D. (2003) Modeling Survival Data in Medical Research. 2nd Edition, Chapman @ Hall/CRS A CRC Press Company, London, New York, Washington, D.C.
18. Filus, J.K., Filus, L.Z. and Krysiak, Z. (2013) Analytical Statistical and Simulation Models Utilized in Modeling the Risk in Finance. 5th International Conference on Risk Analysis, Tomar, Portugal, 30th May-1st June.
19. Filus, J.K. (1986) A Problem in Reliability Optimization. Journal of the Operational Research Society, 37, 407-412.
20. Filus, J.K. (1987) The Load Optimization of a Repairable System with Gamma Distributed Time-to-Failure. Reliability Engineering, 18, 275-284.
21. Levitin, G. and Amari, S.V. (2009) Optimal Load Distribution in Series-Parallel Systems. Reliability Engineering and System Safety, 94, 254-260.
22. Nourelfath, M. and Yalaoui, F. (2012) Integrated Load Distribution and Production Planning in Series-Parallel Multistate Systems. Reliability Engineering and System Safety, 106, 138-145.
23. Freund, J.E. (1961) A Bivariate Extension of the Exponential Distribution. Journal of the American Statistical Association, 56, 971-977.
24. Lu, J. (1989) Weibull Extensions of the Freund and Marshal-Olkins Bivariate Exponential Models. IEEE Transactions on Reliability, 38, 615-619.
25. Filus, J.K. (1991) On a Type of Dependencies between Weibull Life times of System Components. Reliability Engineering and System Safety, 31s, 267-280.

CHAPTER 14

A Stochastic Correlation Model with Time Change for Pricing Credit Spread Options

Zhigang Tong[1,2], Allen Liu[2]

[1]*Department of Mathematics and Statistics, University of Ottawa, Ottawa, Canada*

[2]*Model Validation, Enterprise Risk and Portfolio Management, Bank of Montreal, Toronto, Canada*

ABSTRACT

In this paper, we introduce the stochastic correlation processes for modeling the credit spread. We first model the components of spread process as correlated Ornstein-Uhlenbeck processes and correlation as Jacobi process. Using the properties of Jacobi process, we are able to obtain the analytical solutions for the credit spread option prices. To further enhance the model's ability to capture the abrupt changes in the observed correlation time series, we construct a new model where the correlation is modeled by a Jacobi process time change by Lévy subordinators. We employ the eigenfunction expansion me- thods to obtain the closed-form solutions for the option prices. Our empirical study indicates the time changed Jacobi process fits the correlation series significantly better than the Jacobi process.

Keywords: Stochastic Correlation, Jacobi Process, Stochastic Time Change, Eigenfunction Expansion, Credit Spread Options

1. INTRODUCTION

A credit spread option is an option on a particular borrower's credit spread. The credit spread is the difference between the yield on the borrower's debt and the yield on Treasury debt of the same maturity. The credit spread represents the risk premium that the market demands for holding the borrower's debt. Unlike the credit default swaps, the payoff of credit spread options depends solely on the measurement of the credit spread rather than a specific credit event. Credit spread options protect the holder of the borrower's debt from the loss of the rising yield spread due to default or credit rating downgrade.

Most of the models used in the pricing of credit spread options belong to the multi-factor models where state variables are modeled as correlated stochastic processes. [1] proposes a valuation model for pricing credit spread options by assuming that the risk-free interest rate and logarithm of the credit spread follow correlateed mean reverting diffusion processes. [2] argues that the spread should be modeled using its two components instead of the spread itself and he extends [1] by assuming that the risk-free rate and the two components of credit spread follow correlated Ornstein-Uhlenbeck (OU) processes. [3] proposes a very general mean reverting process for the credit spread and two stochastic volatility processes, the square-root process and the OU process. The volatility process is assumed to be correlated with the spread process. There are many other examples that employ correlated processes for modeling the credit spread and pricing credit spread options, see e.g. [4] - [9] .

Although most of the models stress the role played by the correlation between state variables for pricing credit spread options, a common assumption is that the correlation is constant. There is a growing literature that documents that the correlation is far from being constant and it should be modeled as a random process, just like interest rate, stock price and volatility. [10] provides analytical properties of the correlation process and provides a new approach of modeling correlation as a stochastic process. [11] and [12] propose a modified Jacobi process to evaluate risk premium of the stochastic correlation and develop a series solution for pricing options under the correlation risk. [13] provides a closed-form approximation for pricing of several two-dimensional derivatives under the assumptions of stochastic correlation. [14] and [15] propose to generate the stochastic correlation process through the hyperbolic transformation of any mean-reverting process with positive and negative values.

In this paper, the main problem we are going to solve is how to model the credit spread and price the credit spread options with the stochastic correlation process. We stress that so far most of stochastic correlation models are developed for the equity and foreign exchange derivatives, where the underlying state variables are usually modeled as correlated geometric Brownian motions. To model the credit spread, we need to take the mean-reversion in the state variables into consideration. In particular, we model the components of the spread process as correlated OU processes and stochastic correlation as Jacobi process which guarantees the correlation between state variables is a bounded process. Using the properties of Jacobi process, we are able to derive the analytical formula for the moments of integrated correlation process required for calculating the credit spread options prices. From the data we employ for the empirical study, we observe that the correlation process changes abruptly and a diffusion model may not fully capture its dynamics. We therefore develop a time-changed model for correlation process where the Jacobi process is time changed by Lévy subordinators to yield state-dependent jumps. To the best of our knowledge, the time-changed Jacobi process (TC-Jacobi) has not been studied in the literature for modeling correlation. We notice that the TC-Jacobi process remains analytically tractable and the closed-form solutions for the credit spread options we derive under the Jacobi process require only minor changes. Our empirical results demonstrate that indeed the TC-Jacobi process improves the fit of correlation process significantly compared to the Jacobi process.

The structure of the paper is as follows. In Section 2, we introduce the credit spread model with constant correlation. In Section 3 we extend the constant correlation model to stochastic correlation model. Using the properties of Jacobi process, we derive the closed-form solutions for credit spread option prices. We introduce the time change to Jacobi process in Section 4 and demonstrate that the analytical formulas remain little changed by employing the eigenfunction expansion method. We compare the performance of Jacobi process with the time changed process using the correlation time series in Section 5. We also numerically study the implications on the derivatives for both processes.

2. A CREDIT SPREAD MODEL WITH CONSTANT CORRELATION

Let $(\Omega, \mathcal{F}, P)$ denote a probability space with an information filtration ($\mathcal{F}_t$). We assume that under physical measure P , the dynamics of the two yields included in the credit spread, denoted by X_1 and X_2, follow the correlated OU processes. Thus,

$$dX_1(t) = \kappa_1(\theta_1 - X_1(t))dt + \sigma_1 dB_1(t), \quad (1)$$

$$dX_2(t) = \kappa_2(\theta_2 - X_2(t))dt + \sigma_2 dB_2(t), \quad (2)$$

and

$$dB_1(t)dB_2(t) = \rho dt, \quad (3)$$

where $\kappa_1, \theta_1, \sigma_1, \kappa_2, \theta_2, \sigma_2 \in \mathbb{R}$. $B_1(t)$ and $B_2(t)$ are two dependent Brownian motions with constant correlation coefficient ρ.

To price the credit spread options, we need to work under the risk-neutral measure Q. We assume that under the measure Q, the stochastic processes of $X_1(t)$ and $X_2(t)$ are still correlated Ornstein-Uhlenbeck processes, but with different coefficients to highlight the risk premium. Therefore,

$$dX_1(t) = \kappa_1^Q(\theta_1^Q - X_1(t))dt + \sigma_1 dB_1^Q(t), \quad (4)$$

$$dX_2(t) = \kappa_2^Q(\theta_2^Q - X_2(t))dt + \sigma_2 dB_2^Q(t), \quad (5)$$

and

$$dB_1^Q(t)dB_2^Q(t) = \rho dt. \quad (6)$$

Under the above assumptions, we know that at time T, under the measure Q the vector $\boldsymbol{X}(T) = (X_1(T), X_2(T))$ has a bivariate normal distribution with mean $\boldsymbol{\mu}$ and covariance matrix $\boldsymbol{\Sigma}$ given by

$$\mu = \begin{pmatrix} \mu_1 \\ \mu_2 \end{pmatrix}, \ \Sigma = \begin{pmatrix} \eta_1^2 & \eta_{12} \\ \eta_{12} & \eta_2^2 \end{pmatrix}, \tag{7}$$

where

$$\mu_1 = x_1(0)\exp\left(-\kappa_1^Q T\right) + \theta_1^Q\left(1-\exp\left(-\kappa_1^Q T\right)\right), \tag{8}$$

$$\mu_2 = x_2(0)\exp\left(-\kappa_2^Q T\right) + \theta_2^Q\left(1-\exp\left(-\kappa_2^Q T\right)\right), \tag{9}$$

$$\eta_1^2 = \frac{\sigma_1^2}{2\kappa_1^Q}\left(1-\exp\left(-2\kappa_1^Q T\right)\right), \ \eta_2^2 = \frac{\sigma_2^2}{2\kappa_2^Q}\left(1-\exp\left(-2\kappa_2^Q T\right)\right), \tag{10}$$

and

$$\begin{aligned} \eta_{12} &= \sigma_1\sigma_2\int_0^T \exp\left(-\left(\kappa_1^Q+\kappa_2^Q\right)(T-s)\right)\mathrm{d}s \\ &= \frac{\sigma_1\sigma_2}{\kappa_1^Q+\kappa_2^Q}\left(1-\exp\left(-\left(\kappa_1^Q+\kappa_2^Q\right)T\right)\right). \end{aligned} \tag{11}$$

Let $C(T,K,\rho)$ denote the price of a call option on credit spread with strike price K and maturity T. The option price can be calculated from

$$C(T,K,\rho) = \exp(-rT)E^Q\left[\left(X_1(T)-X_2(T)-K\right)1_{\{X_1(T)-X_2(T)>K\}}\right], \tag{12}$$

where ρ is included in C to highlight that option price depends on the correlation coefficient ρ. r is the risk-free short rate and for simplicity, we assume it is a constant.

Since $(X_1(T), X_2(T))$ follows a bivariate normal distribution, we know that $X_1(T)-X_2(T)$ is a Gaussian process with mean $\mu_1 - \mu_2$ and variance

$$\eta_s^2 = \eta_1^2 + \eta_2^2 - 2\rho\eta_{12}. \tag{13}$$

It becomes straightforward to compute the call option prices. We summarize the result in the following proposition (see also [2] and [12]).

Proposition 1. Assume the dynamics of X_1 and X_2 are given by (4), (5) and (6) then the option price $C(T,K,\rho)$ can be calculated by

$$C(T,K,\rho)=\exp(-rT)\left[(\mu_1-\mu_2-K)N(d)+\frac{\eta_s}{\sqrt{2\pi}}\exp\left(-\frac{d^2}{2}\right)\right], \quad (14)$$

where $d=\frac{\mu_1-\mu_2-K}{\eta_s}$. $N(\cdot)$ is the CDF of standard normal distribution.

3. STOCHASTIC CORRELATION WITH JACOBI PROCESS

3.1. Motivation for Stochastic Correlation Process

In this section, we extend the constant correlation model in Section 2 by modeling the correlation coefficient ρ as a stochastic process. To motivate our model, we first take a look at the data we use in the empirical study. Our data contain daily rates for 10-year constant maturity Treasury bonds and Moody's Aaa and Baa seasoned bond indices. The data cover the period January 2, 1996 to July 25, 2016, for a total of 5146 observations. In Figure 1, we plot the time series of three yields and it is clear that all of them display mean reversion. We also construct the time series of correlation by the following procedure (similar to [10] [14] and [15]):

1) At time t, we regress daily changes in the values of $X_1(t)$ and $X_2(t)$ on the values of $X_1(t)$ and $X_2(t)$ a day before for a time window n_T:

$$\Delta X_1(t)=X_1(t)-X_1(t-\Delta t)=\alpha_1+\beta_1 X_1(t-\Delta t)+\varepsilon_1(t) \quad (15)$$

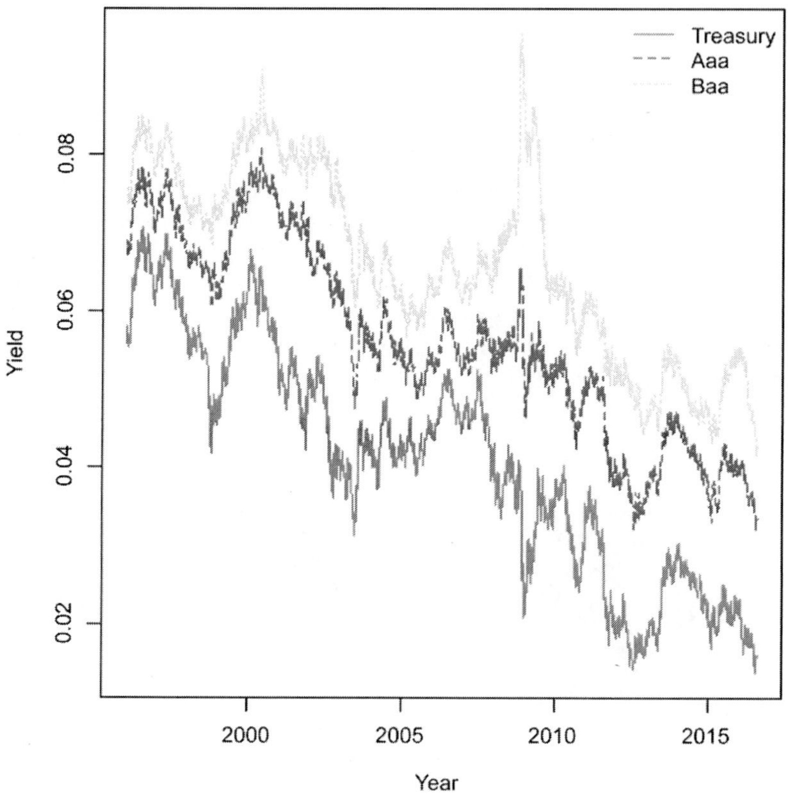

Figure 1. Time series of the yields of Treasury bonds and Moody's Aaa and Baa seasoned bond indices.

and

$$\Delta X_2(t) = X_2(t) - X_2(t-\Delta t) = \alpha_2 + \beta_2 X_2(t-\Delta t) + \varepsilon_2(t). \quad (16)$$

2) The correlation $\rho(t)$ is calculated by the correlation coefficient between $\varepsilon_1(t)$ and $\varepsilon_2(t)$ for a time window n_T, that is

$$\rho(t) = \frac{\sum_{i=1}^{n_T}(\varepsilon_1(t-i\Delta t) - \bar{\varepsilon}_1(t))(\varepsilon_2(t-i\Delta t) - \bar{\varepsilon}_2(t))}{\sqrt{\sum_{i=1}^{n_T}(\varepsilon_1(t-i\Delta t) - \bar{\varepsilon}_1(t))^2 \sum_{i=1}^{n_T}(\varepsilon_2(t-i\Delta t) - \bar{\varepsilon}_2(t))^2}}, \quad (17)$$

where $\bar{\varepsilon}_1(t) = \frac{1}{n_T}\sum_{i=1}^{n_T}\varepsilon_1(t-i\Delta t)$ and $\bar{\varepsilon}_2(t) = \frac{1}{n_T}\sum_{i=1}^{n_T}\varepsilon_2(t-i\Delta t)$.

We then roll over to time $t+\Delta t$ and so on to obtain a series of correlations through the time.

In Figure 2 we plot the dynamics of $\rho(t)$ with $n_T = 30$. We can see clearly that the correlation is far from being constant and it varies very abruptly. The correlation is supposed to be bounded between −1 and 1, but in our dataset, it is concentrated on [0, 1]. In addition, the correlation process displays strong mean-reverting property. In Table 1, we provide the summary statistics for $\rho(t)$ calculated from the two defaultable bond indices. The average correlations between the yields of Treasury bonds and Aaa and Baa bonds are around

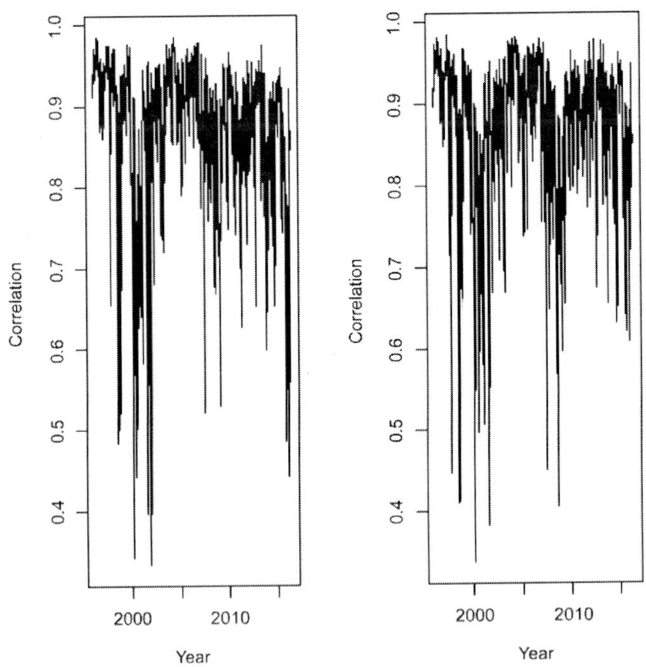

Figure 2. Correlations between the yields for 10-year constant maturity Treasury bonds and Moody's Aaa and Baa bond indices. Left panel: Aaa bond indices; right panel: Baa bond indices.

Table 1. Summary statistics of correlation series.

Statistics	Aaa	Baa
Mean	0.8667	0.8700
Std. Dev.	0.0986	0.0927
Skewness	−1.9326	−1.9076
Kurtosis	4.5146	4.5088
Minimum	0.3324	0.3374
Maximum	0.9842	0.9836

0.87, but the correlations fluctuate between 0.33 and 0.99. The density functions for the correlation series have negative skewness and fat tail.

To satisfy the properties of the correlation series displayed in our data, we model the correlation as a Jacobi process, which takes the value between ρ_m and ρ_M, i.e.

$$\rho(t) = \rho_m + (\rho_M - \rho_m)Y(t), \tag{18}$$

and

$$\mathrm{d}Y(t) = \kappa(\theta - Y(t))\mathrm{d}t + \sigma\sqrt{Y(t)(1 - Y(t))}\mathrm{d}B(t), \tag{19}$$

where $\kappa > 0$, $\sigma > 0$ and $0 < \theta < 1$. $B(t)$ is a standard Brownian motion and independent of $B_1(t)$ and $B_2(t)$ in (1) and (2).

Under the martingale measure Q, we assume the stochastic process of $Y(t)$ is

$$\mathrm{d}Y(t) = \kappa^Q(\theta^Q - Y(t))\mathrm{d}t + \sigma\sqrt{Y(t)(1 - Y(t))}\mathrm{d}B^Q(t), \tag{20}$$

where $\kappa^Q > 0$, $\kappa^Q > 0$ and $0 < \theta^Q < 1$. $B^Q(t)$ is a standard Brownian motion under the measure Q and independent of $B_1^Q(t)$ and $B_2^Q(t)$ in (4) and (5).

The standard Jacobi process Y takes values between 0 and 1. Following [16] , we also impose the following conditions to ensure that the boundaries 0 and 1 are inaccessible to the process $Y(t)$:

$$\frac{\sigma^2}{2\kappa} \leq \theta \leq 1 - \frac{\sigma^2}{2\kappa}, \quad \frac{\sigma^2}{2\kappa^{\varrho}} \leq \theta^{\varrho} \leq 1 - \frac{\sigma^2}{2\kappa^{\varrho}}. \tag{21}$$

3.2. Properties of Jacobi Process

It is well known that there is no closed-form expression for the transition density for the Jacobi process (see e.g. [17]). However, the transition density of Jacobi process can be represented in terms of eigenfunction expansions.

From (19), we know the infinitesimal generator $\mathcal{L}$ of Jacobi process is

$$\mathcal{L}f(x) = \kappa(\theta - x) f'(x) + \frac{1}{2} x(1-x) f''(x), \tag{22}$$

where f is transformation function. f' and f'' are first- and second-order derivatives of f, respectively.

Let $\{\lambda_n\}$ be the eigenvalues of $-\mathcal{L}$ and $\{\psi_n\}$ be the corresponding eigenfunctions, i.e.

$$-\mathcal{L}\psi_n = \lambda_n \psi_n. \tag{23}$$

We know for the Jacobi process Y defined on $I = (0,1)$, the eigenvalues and eigenfunctions can be written as (see e.g. [16] and [17])

$$\lambda_n = \frac{\sigma^2}{2} n \left(n - 1 + \frac{2\kappa}{\sigma^2} \right), \tag{24}$$

and

$$\psi_n(x) = N_n P_n^{(\alpha,\beta)}(2x - 1), \tag{25}$$

where $\alpha = \frac{2\kappa(1-\theta)}{\sigma^2} - 1,\ \beta = \frac{2\kappa\theta}{\sigma^2} - 1$ and

$$N_n^2 = \frac{(\alpha+\beta+2)_{n-1}(2n+\alpha+\beta+1)n!}{(\alpha+1)_n(\beta+1)_n}, \tag{26}$$

where $(a)_0 = 1$, $(a)_n = a(a+1),\cdots,(a+n-1)$, $n>0$. $P_n^{(\alpha,\beta)}(x)$ is the Jacobi polynomials defined as

$$P_n^{(\alpha,\beta)}(x) = \frac{(\alpha+1)_n}{n!}\,{}_2F_1\left(-n, 1+\alpha+\beta+n; \alpha+1; \frac{1}{2}(1-x)\right), \tag{27}$$

where ${}_2F_1(a,b;c;x)$ is the hypergeometric function defined by

$${}_2F_1(a,b;c;x) = \sum_{n=0}^{\infty} \frac{(a)_n(b)_n}{(c)_n}\frac{x^n}{n!}. \tag{28}$$

The spectral decomposition allows us to find the transition density function of Jacobi process Y as

$$p_Y(t, y(0), y(t)) = \pi(y(t))\sum_{n=0}^{\infty}\exp(-\lambda_n t)\psi_n(y(0))\psi_n(y(t)), \tag{29}$$

where $\pi(\cdot)$ is the stationary density function of Jacobi process Y,

$$\pi(y) = \frac{y^\beta(1-y)^\alpha}{B(\alpha+1,\beta+1)}, \tag{30}$$

where $B(a,b)$ is the Beta function defined by

$$B(a,b) = \frac{\Gamma(a)\Gamma(b)}{\Gamma(a+b)}, \tag{31}$$

where Γ is the standard Gamma function.

The most important property of Jacobi polynomials is that they satisfy the following orthogonality condition (see e.g. [16]):

Proposition 2.

$$\int_0^1 P_n^{(\alpha,\beta)}(2x-1)P_m^{(\alpha,\beta)}(2x-1)\pi(x)\mathrm{d}x = \begin{cases} \frac{1}{N_n^2}, & n=m \\ 0, & n\neq m \end{cases} \tag{32}$$

Using this property, we can immediately have the following results:

Corollary 1. Let k , k_1 and k_2 be nonnegative integers.

- If $k<n$, then $\int_{-1}^1 (1+x)^{\beta+k}(1-x)^{\alpha} P_n^{(\alpha,\beta)}(x)\mathrm{d}x = 0.$
- If $k<n$, then $\int_{-1}^1 (1+x)^{\beta}(1-x)^{\alpha+k} P_n^{(\alpha,\beta)}(x)\mathrm{d}x = 0.$
- If $k_1+k_2<n$, then $\int_{-1}^1 (1+x)^{\beta+k_1}(1-x)^{\alpha+k_2} P_n^{(\alpha,\beta)}(x)\mathrm{d}x = 0.$

In this paper, we need to compute the following two integrals:

$$\int_{-1}^1 (1-x)^{\alpha+k_1}(1+x)^{\beta+k_2} P_n^{(\alpha,\beta)}(x)\mathrm{d}x,$$

and

$$\int_{-1}^1 (1-x)^{\alpha}(1+x)^{\beta+1} P_n^{(\alpha,\beta)}(x)P_m^{(\alpha,\beta)}(x)\mathrm{d}x,$$

where k_1 and k_2 are two nonnegative integers.

We can prove the following results:

Lemma 1.

(i)

$$\int_{-1}^1 (1-x)^{\alpha+k_1}(1+x)^{\beta+k_2} P_n^{(\alpha,\beta)}(x)\mathrm{d}x = \begin{cases} 0, & k_1+k_2<n \\ a_n(\alpha,\beta,k_1,k_2), & k_1+k_2\geq n \end{cases} \tag{33}$$

where

$$a_n(\alpha,\beta,k_1,k_2)$$
$$=\frac{2^{\alpha+k_1+\beta+k_2+1}\Gamma(\alpha+k_1+1)\Gamma(\beta+k_2+1)\Gamma(n+1+\alpha)}{n!\Gamma(\alpha+k_1+\beta+k_2+2)\Gamma(\alpha+1)}$$
$$\times {}_3F_2(-n,\alpha+\beta+n+1,\alpha+k_1+1;\alpha+1,\alpha+k_1+\beta+k_2+2;1), \quad (34)$$

where ${}_3F_2(a,b,c;d,e;x)$ is the hypergeometric function defined by

$${}_3F_2(a,b,c;d,e;x)=\sum_{n=0}^{\infty}\frac{(a)_n(b)_n(c)_n}{(d)_n(e)_n}\frac{x^n}{n!}. \quad (35)$$

(ii)

$$\int_{-1}^{1}(1-x)^{\alpha}(1+x)^{\beta+1}P_n^{(\alpha,\beta)}(x)P_m^{(\alpha,\beta)}(x)\mathrm{d}x$$
$$=\begin{cases}0, & |n-m|\geq 2\\ b_{n,m}(\alpha,\beta), & |n-m|<2\end{cases} \quad (36)$$

where

$$b_{n,m}(\alpha,\beta)=\sum_{s=0}^{n}\binom{n+\alpha}{s}\binom{n+\beta}{n-s}(-1)^{n-s}\frac{1}{2^n}a_m(\alpha,\beta,n-s,s+1). \quad (37)$$

Proof. To prove (i), from Corollary 1, we immediately know that for $k_1+k_2<n$, the integral is clearly zero. For $k_1+k_2\geq n$, we can use the following formula (see e.g. [18]):

$$\int_{-1}^{1}(1-x)^{k_1}(1+x)^{k_2}P_n^{(\alpha,\beta)}(x)\mathrm{d}x$$
$$=\frac{2^{k_1+k_2+1}\Gamma(k_1+1)\Gamma(k_2+1)\Gamma(n+1+\alpha)}{n!\Gamma(k_1+k_2+2)\Gamma(1+\alpha)}$$
$$\times {}_3F_2(-n,\alpha+\beta+n+1,k_1+1;\alpha+1,k_1+k_2+2;1). \quad (38)$$

To prove (ii), we first use another representation of the Jacobi polynomials,

$$P_n^{(\alpha,\beta)}(x)=\sum_{s=0}^{n}\binom{n+\alpha}{s}\binom{n+\beta}{n-s}\left(\frac{x-1}{2}\right)^{n-s}\left(\frac{x+1}{2}\right)^{s}. \tag{39}$$

Then, we have

$$\begin{aligned}&\int_{-1}^{1}(1-x)^{\alpha}(1+x)^{\beta+1}P_n^{(\alpha,\beta)}(x)P_m^{(\alpha,\beta)}(x)\mathrm{d}x\\&=\sum_{s=0}^{n}\binom{n+\alpha}{s}\binom{n+\beta}{n-s}\int_{-1}^{1}\left(\frac{x-1}{2}\right)^{n-s}\left(\frac{x+1}{2}\right)^{s}(x+1)^{\beta+1}(1-x)^{\alpha}P_m^{(\alpha,\beta)}(x)\mathrm{d}x.\end{aligned} \tag{40}$$

Using Corollary 1, we know that for $|n-m|\geq 2$, the above integral will be zero. If $|n-m|<2$, we can compute the integral using the definition of function a_n. □

In order to obtain the closed-form solution for the credit spread option prices, we are interested in computing the moments of the following integral

$$\bar{Y}(T)=\int_0^T Y(s)\exp\left(\left(\kappa_1^Q+\kappa_2^Q\right)s\right)\mathrm{d}s. \tag{41}$$

We can prove the following result:

Theorem 1.

$$\begin{aligned}E^Q\left[\bar{Y}(T)^n\right]=&\frac{n!}{B^n(\alpha+1,\beta+1)}\frac{1}{2^{n(\alpha+\beta+2)}}\sum_{(k_1,\cdots,k_n)\in I^n}\prod_{j=1}^{n}\left(N_{k_j}\right)^2\\&\times P_{k_1}^{(\alpha,\beta)}(2y(0)-1)\prod_{j=1}^{n-1}b_{k_j,k_{j+1}}(\alpha,\beta)a_{k_n}(\alpha,\beta,0,1)\\&\times\int_0^T\int_{s_1}^T\cdots\int_{s_n}^T\prod_{j=1}^{n}\exp\left(-\lambda_{k_j}\left(s_j-s_{j-1}\right)+\left(\kappa_1^Q+\kappa_2^Q\right)s_j\right)\mathrm{d}s_n\cdots\mathrm{d}s_1,\end{aligned} \tag{42}$$

where $s_0:=0$,

$$I^n=\left\{(k_1\cdots k_n)\in\mathbb{Z}_+^n:k_n\in\{0,1\},\left|k_j-k_{j+1}\right|\leq 1, j=1,2,\cdots,n-1\right\}, \tag{43}$$

α, β, N_k and λ_k are assumed to be calculated using the coefficients under the measure Q.

Proof. First, we know

$$E^Q\left[\bar{Y}(T)^n\right]=E^Q\left[\left(\int_0^T Y(s)\exp\left(\left(\kappa_1^Q+\kappa_2^Q\right)s\right)\mathrm{d}s\right)^n\right]. \tag{44}$$

We have

$$\begin{aligned}&E^Q\left[\left(\int_0^T Y(s)\exp\left(\left(\kappa_1^Q+\kappa_2^Q\right)s\right)\mathrm{d}s\right)^n\right]\\&=n!\int_0^T\int_{s_1}^T\cdots\int_{s_n}^T E^Q\left[\prod_{j=1}^n Y(s_j)\exp\left(\left(\kappa_1^Q+\kappa_2^Q\right)s_j\right)\right]\mathrm{d}s_n\cdots\mathrm{d}s_1.\end{aligned} \tag{45}$$

Using the eigenfunction expansion for the transition density of Jacobi process Y, we have

$$\begin{aligned}&E^Q\left[\prod_{j=1}^n Y(s_j)\exp\left(\left(\kappa_1^Q+\kappa_2^Q\right)s_j\right)\right]\\&=\int_0^1\cdots\int_0^1\prod_{j=1}^n y(s_j)\exp\left(\left(\kappa_1^Q+\kappa_2^Q\right)s_j\right)\\&\quad\times p_Y\left(s_{j-1},s_j,y(s_{j-1}),y(s_j)\right)\mathrm{d}y(s_1)\cdots\mathrm{d}y(s_n)\\&=\sum_{k_1=0}^\infty\cdots\sum_{k_n=0}^\infty\prod_{j=1}^n\exp\left(-\lambda_{k_j}\left(s_j-s_{j-1}\right)+\left(\kappa_1^Q+\kappa_2^Q\right)s_j\right)\psi_{k_1}\left(y(0)\right)\\&\quad\times\prod_{j=1}^{n-1}\int_0^1 x\pi(x)\psi_{k_j}(x)\psi_{k_{j+1}}(x)\mathrm{d}x\int_0^1 x\pi(x)\psi_{k_n}(x)\mathrm{d}x.\end{aligned} \tag{46}$$

For the first integral in the above equation,

$$\begin{aligned}&\int_0^1 x\pi(x)\psi_{k_j}(x)\psi_{k_{j+1}}(x)\mathrm{d}x\\&=\frac{N_{k_j}N_{k_{j+1}}}{B(\alpha+1,\beta+1)}\frac{1}{2^{\alpha+\beta+2}}\int_{-1}^1(1-x)^\alpha(1+x)^{\beta+1}P_{k_j}^{(\alpha,\beta)}(x)P_{k_{j+1}}^{(\alpha,\beta)}(x)\mathrm{d}x.\end{aligned} \tag{47}$$

For the second integral,

$$\begin{aligned}&\int_0^1 x\pi(x)\psi_{k_n}(x)\mathrm{d}x\\&=\frac{N_{k_n}}{B(\alpha+1,\beta+1)}\frac{1}{2^{\alpha+\beta+2}}\int_{-1}^1(1-x)^\alpha(1+x)^{\beta+1}P_{k_n}^{(\alpha,\beta)}(x)\mathrm{d}x.\end{aligned} \tag{48}$$

Using Lemma 1, we can see that $E^Q\left[\prod_{j=1}^n Y(s_j)\exp\left(\left(\kappa_1^Q+\kappa_2^Q\right)s_j\right)\right]=0$ if $(k_1\cdots k_n)\notin I^n$. Otherwise, we have

$$E^{Q}\left[\prod_{j=1}^{n} Y(s_j)\exp\left(\left(\kappa_1^Q+\kappa_2^Q\right)s_j\right)\right]$$
$$=\frac{1}{B^n(\alpha+1,\beta+1)}\frac{1}{2^{n(\alpha+\beta+2)}}\sum_{(k_1\cdots k_n)\in I^n}\psi_{k_1}\left(y(0)\right)$$
$$\times\prod_{j=1}^{n-1} N_{k_j}N_{k_{j+1}}b_{k_j,k_{j+1}}(\alpha,\beta)N_{k_n}a_{k_n}(\alpha,\beta,0,1)$$
$$\times\prod_{j=1}^{n}\exp\left(-\lambda_{k_j}\left(s_j-s_{j-1}\right)+\left(\kappa_1^Q+\kappa_2^Q\right)s_j\right). \tag{49}$$

Finally, using the definition of ψ_k, we obtain the final result. □

We are also interested in calculating the moments of the following quantity

$$\bar{\rho}(T)=\frac{1}{\int_0^T \exp\left(-\left(\kappa_1^Q+\kappa_2^Q\right)(T-s)\right)ds}$$
$$\times\int_0^T \rho(s)\exp\left(-\left(\kappa_1^Q+\kappa_2^Q\right)(T-s)\right)ds$$
$$=\rho_m+\frac{(\rho_M-\rho_m)\left(\kappa_1^Q+\kappa_2^Q\right)\exp\left(-\left(\kappa_1^Q+\kappa_2^Q\right)T\right)}{1-\exp\left(-\left(\kappa_1^Q+\kappa_2^Q\right)T\right)}$$
$$\times\int_0^T Y(s)\exp\left(\left(\kappa_1^Q+\kappa_2^Q\right)s\right)ds. \tag{50}$$

Using binomial expansion, we immediately have

$$E^{Q}\left[\bar{\rho}(T)^n\right]$$
$$=\sum_{j=0}^{n}\binom{n}{j}\rho_m^{n-j}\left(\frac{(\rho_M-\rho_m)\left(\kappa_1^Q+\kappa_2^Q\right)\exp\left(-\left(\kappa_1^Q+\kappa_2^Q\right)T\right)}{1-\exp\left(-\left(\kappa_1^Q+\kappa_2^Q\right)T\right)}\right)^j E^{Q}\left[\bar{Y}(T)^j\right]. \tag{51}$$

It is also easy to obtain the centered moments of $\bar{\rho}(T)$ from the uncentered moments. The general equation for converting the nth uncentered moment to the centered moment is

$$E^Q\left[\left(\bar{\rho}(T)-E^Q\left[\bar{\rho}(T)\right]\right)^n\right]$$
$$=\sum_{j=0}^{n}\binom{n}{j}(-1)^{n-j}E^Q\left[\bar{\rho}(T)^j\right]\left(E^Q\left[\bar{\rho}(T)\right]\right)^{n-j}. \tag{52}$$

We provide the first two centered moments for $\bar{\rho}(T)$ as follows:

Corollary 2.

$$E^Q\left[\bar{\rho}(T)\right]$$
$$=\rho_m+\frac{(\rho_M-\rho_m)\left(\kappa_1^Q+\kappa_2^Q\right)\exp\left(-\left(\kappa_1^Q+\kappa_2^Q\right)T\right)}{1-\exp\left(-\left(\kappa_1^Q+\kappa_2^Q\right)T\right)}$$
$$\times\frac{1}{B(\alpha+1,\beta+1)}\frac{1}{2^{\alpha+\beta+2}}\times\sum_{k=0}^{1}N_k^2P_k^{(\alpha,\beta)}\left(2y(0)-1\right)a_k(\alpha,\beta,0,1)$$
$$\times\frac{\exp\left(\left(-\lambda_k+\kappa_1^Q+\kappa_2^Q\right)T\right)-1}{-\lambda_k+\kappa_1^Q+\kappa_2^Q}. \tag{53}$$

$$Var^Q\left[\bar{\rho}(T)\right]$$
$$=\left(\frac{(\rho_M-\rho_m)\left(\kappa_1^Q+\kappa_2^Q\right)\exp\left(-\left(\kappa_1^Q+\kappa_2^Q\right)T\right)}{1-\exp\left(-\left(\kappa_1^Q+\kappa_2^Q\right)T\right)}\right)^2\frac{2}{B^2(\alpha+1,\beta+1)}\frac{1}{2^{2(\alpha+\beta+2)}}$$
$$\times(U_1+U_2+U_3+U_4+U_5)-\left(E^Q\left[\bar{\rho}(T)\right]-\rho_m\right)^2, \tag{54}$$

where

$$U_1=N_0^4P_0^{(\alpha,\beta)}\left(2y(0)-1\right)b_{0,0}(\alpha,\beta)a_0(\alpha,\beta,0,1)\frac{1}{\kappa_1^Q+\kappa_2^Q}$$
$$\times\left[\frac{\exp\left(2\left(\kappa_1^Q+\kappa_2^Q\right)T\right)-\exp\left(\left(\kappa_1^Q+\kappa_2^Q\right)T\right)}{\kappa_1^Q+\kappa_2^Q}-\frac{\exp\left(2\left(\kappa_1^Q+\kappa_2^Q\right)T\right)-1}{2\left(\kappa_1^Q+\kappa_2^Q\right)}\right], \tag{55}$$

$$U_2 = N_0^2 N_1^2 P_1^{(\alpha,\beta)}\left(2y(0)-1\right) b_{1,0}(\alpha,\beta) a_0(\alpha,\beta,0,1) \frac{1}{\kappa_1^Q+\kappa_2^Q} \times\left[\frac{\exp\left(\left(-\lambda_1+2\left(\kappa_1^Q+\kappa_2^Q\right)\right)T\right)-\exp\left(\left(\kappa_1^Q+\kappa_2^Q\right)T\right)}{-\lambda_1+\kappa_1^Q+\kappa_2^Q} - \frac{\exp\left(\left(-\lambda_1+2\left(\kappa_1^Q+\kappa_2^Q\right)\right)T\right)-1}{-\lambda_1+2\left(\kappa_1^Q+\kappa_2^Q\right)}\right], \tag{56}$$

$$U_3 = N_0^2 N_1^2 P_0^{(\alpha,\beta)}\left(2y(0)-1\right) b_{0,1}(\alpha,\beta) a_1(\alpha,\beta,0,1) \frac{1}{-\lambda_1+\kappa_1^Q+\kappa_2^Q} \times\left[\frac{\exp\left(2\left(\kappa_1^Q+\kappa_2^Q\right)T\right)-\exp\left(\left(-\lambda_1+\kappa_1^Q+\kappa_2^Q\right)T\right)}{\lambda_1+\kappa_1^Q+\kappa_2^Q} - \frac{\exp\left(2\left(\kappa_1^Q+\kappa_2^Q\right)T\right)-1}{2(\kappa_1^Q+\kappa_2^Q)}\right], \tag{57}$$

$$U_4 = N_1^4 P_1^{(\alpha,\beta)}\left(2y(0)-1\right) b_{1,1}(\alpha,\beta) a_1(\alpha,\beta,0,1) \frac{1}{-\lambda_1+\kappa_1^Q+\kappa_2^Q} \times\left[\frac{\exp\left(\left(-\lambda_1+2\left(\kappa_1^Q+\kappa_2^Q\right)\right)T\right)-\exp\left(\left(-\lambda_1+\kappa_1^Q+\kappa_2^Q\right)T\right)}{\kappa_1^Q+\kappa_2^Q} - \frac{\exp\left(\left(-\lambda_1+2\left(\kappa_1^Q+\kappa_2^Q\right)\right)T\right)-1}{-\lambda_1+2\left(\kappa_1^Q+\kappa_2^Q\right)}\right], \tag{58}$$

$$U_5 = N_1^2 N_2^2 P_2^{(\alpha,\beta)} \left(2y(0)-1\right) b_{2,1}(\alpha,\beta) a_1(\alpha,\beta,0,1) \frac{1}{-\lambda_1+\kappa_1^Q+\kappa_2^Q}$$
$$\times \left[\frac{\exp\left(\left(-\lambda_2+2\left(\kappa_1^Q+\kappa_2^Q\right)\right)T\right)-\exp\left(\left(-\lambda_1+\kappa_1^Q+\kappa_2^Q\right)T\right)}{\lambda_1-\lambda_2+\kappa_1^Q+\kappa_2^Q} \right.$$
$$\left. -\frac{\exp\left(\left(-\lambda_2+2\left(\kappa_1^Q+\kappa_2^Q\right)\right)T\right)-1}{-\lambda_2+2\left(\kappa_1^Q+\kappa_2^Q\right)} \right]. \tag{59}$$

3.3. Credit Spread Option Prices under Jacobi Process

If $\rho(t)$ is not a constant, but deterministic, the credit spread $X_1(T)-X_2(T)$ defined in Section 2 will still be a Gaussian process with same mean $\mu_1-\mu_2$ but different variance

$$\bar{\eta}_s^2 = \eta_1^2 + \eta_2^2 - 2\bar{\rho}(T)\eta_{12}, \tag{60}$$

where $\bar{\rho}(T)$ is defined in (50).

The pricing formula in (14) will be changed to

$$C(T,K,\bar{\rho}) = \exp(-rT)\left[(\mu_1-\mu_2-K)N(\bar{d}) + \frac{\bar{\eta}_s}{\sqrt{2\pi}}\exp\left(-\frac{\bar{d}^2}{2}\right)\right], \tag{61}$$

where $\bar{d} = \frac{\mu_1-\mu_2-K}{\bar{\eta}_s}$.

If the correlation coefficient $\rho(t)$ is a stochastic process, but independent of underlying processes for the two yields $X_1(t)$ and $X_2(t)$, we can employ Taylor series expansion method to obtain the analytical formula for the spread credit spread option prices (see e.g. [11] [12] [13])

$$\bar{C}(T,K,\bar{\rho}) = E^{Q}\left[C(T,K,\bar{\rho})\right]$$
$$= E^{Q}\left[\sum_{k=0}^{\infty}\frac{C^{(k)}\left(T,K,E^{Q}[\bar{\rho}]\right)}{k!}\left(\bar{\rho}-E^{Q}[\bar{\rho}]\right)^{k}\right]$$
$$= \sum_{k=0}^{\infty}\frac{C^{(k)}\left(T,K,E^{Q}[\bar{\rho}]\right)}{k!}E^{Q}\left[\left(\bar{\rho}-E^{Q}[\bar{\rho}]\right)^{k}\right], \tag{62}$$

where $C^{(k)}\left(T,K,E^{Q}[\bar{\rho}]\right)$ denotes the kth derivative of $C(T,K,\rho)$ with respect to ρ evaluated at the point $\rho = E^{Q}[\bar{\rho}]$.

In practice, we need to approximate the series using the first n terms to obtain the option prices

$$\hat{C}_{n}(T,K,\bar{\rho}) = \sum_{k=0}^{n}\frac{C^{(k)}\left(T,K,E^{Q}[\bar{\rho}]\right)}{k!}E^{Q}\left[\left(\bar{\rho}-E^{Q}[\bar{\rho}]\right)^{k}\right]. \tag{63}$$

To estimate the truncation error, we use the remainder of Taylor's formula. We know there exists $\bar{\rho}^{*}$ such that

$$C(T,K,\bar{\rho}) = \sum_{k=0}^{n}\frac{C^{(k)}\left(T,K,E^{Q}[\bar{\rho}]\right)}{k!}\left(\bar{\rho}-E^{Q}[\bar{\rho}]\right)^{k}$$
$$+\frac{C^{(n+1)}\left(T,K,\bar{\rho}^{*}\right)}{(n+1)!}\left(\bar{\rho}-E^{Q}[\bar{\rho}]\right)^{n+1}. \tag{64}$$

Then, the bound for the error of $\hat{C}_{n}(T,K,\bar{\rho})$ is

$$\left|\bar{C}(T,K,\bar{\rho})-\hat{C}_{n}(T,K,\bar{\rho})\right|$$
$$= \left|E^{Q}\left[\frac{C^{(n+1)}\left(T,K,\bar{\rho}^{*}\right)}{(n+1)!}\left(\bar{\rho}-E^{Q}[\bar{\rho}]\right)^{n+1}\right]\right|$$
$$\le \frac{\sup_{0<\bar{\rho}^{*}<1}\left|C^{(n+1)}\left(T,K,\bar{\rho}^{*}\right)\right|}{(n+1)!}\left|E^{Q}\left[\left(\bar{\rho}-E^{Q}[\bar{\rho}]\right)^{n+1}\right]\right|. \tag{65}$$

For example, the second-order approximation to the credit spread option price is

$$\hat{C}_2(T,K,\bar{\rho}) = C\left(T,K,E^Q[\bar{\rho}]\right) + \frac{1}{2}C''\left(T,K,E^Q[\bar{\rho}]\right)Var^Q(\bar{\rho}), \tag{66}$$

where $E^Q[\bar{\rho}]$ and $Var^Q[\bar{\rho}]$ can be found from Corollary 2. The second-order derivative of C is

$$\begin{aligned} C''(T,K,\rho) &= \exp(-rT)\left(\frac{\sigma_1\sigma_2}{\kappa_1^Q+\kappa_2^Q}\right)^2\left(1-\exp\left(-\left(\kappa_1^Q+\kappa_2^Q\right)T\right)\right)^2\frac{\exp\left(\frac{-d^2}{2}\right)}{2\pi} \\ &\times\left\{-d(\mu_1-\mu_2-K)\left[(\mu_1-\mu_2-K)^2\eta_{12}^{-3}-\eta_{12}^{-2}-d(\mu_1-\mu_2-K)\eta_{12}^{-\frac{5}{2}}\right]\right. \\ &\left.+\left[2(\mu_1-\mu_2-K)^2\eta_{12}^{-\frac{5}{2}}-\eta_{12}^{-\frac{3}{2}}-2d(\mu_1-\mu_2-K)\eta_{12}^{-2}\right]\right\}. \end{aligned} \tag{67}$$

4. STOCHASTIC CORRELATION WITH TIME CHANGED JACOBI PROCESS

From Figure 2, it is clear that the correlation process changes abruptly through the time. We also know that the density function of correlation process has fat tail. These indicate that a diffusion process such as Jacobi process may not fully capture the main features of the correlation process and it is necessary to include the jumps into the correlation process. One way is to add the jump process directly into the correlation process in (19), which is the approach taken by [19] . The shortcoming of this approach is that it is difficult to ensure the correlation stays between ρ_m and ρ_M.

In this section, we model the correlation with Jacobi process time changed by Lévy subordinators to yield state-dependent jumps. The models with stochastic time changes have been studied in [20] and [21] for equity markets, [22] and [23] for credit markets and [24] and [25] for commodity markets. Here we model the stochastic correlation with time changed process.

4.1. Lévy Subordinator

A Lévy subordinator $T(t)$ is a non-decreasing Lévy process with jumps and non-negative drift (see e.g. [26]). The Laplace transform of a Lévy subordinator can be written as

$$E\left[\exp\left(-\lambda T(t)\right)\right]=\exp\left(-\phi(\lambda)t\right), \tag{68}$$

where function $\phi(\lambda)$ is known as the Lévy exponent of the subordinator $T(t)$ and is given by the Lévy-Khintchine formula (see e.g. [26])

$$\phi(\lambda)=\gamma\lambda+\int_0^{\infty}\left(1-\exp(-\lambda s)\right)\nu(\mathrm{d}s), \tag{69}$$

where $\gamma\geq 0$ is the drift term of $T(t)$. ν is the Lévy measure which must satisfy

$$\int_0^{\infty}(1\wedge s)\nu(\mathrm{d}s)<\infty. \tag{70}$$

An important sub-class of Lévy subordinators are the tempered stable subordinators. For such subordinators, the Lévy measure $\nu(s)$ is given by

$$\nu(s)=\delta s^{-1-p}\exp(-\eta s), \tag{71}$$

where $\delta>0$ and $\eta>0$. Important special cases are the Gamma subordinator

with $p=0$, the IG subordinator with $p=\frac{1}{2}$ and the compound Poisson sub-ordinator with $p=-1$ and $\eta>0$. For such subordinators, the Lévy exponent is given by

$$\phi(\lambda)=\begin{cases}\gamma\lambda-\delta\Gamma(-p)\left[(\lambda+\eta)^p-\eta^p\right], & p\neq 0\\ \gamma\lambda+\delta\log\left(1+\dfrac{\lambda}{\eta}\right), & p=0\end{cases}. \tag{72}$$

4.2. The TC-Jacobi Process

We model the correlation as the TC-Jacobi process, i.e.

$$\rho(t) = \rho_m + (\rho_M - \rho_m) Y_\phi(t), \quad (73)$$

$$Y_\phi(t) = Y(T(t)), \quad (74)$$

$$\mathrm{d}Y(t) = \kappa(\theta - Y(t))\mathrm{d}t + \sigma\sqrt{Y(t)(1 - Y(t))}\mathrm{d}B(t), \quad (75)$$

where $T(t)$ is a Lévy subordinator and $Y_\phi(t)$ is a time-changed Jacobi process. Under the martingale measure Q, we assume the stochastic process of $Y(t)$ is

$$\mathrm{d}Y(t) = \kappa^Q(\theta^Q - Y(t))\mathrm{d}t + \sigma\sqrt{Y(t)(1 - Y(t))}\mathrm{d}B^Q(t). \quad (76)$$

Therefore, the original Jacobi process $\rho(t)$ is time changed by Lévy subordinator ϕ to generate jumps. With these specifications, the new correlation pro- cess $\rho(t)$ is guaranteed to lie between ρ_m and ρ_M.

An important advantage of modeling the correlation as Jacobi process time changed by Lévy subordinator is that we can express the transition density of Y_ϕ via eigenfunction expansion (see e.g. [24] and [25]), i.e.

$$\begin{aligned} & p_{Y_\phi}(t, y_\phi(0), y_\phi(t)) \\ & = \pi(y_\phi(t)) \sum_{n=0}^{\infty} \exp(-\phi(\lambda_n)t)\psi_n(y_\phi(0))\psi_n(y_\phi(t)), \end{aligned} \quad (77)$$

where λ_n, ψ_n and $\pi(y)$ are the eigenvalues, eigenfunctions and steady-state density of Jacobi process, respectively, which can be found in (24), (25) and (30). Thus, the eigenfunction expansion of Y_ϕ remains the same form as Y , but with $\exp(-\lambda_n t)$ replaced by $\exp(-\phi(\lambda_n)t)$.

In order to obtain the closed-form solution for the credit spread option prices under the TC-Jacobi process, we need to calculate the moments of the following integral

$$\bar{Y}_\phi(T)=\int_0^T Y_\phi(s)\exp\left(\left(\kappa_1^Q+\kappa_2^Q\right)s\right)\mathrm{d}s. \tag{78}$$

Using the eigenfunction expansion of the transition density of Y_ϕ, we see that

the moments of $E^Q\left[\left(\bar{Y}_\phi(T)\right)^n\right]$ can be computed exactly same as $E^Q\left[\left(\bar{Y}(T)\right)^n\right]$

by replacing λ_j with $\phi(\lambda_j)$. We immediately obtain the following result:

Theorem 2.

$$\begin{aligned}
&E^Q\left[\bar{Y}_\phi(T)^n\right]\\
&=\frac{n!}{B^n(\alpha+1,\beta+1)}\frac{1}{2^{n(\alpha+\beta+2)}}\sum_{(k_1,\cdots,k_n)\in I^n}\prod_{j=1}^n\left(N_{k_j}\right)^2\\
&\times P_{k_1}^{(\alpha,\beta)}(2y(0)-1)\prod_{j=1}^{n-1}b_{k_j,k_{j+1}}(\alpha,\beta)a_{k_n}(\alpha,\beta,0,1)\\
&\times\int_0^T\int_{s_1}^T\cdots\int_{s_n}^T\prod_{j=1}^n\exp\left(-\phi\left(\lambda_{k_j}\right)\left(s_j-s_{j-1}\right)+\left(\kappa_1^Q+\kappa_2^Q\right)s_j\right)\mathrm{d}s_n\cdots\mathrm{d}s_1,
\end{aligned} \tag{79}$$

where $s_0:=0$

$$I^n=\left\{\left(k_1\cdots k_n\right)\in\mathbb{Z}_+^n:k_n\in\{0,1\},\left|k_j-k_{j+1}\right|\le 1,j=1,2,\cdots,n-1\right\}. \tag{80}$$

The credit spread options prices under the TC-Jacobi process can then be computed by employing exactly the same technique as Jacobi process.

5. EMPIRICAL STUDY AND NUMERICAL ANALYSIS

5.1. Estimation of Stochastic Correlation

We estimate the Jacobi and TC-Jacobi processes using daily correlation series constructed as in Section 3. We choose maximum likelihood estimation (MLE) method for both models and the estimation is carried out by using R software. For a sample of size N of $\{\rho_t\}_{t=1}^{N}$ the log conditional likelihood function is given by

$$\log\left(L(\Theta)\right)=\sum_{t=1}^{N}\log\left(p\left(t,t+\Delta t,\rho(t),\rho(t+\Delta t);\Theta\right)\right), \tag{81}$$

where Θ is the set of parameters to be estimated. $p\left(t,t+\Delta t,\rho(t),\rho(t+\Delta t);\Theta\right)$ is the transition density function of ρ. The transition density functions are

$$\begin{aligned}&p\left(t,t+\Delta t,\rho(t),\rho(t+\Delta t)\right)\\&=\frac{1}{\rho_M-\rho_m}\pi\left(\frac{\rho(t+\Delta t)-\rho_m}{\rho_M-\rho_m}\right)\\&\times\sum_{n=0}^{\infty}\exp\left(-\lambda_n\Delta t\right)\psi_n\left(\frac{\rho(t)-\rho_m}{\rho_M-\rho_m}\right)\psi_n\left(\frac{\rho(t+\Delta t)-\rho_m}{\rho_M-\rho_m}\right),\end{aligned} \tag{82}$$

for Jacobi process and

$$\begin{aligned}&p\left(t,t+\Delta t,\rho(t),\rho(t+\Delta t)\right)\\&=\frac{1}{\rho_M-\rho_m}\pi\left(\frac{\rho(t+\Delta t)-\rho_m}{\rho_M-\rho_m}\right)\\&\times\sum_{n=0}^{\infty}\exp\left(-\phi(\lambda_n)\Delta t\right)\psi_n\left(\frac{\rho(t)-\rho_m}{\rho_M-\rho_m}\right)\psi_n\left(\frac{\rho(t+\Delta t)-\rho_m}{\rho_M-\rho_m}\right),\end{aligned} \tag{83}$$

for TC-Jacobi process.

The MLE approach based on eigenfunctions for Jacobi process has been investigated in [17] . They also compare the performance of MLE estimator with other estimators proposed in the literature, such as the estimator in [27] , which is a method of moments based on an approximation of score function or with a generalized method of moments (GMM) estimator. The MLE estimator is further compared with computer-intensive simulation-based estimator, such as the simulated method of moments (SMM). They show that MLE outperforms the other estimators relative to bias and variance, while being easy to implement.

In the estimation, we set $\rho_m = 0$ and $\rho_M = 1$ so that the correlation is bounded by 0 and 1. The estimation results are presented in Table 2. We notice that all the parameters for both Jacobi and TC-Jacobi processes are highly significant. For Jacobi process, the mean parameters are very close for the correlations calculated from Aaa and Baa bonds. The correlations for Aaa bonds have higher mean reversion and smaller variance compared to Baa bonds. Whence Lévy subordinator is included, we clearly see that mean reversion and variance parameters drop significantly. This is as expected since the time-change process introduces the mean-reverting jumps into the correlation process, which is contrary to Jacobi process where the only source to generate mean reversion is through diffusion process.

We also report the log-likelihood function values, Akaike Information Criterion (AIC) and the Bayes Information Criterion (BIC) in Table 2. It is clear for the correlation calculated from both Aaa and Baa bonds, the TC-Jacobi model produces much larger likelihood values than Jacobi model, which indicates that it is necessary to include the jumps into the Jacobi process to improve its performance. We also employ AIC and BIC to compare the relative performance of Jacobi and TC-Jacobi processes. Again, it is clear that TC-Jacobi process captures the dynamics of the correlation much better than the Jacobi process.

5.2. Numerical Analysis of Credit Spread Options

We compute the credit spread option prices using second-order approximation that is provided in sections 3 and 4. The higher-order approximation yields similar results.

First, we test if stochastic correlation leads to different credit spread option prices compared to the constant correlation case. For the Jacobi process, we take the parameters from those reported in Table 2 based on the Aaa bonds. From Figure 3, we observe when the correlation process starts from the value lower/higher than the long-run mean, the prices under constant correlation will be higher/lower than those under the stochastic correlation. When starting value is close to the long-run mean, there is no decisive difference between the two prices. These are due to the fact that under the Jacobi process, the correlation will converge to the long-run mean due to the strong mean-reversion. When the correlation is higher, the variance of credit spread will be lower, which will lead to the lower option prices and vice versa.

Second, we test if there is significant difference in the prices from the Jacobi and TC-Jacobi processes. We use the parameters from those in Table 2 for Aaa bonds for both correlation processes. From Figure 4 it is clear the prices from the TC-Jacobi process are higher than those from Jacobi process. This can be explained by the fact that the long-run mean from the TC-Jacobi process is lower than the Jacobi process.

In Figure 5, we also illustrate the role of different parameters of TC-Jacobi process played in the pricing of credit spread options. We vary each parameter while keeping others at values reported in Table 2. It seems that higher mean reversion parameter κ will lead to higher prices. As expected, increase in long-run mean will decrease the option prices. We also note

Table 2. Estimation results for the correlation processes.

Parameters	Aaa		Baa	
	Jacobi	TC-Jacobi	Jacobi	TC-Jacobi
κ	4.6027	0.0790	4.4080	0.0668
	(0.6217)	(0.0180)	(0.5546)	(0.0176)
θ	0.8717	0.8180	0.8782	0.8394
	(0.0156)	(0.0173)	(0.0141)	(0.0160)
σ	0.9054	0.1287	0.8540	0.1106
	(0.0022)	(0.0098)	(0.0028)	(0.0120)
δ		95.4745		114.7419
		(4.7448)		(5.7797)
η		2.7278		2.2464
		(0.3720)		(0.4398)
Log likelihood	13549.46	15491.76	13859.33	15186.61
AIC	−27092.91	−30973.53	−27712.66	−30363.22
BIC	−27073.29	−30940.83	−27693.04	−30330.75

The standard errors are in the parentheses.

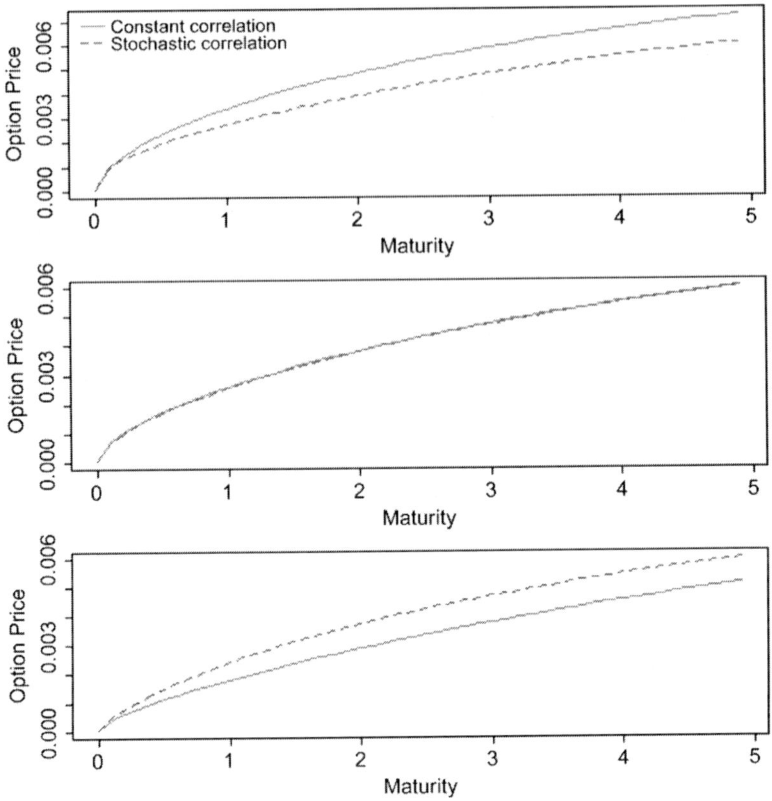

Figure 3. Comparison of credit spread option prices between stochastic and constant correlation. Top panel: $y(0)=0.7$; middle panel: $y(0)=0.87$; bottom panel: $y(0)=0.98$. The parameters for correlation process are obtained from Table 2. $x_1(0)=0.05$ $K=0.03$, $r=0.01$, $\sigma_2=0.01$, $\theta_2=0.02$, $\kappa_2=0.17$, $\sigma_1=0.008$, $\theta_1=0.06$, $\kappa_1=0.15$, and $x_2(0)=0.02$.

there is no significant effect on the prices for different volatility parameters. Finally, we find the higher δ, the lower η in the Lévy process, the higher the price is.

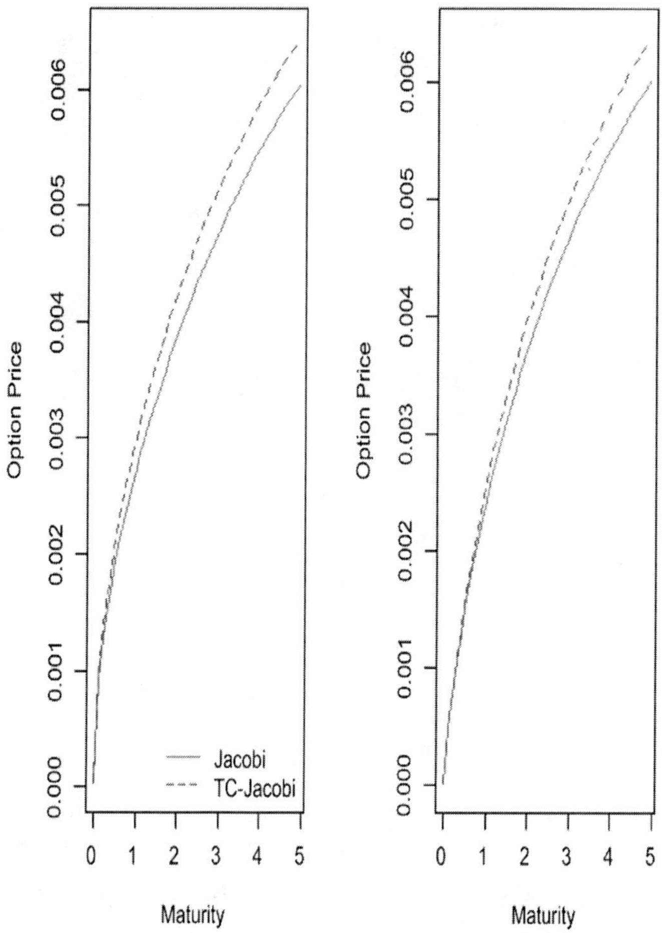

Figure 4. Comparison of credit spread option prices from Jacobi and TC-Jacobi processes. Left panel: $y(0)=0.7$; right panel: $y(0)=0.98$. The parameters for correlation process are obtained from Table 2. $x_1(0)=0.05$ $K=0.03$, $r=0.01$, $\sigma_2=0.01$, $\theta_2=0.02$, $\kappa_2=0.17$, $\sigma_1=0.008$, $\theta_1=0.06$, $\kappa_1=0.15$, and $x_2(0)=0.02$.

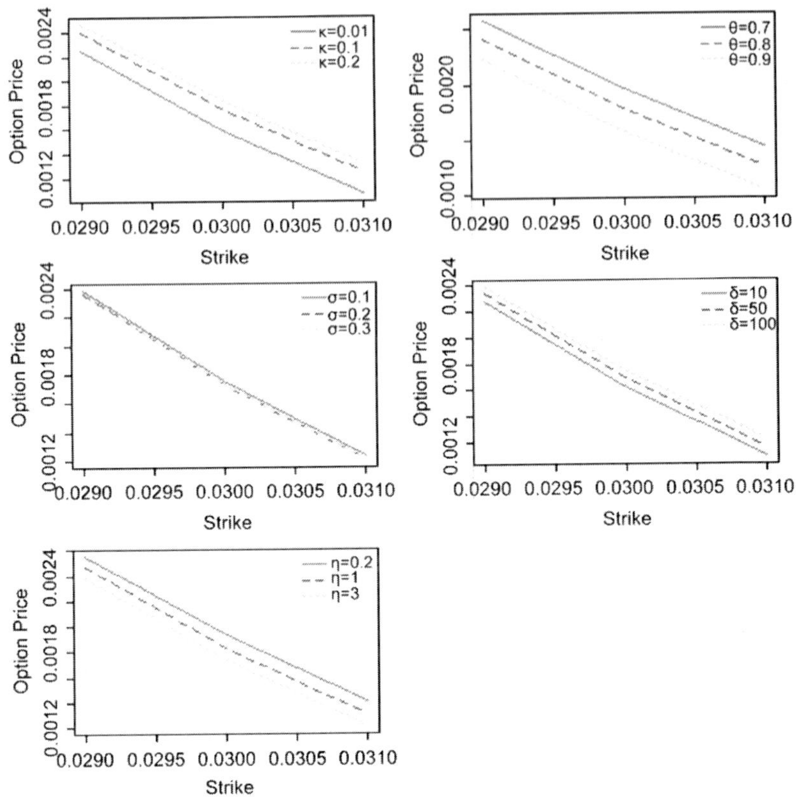

Figure 5. Credit spread option prices with different parameters from TC-Jacobi process. The parameters for correlation process are obtained from Table 2. $x_2(0)=0.02$ $x_1(0)=0.05$, $T=0.5$, $r=0.01$, $\sigma_2=0.01$, $\theta_2=0.02$, $\kappa_2=0.17$, $\sigma_1=0.008$, $\theta_1=0.06$, $\kappa_1=0.15$, and $y(0)=0.9$.

6. CONCLUSIONS

In this paper, we introduce the stochastic correlation process for modeling the credit spread. We model the components of spread process as correlated OU processes and correlation as Jacobi process. We study the properties of Jacobi process and show how to obtain the analytical solutions for credit spread options. To enhance the model's ability to capture the main features of correlation process, we extend the Jacobi process to TC-Jacobi process where the correlation is modeled by a Jacobi process time changed by Lévy subordinators. We demonstrate that the TC-

Jacobi process remains analytically tractable and the closed-form solutions for the credit spread options derived under the Jacobi process requires only minor changes. From an empirical study using the correlation series constructed from the yields on Treasury bonds and Aaa/Baa bonds, we demonstrate the superiority of TC-Jacobi compared to Jacobi process for modeling the correlation. We also numerically illustrate that modeling the correlation as a stochastic process will have an effect on the option prices compared to the constant correlation model. Furthermore, there are differences in option prices from Jacobi and TC-Jacobi processes.

To keep our model as parsimonious as possible, we assume that both risk-free short rate and volatility are constants. It will be an interesting extension to allow them to be stochastic. As long as the stochastic processes for the short rate and volatility are independent of correlation process, our method of obtaining the closed-form solutions for option prices will continue to work. We would also like to apply the TC-Jacobi process to the modeling of correlation in equity, foreign exchange and commodity markets.

ACKNOWLEDGEMENTS

Zhigang Tong would like to thank NSERC for partially funding this work. The opinions in this article are solely those of the authors and do not reflect the official opinion of Bank of Montreal.

CITE THIS PAPER

Tong, Z.G. and Liu, A. (2017) A Stochastic Correlation Model with Time Change for Pricing Credit Spread Options. Journal of Mathematical Finance, 7, 445-466. https://doi.org/10.4236/jmf.2017.72024

REFERENCES

1. Longstaff, F.A. and Schwartz, E.S. (1995) Valuing Credit Derivatives. The Journal of Fixed Income, 5, 6-12.
2. Mougeot, N. (2000) Credit Spread Specification and the Pricing of Spread Options. Unpublished, Ecole des HEC.
3. Tahani, N. (2004) Valuing Credit Derivatives Using Gaussian Quadrature: A Stochastic Volatility Framework. Journal of Futures Markets, 24, 3-35.
4. Chu, C.C. and Kwok, Y.K. (2003) No-Arbitrage Approach to Pricing Credit Spread Derivatives. The Journal of Derivatives, 10, 51-64.

5. Giacometti, R. and Teocchi, M. (2005) On Pricing of Credit Spread Options. European Journal of Operational Research, 163, 52-64.
6. Tchuindjo, L. (2007) Pricing of Multi-Defaultable Bonds with a Two-Correlated-Factor Hull-White Model. Applied Mathematical Finance, 14, 19-39.
7. Tchuindjo, L. (2011) Closed-Form Solutions for Pricing Credit-Risky Bonds and Bond Options. Applied Mathematics and Computation, 217, 6133-6143.
8. Romo, J.M. (2014) Modeling Credit Spreads under Multifactor Stochastic Volatility. The Spanish Review of Financial Economics, 12, 40-45.
9. Chen, S.N., Hsu, P.P. and Li, C.Y. (2016) Pricing Credit-Risky Bonds and Spread Options Modelling Credit-Spread Term Structures with Two-Dimensional Markov-Modulated Jump-Diffusion. Quantitative Finance, 16, 573-592.
10. van Emmerich, C. (2006) Modelling Correlation as a Stochastic Process. Unpublished, University of Wuppertal, Wuppertal.
11. Ma, J. (2009) Pricing Foreign Equity Options with Stochastic Correlation and Volatility. Annals of Economics and Finance, 10, 303-327.
12. Ma, J. (2010) Multi-Factor Models for Pricing Correlation-Dependent Interest-Rate Contingent Claim. International Review of Applied Financial Issues and Economics, 2, 359-378.
13. Alvarez, A., Escobar, M. and Olivares, P. (2011) Pricing Two Dimensional Derivatives under Stochastic Correlation. International Journal of Financial Markets and Derivatives, 2, 265-287.
14. Teng, L., Ehrhardt, M. and Günther, M. (2016) A Versatile Approach for Stochastic Correlation Using Hyperbolic Functions. International Journal of Computer Mathematics, 93, 524-539.
15. Teng, L., Ehrhardt, M. and Günther, M. (2016b) Modelling Stochastic Correlation. Journal of Mathematics in Industry, 6, 1-18.
16. Delbaen, F. and Shirakawa, H. (2002) An Interest Rate Model with Upper and Lower Bounds. Asia-Pacific Financial Markets, 9, 191-209.
17. Valéry, P. and Gouriéroux, C. (2011) A Quasi-Likelihood Approach Based on Eigenfunctions for a Bounded-Valued Jacobi Process. Unpublished, HEC, Montréal.
18. Prudnikov, A.P., Brychkov, Y.A. and Marichev, O.I. (1986) Integrals and Series. Vol. 2, Gordon and Breach Science Publishers.
19. Zetocha, V. (2015) Jumping off the Bandwagon: Introducing Jumps to Equity Correlation. Unpublished, SSRN.
20. Carr, P., Geman, H., Madan, D.B. and Yor, M. (2003) Stochastic Volatility for Lévy Processes. Mathematical Finance, 13, 345-382.
21. Carr, P. and Wu L. (2004) Time Changed Lévy Processes and Option Pricing. Journal of Financial Economics, 71, 113-141.
22. Mendoza-Arriaga, R., Carr, P. and Linetsky, V. (2010) Time Changed Markov Processes in Unified Credit-Equity Modeling. Mathematical Finance, 20, 527-569.
23. Mendoza-Arriaga, R. and Linetsky, V. (2013) Time-Changed CIR Default Intensities with Two-Sided Mean-Reverting Jumps. The Annals of Applied Probability, 24, 811-856.
24. Li, L. and Linetsky, V. (2014) Time-Changed Ornstein-Uhlenbeck Processes and Their Applications in Commodity Derivative Models. Mathematical Finance, 24, 289-330.
25. Li, L., Mendoza-Arriaga, R., Mo, Z. and Mitchell, D. (2016) Modelling Electricity Prices: A Time Change Approach. Quantitative Finance, 16, 1089-1109.
26. Sato, K. (1999) Lévy Processes and Infinitely Divisible Distribution. Cambridge, UK.
27. Kessler, M. and Sorensen, M. (1999) Estimating Equations Based on Eigenfunctions for a Discretely Observed Diffusion Process. Bernoulli, 5, 299-314.

Index